Cybersecurity for Accounting & Business

Nancy A. Bagranoff

University of Richmond

Scott R. Boss

Bentley University

Prospect Press

Founded in 2014, Prospect Press serves the academic disciplines of Information Systems and Business Analytics by publishing essential and innovative textbooks across the curriculum, including introductory, emerging, and upper-level courses. Based in Burlington, Vermont, Prospect Press distributes titles worldwide. We welcome new authors to send proposals or inquiries to Beth.golub@prospectpressvt.com.

Editor: Beth Lang Golub
Associate Editor: Dave Williams
Publishing Coordinator: Giulia Bronzi
Production Management: Peter Holm, Sterling Hill Productions
Copyeditor: Carl Quesnel, Assurance Editorial
Cover Design: Annie Clark
Illustrations: Marcia Blanco

eTextbook:
Edition 1.0
ISBN: 978-1-958303-43-6
Available from VitalSource, Kortext, Perusall, and BibliU

Printed Paperback:
Edition 1.0
ISBN: 978-1-958303-44-3

For more information, visit https://www.prospectpressvt.com/textbooks/bagranoff-cybersecurity.

CONTENTS

Preface

Why We Wrote this Book

Cybersecurity impacts everyone. You may have had your identity stolen, been unable to access a website, or been denied access to a service. Each of these situations may come down to a failure of cybersecurity to protect a person, organization, or nation from a cyber-attack. Hardly a day goes by that we do not learn of a business or other entity having suffered an adverse cyber incident.

As this book's authors, we are passionate about cybersecurity. We worked during the 1990's as professionals in the cybersecurity industry, hacking into computer systems as penetration testers, participating in cybersecurity implementations, and auditing organizational security measures. We are also academics with long careers as accounting faculty who teach courses in cybersecurity, information systems (IS), and accounting information systems (AIS). We know that our students are not going to be computer scientists or engineers but they will need a basic understanding of cybersecurity and be able to intelligently talk to the cybersecurity experts they will work with throughout their careers.

We wrote this book because both of us realized that there just was no cybersecurity textbook available for accounting and general business students and professionals. As we listened to other faculty teaching this type of course, we realized that we were not alone. Existing cybersecurity texts tend to focus heavily on the extremes: "Technical" (what one of our colleagues calls a "fingers on the keyboard" focus) with detailed explanations of exploits and solutions, or "Strategy" where cybersecurity is approached from a strategic perspective or C-Suite point of view.

Teaching cybersecurity in accounting and business programs today largely means creating your own materials. We see this as an impediment to accounting and business degree programs adding cybersecurity content to meet our students' needs. In the absence of a textbook, faculty who might be interested in teaching the subject may consider it too much of an investment in time or not know where to start. So we decided to remedy this by using the materials we had created to write a book.

Bridging the Cybersecurity Communication Gap

Cybersecurity is a subject that is constantly evolving. There are new threats and risks every day. That said, the basic underpinnings are somewhat constant. When we were doing cybersecurity consulting for large public accounting firms in the 1990's, that type of consulting was just getting started, but we were able to "hack" into computer networks because of human behavior. Individuals failed to change their default passwords, posted passwords written on sticky-notes on the underside of keyboards (or on their monitor!), and allowed easily guessed passwords on the system.

Cybersecurity is about people and their behaviors as much as it is about building the programmed defense structures. CrowdStrike's 2025 Global Threat Report calls out social engineering as a major way that threat actors (hackers) are gaining access. So sometimes, things do not change so much as it is just different people doing them. That said, we do know that AI is a game changer that will impact cybersecurity in profound ways. We will continue to assess AI's use by both the good guys and the bad actors as it evolves and posting updates at our Supplemental Instructional Resources site.

Changes in the CPA Exam

While this book is relevant for all business professionals, accounting students in particular will benefit from its content coverage of the Information Systems and Control (ISC) discipline material that is now in the CPA examination. We have paid careful attention to the topics in the CPA Evolution Model, and a large majority of this subject matter is described in this book. For example, the subject of System and Organization Controls (SOC), comprises between 15 and 25 percent of the ISC discipline content. Chapter 8 of this book is devoted to that topic and includes coverage of all SOC reports, with a focus on SOC 2 reporting. The Supplemental Resource website has a mapping of ISC topics to chapter coverage in this book.

Target Audience

Our book recognizes that there are fundamentals of cybersecurity that should be learned by everyone, but that the understanding does not necessarily need to be at the level of how to physically or programmatically change a poorly protected system. Instructors who teach AIS, IT Audit or Advanced Auditing, Information Systems Security, or Cybersecurity will find this book applicable in their course. In addition, accounting and business professionals seeking an interesting-to-read cybersecurity primer, may also find the book useful.

Book Features

- **Bridges the Communication Gap.** There is often a gap between how IT personnel and other functional staff talk about cybersecurity. This book bridges that space. It provides non-technical individuals with a general knowledge of cybersecurity that will allow them to intelligently describe cybersecurity problems with technical personnel.
- **Emphasis on business and organizational cyber risk.** We provide some technical coverage in our chapter on IT infrastructure for those educators in programs that have little coverage of this content and as a way to provide the vocabulary of cybersecurity. Otherwise, we focus on the concepts that accounting and business students need to know.
- **Accounting topics**: The book addresses many of the topics that are covered in the Certified Public Accountant examination. We present most of the topics in the Information Security and Control discipline, along with some of the subject matter in the core Auditing and Attestation section.
- **Coverage of Cyber Risk Elements:** The heart of our book is devoted to the elements associated with cyber incidents, including threats and vulnerabilities, risk assessment and management. Additional topics are loss estimation and legal and regulatory compliance requirements. We conclude with two "bonus" chapters that concern individual cybersecurity protection and cyberwarfare.
- **Ample pedagogical material:**
  - Case in Point descriptions. These short cases inject real world scenarios to show how concepts apply in practice.
  - Examples. To illustrate how a concept might work, an example is given, such as illustrating a cipher or IP address.
  - Table Summaries. Sometimes it is easier to look at a list in a table format rather than in narrative form. These table summaries provide an easier way for students to absorb complex information.
  - Glossary/ Cybersecurity has a language all its own. The Glossary provides definitions for key words.

- Rich end of chapter material including Multiple Choice, Review Questions, Short Cases, and Exercises. To ensure that students have mastered chapter material, an instructor may use this variety of end of chapter material to allow students to practice the material included in each chapter.

Overview of Chapters

Book chapters are organized as follows:

The Foundations of Cybersecurity

Chapter One: Navigating Cybersecurity for Accounting and Business

This first chapter sets the stage by describing the importance of cybersecurity to accounting and business professionals, cybersecurity education, and potential career paths.

Chapter Two: IT Infrastructure

Cybersecurity is about protecting data, and this requires a basic understanding of information technologies, including networks, cloud computing, the Internet, and AI, which are discussed here.

Chapter Three: Threats to Information Security

Bad things can happen in cyberspace. The chapter describes threats and threat actors, along with assets at risk and how to garner threat intelligence.

Chapter Four: Cybersecurity Vulnerabilities

A threat only leads to a cyber incident if there is a vulnerability. Here we describe common cybersecurity vulnerabilities, including human, technical, software, and operational weaknesses that lead to realized threats.

Assessing and Managing Cyber Risk

Chapter Five: Information Security Risk and Risk Assessment

Assessing risk is a first step in building a cybersecurity risk management program. In this chapter, we categorize risks as their ability to impact operations, reputation, and/or compliance. We also describe other features of risk assessment, including frameworks and models.

Chapter Six: Information Security Risk Management

Once risks are assessed, they can be managed with technical, administrative, and physical controls, described here, along with a brief description of popular risk management frameworks. We also describe how to plan for incident response and the potential to transfer risk through insurance.

Chapter Seven: Cybersecurity Governance and Risk Management Frameworks

Many frameworks are available for use in building a cybersecurity risk management program. Chapter Seven describes several of the most popular ones, with a focus on the NIST Cybersecurity Framework (2.0) and ISO 27001.

Chapter Eight: System and Organization Controls

System and Organization Control (SOC) reports are an important part of a cybersecurity risk management program, since third parties are often the weakest link. This chapter describes SOC reports and AICPA Trust Services Criteria, with an emphasis on SOC 2 reporting.

Cybersecurity Laws and Losses

Chapter Nine: Cybersecurity Legal and Regulatory Environment

Prosecuting cyber crimes and enforcing cyber regulations can be a challenge. Here we discuss both US and international laws as well as regulations that protect businesses and individuals in cyberspace.

Chapter Ten: Accounting for Cyber Losses

Here we explain how to assess and report on the potential and actual damage ensuing from a cyber incident. We link these to the risk categories identified in Chapter Five.

Cybersecurity Beyond Business

Chapter Eleven: Protecting Privacy

Privacy is increasingly important in a digital, globally connected environment. This chapter describes elements of privacy policies (that we tend not to read!) and various laws and regulations intended to keep data private.

Chapter Twelve: Geopolitics and Cybersecurity

Cyberwarfare is happening today and everyone is doing it. This chapter describes nation state threats and the potential for cyber incidents affecting national infrastructure, including the power grid, communications, and the water supply.

Chapter Thirteen: Protecting Individuals in Cyberspace

Like the other chapters in this section, this final chapter is a bit of a "bonus." Many individuals, including college students, are seeking ways to stay safe in cyberspace. Here we discuss the traps and tips for general cyber-hygiene.

Supplemental Instructional Resources

Teaching a course in cybersecurity for the first time can be a daunting proposition. Having a book will provide faculty with content and course structure. There are many resources available to supplement a course and make it one of the most interesting and informative classes your students will take.

At the resource website for this textbook, we provide

- End of Chapter Solutions
- Slides
- Test bank of 30 or more multiple choice questions per chapter
- Chapter mapping to CPA exam ISC discipline
- Sample course syllabi
- Sources for newsfeeds (e.g., Recorded Future)
- Websites for cybersecurity reports (e.g., the IBM Data Breach Report)
- A list of cyber movies, tv shows, and podcasts
- Links to cases. There are many cases available at Harvard Business Publishing, *Issues in Accounting*, and EY Academic Resource Center (EYARC) to assign to students.
- Updates (as needed). Because cybersecurity is such an important topic in accounting and business today, there are many publicly available resources to help Instructors create a great course. Given the dynamic nature of cybersecurity, our Instructional Resources will also provide significant updates, such as AI's evolving impact on cybersecurity.

To access Instructional Resources for the book, visit https://www.prospectpressvt.com/textbooks/bagranoff-cybersecurity

- Scroll to the red horizontal menu bar
- Click on Instructor Resources
- Click on the yellow box with Resources Login
- Click on the Instructor Resources Access Request Form and fill out the form. *Only verified faculty will receive access information.*

Thank you

We would like to express our thanks to the many people who helped us make this book possible. We owe a big debt of gratitude to Beth Lang Golub, President and Editor, Prospect Press, for understanding the need for this book and having the confidence in us to write it. Many other individuals have also shared their materials and expertise with us including the folks at Cybercrime, who let us use a few of their creative cybersecurity cartoons (https://cybersecurityventures.com/cybersecurity-cartoon-archives/). Thank you to Scott McAuliffe at Keiter for sharing his experience with SOC reporting. We are grateful to Scott, John Craft, Rebecca Likens, and Peter Neal for allowing us to profile their career paths in Chapter 1. Finally, we are appreciative of our students, who have been testing our book drafts over the past few years.

We also thank our many reviewers listed below and especially two early ones who encouraged the project, Diane Janvrin at Iowa State University and Won No at Rutgers University:

Sidney Askew, CPA, MBA, City University of New York, Borough of Manhattan Community College

Laura Barthel, PhD, CPA, Eastern Kentucky University
David Biros, Oklahoma State University
Marianne Bradford, North Carolina State University
Hermano Jorge De Queiroz, PhD, Lindsey Wilson University
Nusrat Farah, Southern Illinois University Carbondale
Dr. Renee Flasher, CPA, CFE, CMA, Penn State Harrisburg
Joseph Giordano, Western Carolina University
Neil Hwang, Bronx Community College
Sharif Islam, Southern Illinois University Carbondale
Diane J. Janvrin, Iowa State University
Randi Jiang, Grand Valley State University, School of Accounting
Sohee Kim, Missouri State University
Chih-Chen Lee, Northern Illinois University
Lorraine Lee, UNC Wilmington
Don Lux, Creighton University
Dr. Kasra Saad Mohammadi, University of Houston-Downtown
Billy Morehead, Mississippi College
Prof. Waleed Muhanna, Ohio State University
Francis J. Nemia, MBA, CISA, CRISC, CDPSE, Lecturer, Bentley University
Won Gyun No, Rutgers University
Professor Joseph Onyeocha, South Carolina State University, Orangeburg
Uchenna Peters, Augustana University
Matt Reidenbach, James Madison University
Allison Richardson, College of the Holy Cross
Laura K. Rickett, Cleveland State University
Fangjun Sang, University of Wisconsin – Green Bay

Matthew J. Sargent, The University of Southern Mississippi
Dr. Yuxin Shan, University of Wisconsin, Eau Claire
Juergen Sidgman, University of Alaska Anchorage

Nancy Bagranoff and Scott Boss
May 2026

Chapter 1

Navigating Cybersecurity for Accounting and Business

Chapter Outline

What Is Cybersecurity?
Cybersecurity for Business
Cybersecurity for Accounting
Cybersecurity in Education
Cybersecurity Credentials and Career Paths
About This Book

Learning Objectives (LO)

After reading this chapter, you should be able to:

- LO1. Define *cybersecurity*.
- LO2. Know why cybersecurity is important to business.
- LO3. Know why cybersecurity is important to accounting.
- LO4. Understand the importance of cybersecurity to each business function.
- LO5. Know how to obtain a cybersecurity education.
- LO6. Evaluate popular cybersecurity credentials.
- LO7. Compare cybersecurity career paths.

Introduction

Tell someone that you are studying cybersecurity, and you are likely to hear a story from them about how they have fallen prey to a personal cyberattack. If you discuss cybersecurity with a business or other professional leader, they may relate to you a cyber incident that impacted their organization. A discussion with a political leader will include their fears about threats to national security or critical infrastructure. Cybersecurity is a topic on everyone's mind today. Accountants and business professionals are particularly concerned with the subject because cyber threats are widely viewed as one of the most critical risks facing business organizations (Segal, 2023).

This first chapter defines **cybersecurity** and discusses its importance for accounting professionals and business leaders. People are sometimes surprised that cybersecurity is an important topic in accounting and business, but the reality is that accountants are specialists in risk, and when it comes to business, cyber threats are one of the most significant risks they face. This chapter includes discussions about cybersecurity education and career paths as well. We conclude with an overview of this book and its topical coverage. This textbook is a (relatively) nontechnical introduction to cybersecurity for students in any major and is also a useful primer for accounting and business professionals.

Cybersecurity
The practice of protecting systems, networks, and programs from digital attacks, damage, or unauthorized access.

What Is Cybersecurity?

The technological revolution in the 21st century has brought the world together in a new geography—cyberspace. This connectivity has not only created many benefits for humanity but has also incurred some significant costs. One of these costs is increased risk. No longer does even a small business operate with paper files in a metal cabinet that can be physically secured. Data and information are stored on networks and in virtual clouds of data that make up this new cyberspace. Protecting the data and information of individuals, businesses, and governments is the primary goal of cybersecurity professionals.

Cybersecurity both affects everyone and is everyone's responsibility. Cyberattacks can harm our individual wealth, our economy, our national security, and even our health and very lives. Here are a few compelling cyber facts:

- According to Cybersecurity Ventures' *2025 Official Cybercrime Report*, the global cost of cybercrime is expected to be more than $10 trillion by 2025. This is up from $3 trillion in 2015.
- A cyberattack on our critical infrastructure (e.g., water supply or electrical grid) could cripple our economy and cost lives.
- Per the *Cost of a Data Breach Report 2025* from the Ponemon Institute and IBM, in 2024 the average cost of a data breach was $4.44 million.
- The Federal Trade Commission reported that consumers lost more than $12.5 billion to fraud in 2024, with social media being a frequent point of contact for scammers.
- Ransomware payments have escalated. According to Sophos's *The State of Ransomware 2024* report, the median ransom payment was $2 million.
- One of the greatest threats in the Russian invasion of Ukraine was cyberwarfare. Hacktivists, hackers, nation-state cyber experts, and everyday citizens all engaged on both sides of the conflict.
- The Colonial Pipeline hack in 2021 created the longest lines at the gas pump since the 1973 OPEC oil embargo and stemmed from the compromise of a single password.
- Cyberattacks on healthcare systems can lead to diverted emergency services and delays in filling drug prescriptions and have been attributed to patient deaths worldwide (Mogg, 2025).

Figure 1-1. Cyberspace

Cybersecurity is the protection of computer systems, networks, and data from disclosure, disruption, or damage. While it can be a very technical topic, most cyberattacks contain a human element. Consider a phishing attack that introduces malware into a system. That attack was able to occur because someone clicked on a link or opened an attachment when they should not have done so. Cybersecurity is most often a people problem. This is one of many reasons why it is interesting to study.

Cybersecurity is in the news every day. Sometimes the story is about an individual who fell for an Internet scam, while other times it might be about a nation-state attack on a country's infrastructure. Quite often, the story is about a business—perhaps a ransomware attack or a data breach. As you listen to and read these reports, you might feel that you are hearing people talking in another language. Phishing, ransomware, DDoS attacks, hacktivism—all of these are part of the terminology that makes up the language of cybersecurity. Learning the language will make you cyber-literate, which you might think of as the first line of defense in protecting assets from cyber threats.

Cybersecurity for Business

While cyberattacks impact individuals, businesses, and other organizational entities, as well as countries and the global economy, the focus of this book is on cybersecurity for business and accounting, specifically on educating those who are not cybersecurity experts. Business has always been concerned with the security of data and information. Prior to relying on computers for processing data, securing these data required physical locks and keys, along with shredders for destroying sensitive materials. With the advent of computer programs that could process electronic data and send mail and documents, information security took on a new form. The introduction of networked computers—namely, the Advanced Research Projects Agency Network (ARPANET) in the 1970s—made way for computer viruses that could infect multiple devices.

Viruses were largely an annoyance in their early days, until criminals realized their revenue potential. The advent of the Internet exponentially increased threats to information security. The Internet and modern technology have brought us many, many positives, but these advances also have the downside that increased connectivity means more opportunity for widespread harm. Data breaches, denial of service attacks, and ransomware are just a few of the threats that impact business.

> **CASE IN POINT**
>
> **In November 1988, Robert Morris, a graduate student at Cornell University, wrote a program to determine the size of the Internet using existing UNIX operating system vulnerabilities. The resulting program moved from computer to computer and sent back a message to the control server to keep count. Within 24 hours, the program had infected about 10% of the computers on the Internet. It slowed the existing transmission traffic to a crawl and took up to 72 hours to bring under control. The overall cost of the damage was estimated to be in the millions of US dollars. Having created the first Distributed Denial of Service (DDoS) attack, Robert Morris, perhaps despite lack of intent, was the first person successfully prosecuted under the new Computer Fraud and Abuse Act in 1988. He was fined $10,000 and sentenced to three years of probation.**

Although *information security* is sometimes viewed as a broader term than *cybersecurity*, the terminology is used interchangeably in this book. Both are about protecting **data** and **information**, along with computer systems and networks. Data are distinguished from information because they have no meaning. Information is meaningful data.

> **EXAMPLE**
>
> **The numeral 114 is a piece of data. Employee number 114 is information. The numeral 98 is another piece of data. But 98° Fahrenheit is information. The numerals 65, 68, 64, and 72 are raw data elements. Adding that these numerals signify inches in the heights of a group of individuals turns these data into information.**

Both cybersecurity and information security are concerned with the **CIA triad**: the confidentiality, integrity, and availability of data. Confidentiality restricts data access to authorized parties; integrity ensures that the data are complete and accurate; and availability indicates that data may be accessed by authorized parties as needed. These are critical issues for businesses, which run on information. This triad is one of the most important concepts in cybersecurity, and this book will discuss it in depth.

Cybersecurity protection for business has grown significantly as an industry since the early days when viruses were just annoying distractions. The global cybersecurity market is expected

Data Raw facts, figures, or characters that are collected, stored, and processed by computer systems, often in digital format.

Information Data that have been processed or organized in a way that provides meaning or value to the user.

CIA Triad A model designed to guide policies for information security within an organization, consisting of three key principles: confidentiality, integrity, and availability.

to increase from about $172 billion in 2023 to more than $424 billion by 2030 (Insikt Group, 2025). The industry grew significantly during the COVID pandemic, as business employees moved to remote work. Future growth will be propelled by technology, including the Internet of Things (IoT), artificial intelligence (AI), cryptocurrency, and cloud-based solutions (discussed in Chapter 2).

Cyberattacks today are so common that it is no longer a matter of *if* a business will be targeted by a hacker, but rather *when*. Business executives list cybersecurity as one of the top risks their companies face. A special cybersecurity risk factors section of public company annual 10-K reports discusses risk management, strategy, and governance over cybersecurity, as shown in the example in Figure 1-2.

> Item 1C. Cybersecurity. Set forth below is information regarding our cybersecurity risk management, strategy, and governance, along with a related description of our information security and data privacy practices. Securing company systems, business information, and personal information of our guests, team members, vendors, and other third parties is important to us. We have systems in place to: • safely receive, protect, and store that information; • collect, use, and share that information appropriately; and • detect, contain, and respond to information security, cybersecurity, and data privacy incidents. While everyone at Target plays a part in information security, cybersecurity, and data privacy, oversight responsibility is shared by our Board of Directors, its committees, and management.

Figure 1-2. Excerpt from Target 2024 Annual Report

To protect against breaches, businesses are investing in cybersecurity for detection, protection, and response. According to a 2021 Deloitte report, businesses spent more than 10% of their total IT budget or about one-half of a percent of their total revenue on cybersecurity (Fortune Business Insights, 2025). The spending goes to employee salaries and automated security instead of operations, networks, and access controls. Corporate boards have gotten the message and expect briefings from their Chief Information Officer (CIO) and Chief Information Security Officer (CISO) about the risks and risk mitigation efforts. Both large and small businesses are impacted. Cyberattacks reportedly impacted more than 41% of small business enterprises in 2023 (Hiscox Inc., 2023).

All businesses are potential targets for bad actors, but some industry sectors face unique threats. For example, an electric utility might be the target of hackers trying to disrupt power supply, whereas an attack on the education sector might be after intellectual property. Money, of course, is an objective of many adversaries, and they will go after those businesses with an ability to pay or enterprises with valuable data. Financial institutions recognize the escalated threat they face in cyberspace, and that is why they spend more on cybersecurity than any other industry sector (Makortoff, 2025). While any attack on a business is worrisome to management and investors, as the following case illustrates, some industry attacks have ramifications that go beyond the business's direct stakeholders.

CASE IN POINT

In the case of the Colonial Pipeline ransomware attack in 2021, the CEO decided that he had no option but to pay the ransom. However, even though the attackers then provided the company with a tool to decrypt infected systems, this was not enough to enable a speedy, complete recovery and normal operations. The company spent several days trying to restore all systems at a cost of millions of dollars. And even a few weeks after the attack, Colonial could not bill customers because its billing system remained inoperable (Eaton & Volz, 2021).

Every aspect of a business is touched by cybersecurity. Losses can be significant. Target's data breach in 2013 cost the company more than $150 million. A 2023 cyberattack at MGM may have cost as much as $100 million. These costs include both direct and indirect costs, such as reputational damage. Chapter 10 of this book discusses cyber loss valuation.

Cybersecurity for Accounting

An obvious rationale for accounting professionals to understand cybersecurity beyond what you hear in the media is that it impacts their work and offers both challenges and opportunities. Table 1-1 lists some of the specific reasons why accounting professionals should be knowledgeable about cybersecurity.

Table 1-1. Benefits of Cybersecurity Knowledge for Accounting Professionals

Accountants Need Cyber Knowledge to:	Description
Assess and manage risk	Cyber risk is a significant part of the risks businesses face today, and accountants need to be able to assess and manage it for their organizations and clients.
Estimate financial losses	In the event of a cyber incident, accountants will need to estimate losses for inclusion in financial reports.
Prevent cyber incidents	Accountants and auditors implement and evaluate internal controls that protect against cyber incidents.
Protect data	Accounting professionals handle sensitive financial data that they must safeguard against unauthorized access and data breaches.
Comply with laws and regulations	Accountants must follow many laws and regulations governing cybersecurity for their businesses and clients. Some of them concern financial reporting, while others relate to privacy and data protection.
Be technologically literate	Cybersecurity requires some technology knowledge, including understanding emerging technologies. Accountants who understand cybersecurity will be able to apply this knowledge to their own tasks and leverage technology for greater efficiencies in their work.

A few recent developments have increased the importance of cybersecurity knowledge for accounting professionals. An example is the cybersecurity disclosure rule issued by the Securities and Exchange Commission (SEC) in 2023 (Securities and Exchange Commission, 2023). This rule requires disclosure of material cybersecurity incidents and reporting on an entity's policies and procedures with respect to cybersecurity risk. Accountants already consider cybersecurity when assessing business risk, but this new requirement emphasizes that consideration. In addition to working with business leaders to comply with SEC and other regulatory requirements, accounting professionals are well poised to help businesses assess and manage cyber risk. Accountants' skill set includes skepticism, internal control knowledge, measurement expertise, and materiality judgment. Auditors who combine their accounting knowledge with knowledge and experience working with information technology are also well suited to providing consultancy services in cyber risk management.

In 2024, the American Institute of Certified Public Accountants (AICPA) introduced a new format for the Certified Public Accountant (CPA) exam that recognized the evolution of the accounting profession, which includes an Information Systems and Control (ISC) discipline. The new exam format, described later in this chapter, is testament to the need for accounting professionals to be knowledgeable about concepts related to cybersecurity.

The AICPA and the Chartered Institute of Management Accountants (CIMA) recognize the opportunities in cybersecurity available to accountants. They offer several certificates in cybersecurity and many hours of continuing professional education. Professional accountants have been engaged for some time in issuing **Service Organization Controls (SOC) reports**. The SOC 2 report has grown in popularity with the evolution of digital technologies and the need for data security and privacy. This report can only be issued by CPAs.

Service Organization Controls (SOC) Reports Reports that provide information about the controls at a service organization, including SOC 1, SOC 2, and SOC 3; SOC for Cybersecurity; and SOC for Supply Chain reports.

The work of professional accountants is changing with new technologies, such as AI and analytical tools. So, too, are the services offered by traditional accounting firms. Public accounting firms offer many career opportunities in cybersecurity (described in more depth later in the chapter), and the Big Four firms each have cybersecurity professionals numbering in the thousands.

Accountants need to understand both IT and cybersecurity risks and issues so they can connect them to accounting and audit concerns, giving their clients a more holistic view of risk.

> **CASE IN POINT**
> **Deloitte, a Big Four public accounting firm, has a service line in cyber and strategic risk management. A tagline they use is "Cyber is everywhere. So are our services." Deloitte offers clients a range of services, from the highly technical to more strategic risk management. For example, they can help with cybersecurity assessments, benchmarking, and alignment with cybersecurity frameworks. Deloitte also offers services in technical resilience and mobile and endpoint security. As with other Big Four firms, Deloitte hires from a variety of different educational backgrounds and skill sets to do this work (Deloitte, 2025).**

Cybersecurity in Education

As we said at the beginning of this chapter, cybersecurity affects everyone. Everyone who works in an organization needs to understand cyber threats and risks and how to protect against them. Table 1-2 shows the specialized knowledge that each functional business area needs.

Table 1-2. Business Functional Areas and Cyber Expertise

Business Function	Cybersecurity Expertise Needed
Accounting	Accountants are expert at risk management and control. They need cybersecurity knowledge to safeguard business at appropriate risk levels and comply with legal standards. They are engaged with loss valuation, resource allocation, and internal control. They also are charged with all cybersecurity-related disclosures in financial statements.
Data Analytics	Analysis of large data sets can help protect companies against cyber intrusions. For example, sampling user logins for abnormalities can identify unauthorized activity.
Finance	Finance professionals need to understand cybersecurity to safeguard financial assets. They are also engaged in estimating losses and resource allocation. This may mean weighing the cost of protection against possible losses.
Human Resources	Human resources (HR) is concerned with protecting sensitive employee personal data, particularly with regards to confidentiality and privacy. In addition, the HR function may be charged with delivering cybersecurity training.
Information Technology (IT)	IT is the functional area responsible for providing information security. The cybersecurity controls are largely within this unit, which may have a Chief Information Security Officer (CISO) or information security spread across employees, such as network and database administrators.
Management	Managers across all units, including those in the C-suite and board members, need to understand the cyber risks and threats a company faces so they can make informed decisions to protect the entity and its stakeholders.
Marketing	The marketing function needs to be knowledgeable about cybersecurity, as they handle brand reputation and communications. They may be charged proactively with communications and PR about cyber incident preparedness, and in the event of a cyber incident, they will need to manage these activities.

Accounting and business students have many options for developing their skills in cybersecurity and information technology. Core business curricula include coursework in information technology and may include information security topical coverage. Most undergraduate accounting programs have at least one course in Accounting Information Systems (AIS) that includes a component on IT internal controls and risk management. Auditing courses also are likely to include segments on IT auditing, continuous auditing, and automated general and application controls. Undergraduate and graduate information security courses may be offered by Management Information Systems or Information Systems programs within business schools. This book may be used in any of these courses or in a stand-alone cybersecurity or information security course offered in accounting or business programs at either the undergraduate or graduate level.

The impact of technology on accounting has been one of the primary motivators behind a change in the educational model for Certified Public Accountants in the United States The AICPA, together with the National Association of State Boards of Accountancy (NASBA), developed **CPA Evolution** to address the changing educational and licensure needs of professional accountants. CPA Evolution includes a model curriculum and a new structure for the CPA licensing examination.

CPA Evolution An initiative by the AICPA and NASBA to reshape the CPA licensure model to meet the needs of the changing profession, including technology and cybersecurity skills.

A task force made up of more than 40 volunteers, representing a variety of professional accounting stakeholder groups, developed the CPA Model Curriculum. An emphasis of the new curriculum is on new and emerging technologies that CPAs need to begin their professional careers. The model includes content specifications and potential course structuring. The curriculum is intended to support the CPA exam, which now includes a core and three possible paths, as shown in Figure 1-3.

The core itself includes information technologies that impact accounting practice, with an emphasis on data analytics tools. The three components of the core are: Accounting and Data Analytics, Audit and Accounting Information Systems, and Tax. The Model Curriculum core has several content specifications related to cybersecurity. These include the COSO ERM framework, SOC 2 engagements, IT infrastructure, IT governance, cloud computing risks, IT general and application controls, and IT-related risks.

Aspiring CPAs can choose from three disciplines to support their core knowledge. They include Business Analysis and Reporting, Information Systems and Controls (ISC), and Tax Compliance and Planning. The ISC discipline has a heavy emphasis on cybersecurity topics, and the Model Curriculum suggests a few courses that might be included in a college curriculum to address this topical coverage: courses in Information Security and Forensics and one or two in Information Systems Assurance and Advisory. These courses are in addition to a traditional Accounting Information Systems (AIS) course. Cybersecurity topics described in the ISC discipline model curriculum include IT controls, use of IT resources, IT risk assessment, IT control frameworks, cybersecurity risk management, SOC engagements, use and management of data, information asset protection, and business resiliency and continuity.

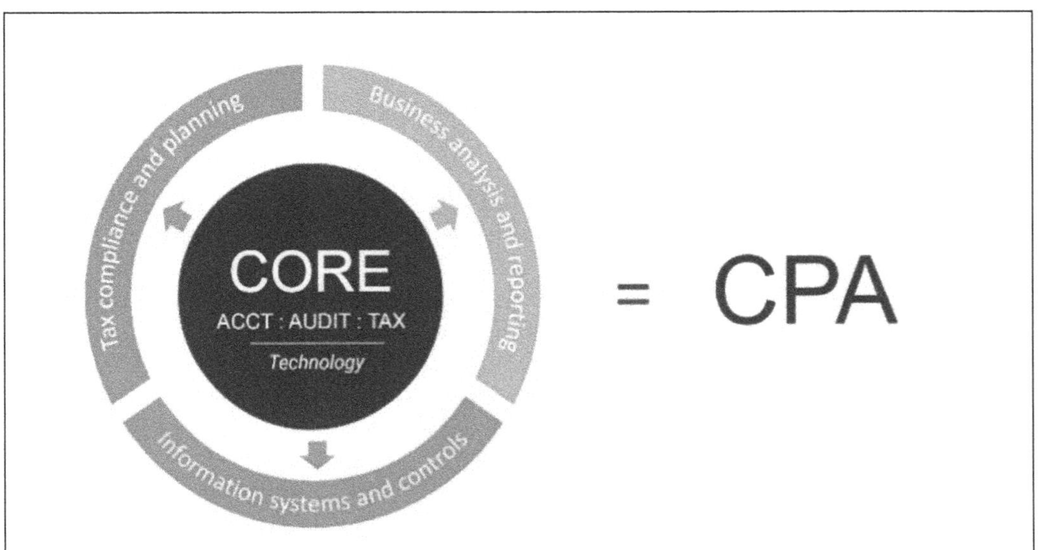

Figure 1-3. CPA Examination Components

Cybersecurity Credentials and Career Paths

Jobs in cybersecurity have shown significant growth over the past decade, and that trend is likely to continue for at least the next. For example, according to the US Bureau of Labor Statistics, jobs for information security analysts are projected to grow at a 33% rate between 2023 and 2033 (US Bureau of Labor Statistics, 2025). Many cybersecurity jobs require engineering, cybersecurity, or computer science education, but accountants and business graduates can also find jobs in the discipline, such as within Big Four public accounting firms, consulting firms, and information technology departments within medium to large businesses. Acquiring a cybersecurity credential or some modicum of technical training, such as in data analytics software, can prepare a business or accounting graduate for work in the field of information security.

Cybersecurity experts may choose to obtain any number of credentials. They include Certified Information Systems Auditor (CISA), Certified Information Security Manager, Certified Ethical Hacker, Certified Information System Security Professional, CompTIA Security+, and add-ons to the CPA.

Credentials

Many different organizations offer a variety of credentials related to cybersecurity that include badges, certifications, and certificates. Individuals wishing to demonstrate that they have knowledge of principles and concepts related to cybersecurity and information systems may obtain certifications in COBIT (multiple), IT Audit Fundamentals, IT Risk Fundamentals, Cloud Auditing Knowledge, Cybersecurity Audit, Computing Fundamentals, Network and Infrastructure Fundamentals, Cybersecurity Fundamentals, AI Fundamentals, and IoT Fundamentals.

Certified Information Systems Auditor (CISA) A certification for professionals responsible for ensuring that an organization's IT systems are effectively monitored, managed, and protected.

CISA

The **Certified Information Systems Auditor (CISA)** credential is offered by ISACA for entry-level or mid-career audit professionals wishing to demonstrate expertise in information systems audit and control. The credential might complement a CPA license or be for an information systems professional in an audit function. Topical expertise is in the information systems audit process; IT governance and management; IS acquisition, development, and implementation; IS operations and business resilience; and the protection of information assets. These topical areas are covered in the required credentialing examination. Candidates for the CISA should have five or more years of experience in IS/IT audit, control, assurance, or security.

Certified Information Systems Manager (CISM) A certification for professionals who manage, design, and oversee an enterprise's information security program.

CISM

The **Certified Information Systems Manager (CISM)** credential is an advanced certification for career professionals who are creating and/or managing an information security program for an enterprise and is also offered by ISACA. These individuals are interested in moving from a technical role into more of a managerial position. Topics covered on the required CISM examination are information security management, information risk management and compliance, information security program development and management, and information security incident management. To obtain the CISM credential, a candidate should have at least five years of experience in information security management.

Certified Risk in Information Systems and Control (CRISC) A certification for professionals who manage enterprise IT risk and implement and maintain information systems controls.

CRISC

ISACA's **Certified Risk in Information Systems and Control (CRISC)** credential is uniquely focused on an enterprise's IT risk management. Risk assessment and management are an important part of the work of almost every accountant, making this an appealing credential for accountants to obtain mid-career, particularly if they are working in IT/IS audit or working with IT auditors. The exam covers governance, IT risk assessment, risk response and reporting, and information technology and security. The CRISC credential requires at least three years of relevant work experience.

CISSP

One of the best-known technical cybersecurity credentials is the **Certified Information Systems Security Professional (CISSP)** credential. This global certification has been awarded by the International Information System Security Certification Consortium (ISC2) since 1994. As with several other credentials, candidates need to have about five years of professional experience and pass a certification examination. Topical coverage for the credential is in the following eight domains: security and risk management, asset security, security architecture and engineering, communication and network security, identity and access management, security assessment and testing, security operations, and software development security.

Certified Information Systems Security Professional (CISSP) A globally recognized certification in the field of IT security, offered by the International Information System Security Certification Consortium (ISC2).

CEH

The **Certified Ethical Hacker (CEH)** credential, granted by the EC-Council, is a more specific credential for information systems professionals interested in working in **penetration testing**. An ethical hacker is one who seeks out an information system's vulnerabilities to help the target organization set up controls for those weaknesses. This credential requires two years of professional experience and an examination, which includes a lab environment penetration simulation test.

Certified Ethical Hacker (CEH) A professional certification provided by the International Council of E-Commerce Consultants (EC-Council) that validates an individual's skills in ethical hacking.

Penetration (PEN) Testing A security exercise where a cybersecurity expert attempts to find and exploit vulnerabilities in a computer system. The purpose of this simulated attack is to identify any weak spots in a system's defenses.

CompTIA Security+

The **CompTIA Security+** credential is a global technical credential for those in early stages of information systems careers or for individuals hoping to change their professional path. There is no experience requirement, though two years of work experience in information technology/security is recommended. Career paths for individuals with this credential include systems administrator, security analyst, IT auditor, and network engineer. Topical areas include attacks, threats, and vulnerabilities; architecture and design; security program implementation; operations and incident response; and governance, risk, and compliance.

CompTIA Security+ A global certification that validates the baseline skills necessary to perform core security functions and pursue an IT security career.

Continuing Education

In addition to degree programs and credentials, individuals interested in pursuing careers in cybersecurity may opt to enroll in an executive education program or a self-study program. Many, many books, podcasts, and videos are available for interested learners, as well as subscriptions to cybersecurity newsletters, such as *Dark Reading*. Cybersecurity is a dynamic discipline and requires continuous learning. It is an ideal field of study for the intellectually curious.

Cybersecurity Career Paths and Opportunities

A wide range of career opportunities are available to those who study cybersecurity, from information security analysts to penetration testers, to threat bounty hunters, to risk consultants, to Chief Information Security Officers. In this section, we describe three "buckets" of cybersecurity careers: auditing, consulting, and IT security.

Auditing

While cybersecurity professionals may not have a desire to become auditors, education in cybersecurity is important for all accountants and auditors. The CPA credential is valuable for doing many types of jobs related to cybersecurity, particularly in auditing. A CPA working in financial audit who has an educational background and/or technical training in information security will be able to prepare SOC 2 reports, become an Information Systems Auditor or interface with them,

Internal Auditor
A professional within an organization who provides independent, objective assurance and consulting services designed to evaluate and improve the effectiveness of risk management, internal controls, and governance processes.

IT General Controls
Foundational controls that apply to all systems, applications, and data across an organization's IT environment. They ensure the reliability, integrity, and security of information systems that support business processes and financial reporting.

IT Application Controls
Automated controls embedded within software applications that ensure the accuracy, completeness, validity, and authorization of data processing specific to a business process (e.g., payroll, invoicing, purchasing). They rely on strong IT general controls to function effectively.

External Auditor
An independent third-party professional or firm hired to examine an organization's financial statements and issue an opinion on whether they are fairly presented in accordance with accounting standards.

Information Technology (IT) Auditor
A professional who evaluates and tests an organization's technology infrastructure, systems, controls, and processes.

Profile 1-1. Risk Advisory Services Partner

Profile: Scott McAuliffe, Risk Advisory Services Partner
Education: Bachelor of Science in Accounting
Certifications/Credentials: CPA, CISA, CFE (Certified Fraud Examiner)
Job History: Accountant at an architecture firm, consultant at Deloitte & Touche, Manager to Partner at Keiter
Current Job Description: Evaluating companies' internal control environments and providing clients with cost-effective solutions to build strong, efficient internal controls systems and practices to support their strategic objectives.

and liaison with cybersecurity teams in work such as risk assessment and risk management and preparing required client disclosures.

Auditors may be internal or external. An **internal auditor** works inside an organization and is responsible for implementing and evaluating internal controls, managing risk, and contributing to and evaluating governance policies and procedures.

With respect to IT and cybersecurity, these internal controls may be general or application controls. **IT general controls** encompass all hardware and software applications. These controls ensure the integrity and security of IT governance, processing, and infrastructure. **IT application controls** ensure data confidentiality, integrity, and availability for specific software applications.

> **EXAMPLE**
> **Two examples of IT general controls are change management and backup and recovery procedures. Change management ensures that software changes are made as needed and carried out in a safe environment. Backup and recovery procedures ensure that accurate data are available as needed. Two examples of application controls are an input control that ensures the completeness of data entered (e.g., user ID) and an authorization check to make sure that data are only entered by an authorized user.**

Internal auditors use these IT internal controls to manage cybersecurity risk. They work with management and IT staff to assess the cybersecurity risk and evaluate the cost of controls versus the risk of loss.

External auditors are hired from outside an organization to provide an independent review of financial statements and to ensure compliance with regulations, including those that govern cybersecurity. They may also conduct other assessments, including SOC reporting (discussed in Chapter 8) and risk analysis. External financial auditors need cybersecurity knowledge as it relates to financial reporting. IT systems process and output the data that underlie this reporting, so external auditors need to ensure IT security. These financial auditors will also be involved in the event of a cyber incident, where they will need to calculate losses.

Profile 1-2. IT Audit Manager

Profile: Peter Neal, IT Audit Manager, PwC
Education: Bachelor of Science in Business Administration, Accountancy major
Certifications/Credentials: CISA
Job History: 6 years at PwC
Current Job Description: Oversees multiple client accounts and relationships. Responsibilities include leading audit teams, mentoring and coaching staff, and managing both internal and external IT audit and compliance work. Focus is on automated control testing, key report testing, IT general control testing, and system transformation, including data conversion, security, and project governance. Also engaged in designing IT controls and conducting risk assessments.

External (and internal) financial auditors may work with specialized **information technology (IT) auditors** to ensure that IT general and application controls are designed and are operating effectively to manage cybersecurity risk. IT auditing is an excellent career path for individuals who have studied both accounting and cybersecurity. Two credentials available for this career path are the CISA offered by ISACA and the Certified Information Technology Professional (CITP), which complements the CPA, offered by the AICPA.

Consulting

Individuals with cybersecurity expertise have many opportunities to work in consulting. A combination of accounting education and experience with cybersecurity also provides a pathway to consulting careers. The Big Four public accounting firms (Deloitte, EY, PwC, KPMG) are all among the 10 largest cybersecurity consulting firms. Other traditional consulting firms, such as Accenture and Booz Allen Hamilton, also have large cybersecurity consulting practices. In addition, some consulting firms are dedicated to general or specialized cyber consulting.

> **CASE IN POINT**
>
> **CrowdStrike is a cybersecurity company that offers both products and services. One of the products they offer clients is Falcon, an endpoint detection and response platform that monitors endpoints (e.g., laptops) for malware and other threats. The product uses AI and other threat intelligence to detect, prevent, and respond to cyber threats. Their cybersecurity consulting services include proactive help that can identify specific threats to an organization. One example would be tabletop exercises to assess clients' cyber incident preparedness. Founded in 2011, the company has, like many cybersecurity companies, a relatively short history but has shown significant growth.**

Cybersecurity consulting companies offer a range of specialized services to help organizations protect their information assets, manage risks, and ensure compliance with relevant standards and regulations. Many of their services are related to risk assessment and management, but they also can conduct penetration testing and tabletop exercises (cyber incident simulations) to help clients find information security vulnerabilities. Some consulting companies offer managed services, whereby a company outsources its information security to the consulting firm. This arrangement benefits clients who may not have the necessary personnel with expertise to effectively and efficiently provide for cybersecurity in-house.

Profile 1-3. Platform Security Engineer Intern

Profile: Rebecca Likens, Platform Security Engineer Intern, CrowdStrike
Education: AAS, Cybersecurity; BAS Cybersecurity (expected graduation May 2026)
Certifications/Credentials: CompTIA Security+ SYO-601, CompTIA CySA+ (Cybersecurity Analyst), Google's Automate Cybersecurity Tasks with Python
Job History: Cybersecurity Operations Intern, Inova Health System; Desktop Risk Mitigation Intern, Navy Federal Credit Union; Penetration Tester and PowerShell Developer Intern, Frost Mountain
Current Job Description: Uses technology tools to enhance event analysis for threat hunting workflows and streamline alerts and incident documentation. Creating repeatable threat hunting stories and workflows to streamline alert escalation and cyber incident documentation. Conducting threat hunting across diverse data sets, identifying anomalous behaviors and correlation indicators of compromise. Developing hypotheses based on emerging threat intelligence.

> **CASE IN POINT**
>
> **KPMG is a Big Four global public accounting firm that offers accounting, tax, and advisory services. Cybersecurity is a service line within advisory services. They offer managed services, products, subscription-based services, and technology-related expertise, such as**

in managing the IoT and cloud security. In addition to these offerings, KPMG can help with incident response if a cyber incident occurs.

If you want to pursue a career in cybersecurity consulting, a relevant educational background, such as in accounting, computer science, consulting, data analytics, engineering, or information systems, provides a good foundation. Business majors might consider obtaining a credential or certificate to demonstrate cybersecurity knowledge and skills. Many consulting firms offer internships in cybersecurity consulting or risk advisory services, which are available to college students with some demonstrated skills, such as expertise in data analysis and IT.

IT Security

The top-level IT security job within a business organization is Chief Information Security Officer (CISO) or Director of Information Security. Moving up the corporate ladder from those positions would mean expanding to a role beyond cybersecurity, perhaps as a Chief Information Officer or a VP of Security, which would include physical as well as digital security.

There are many routes to careers in cybersecurity. Postsecondary degrees in business, engineering, computer science, or even law can all prepare an individual for a job in IT security, though attaining such a job will require at least some technical expertise. Candidates may begin work in consulting or within IT functional areas of organizations. Others might begin their careers with military service or government employment, including the US Federal Bureau of Investigation (FBI).

CASE IN POINT

The FBI is an excellent starting place for a career in cybersecurity. The FBI is the main US federal agency charged with investigating cyberattacks by criminals and nation-states. The bureau maintains expert cyber squads in each field office and has a rapid response team that can be deployed in the event of a major cyber event. Cyber incidents frequently cross national boundaries, and the FBI works with counterparts in other countries to pursue threat actors and seek recovery of stolen assets. The FBI also serves as a collection center for Internet crime reporting through its Internet Complaint Center (ic3.gov).

Depending on the size of an organization and its industry, there may be several roles for information security personnel within the IT function. Table 1-3 describes a few of these jobs that may exist within an organization.

The roles described in Table 1-3 work as a team to create a comprehensive information security framework that safeguards the organization's assets and data from internal and external cyber-

Table 1-3. Sample Cybersecurity Job Titles Within the IT Function

Job Title	Description
Chief Information Security Officer	Develops and maintains cybersecurity strategies and policies. Manages information security for an organization and reports to top management. Together with Information Security Manager, coordinates incident response.
Identity and Access Management Specialist	Designs and manages processes and procedures to verify identities and control access to information resources.
Information Security Manager	Responsible for day-to-day IT security operations. Manages a cybersecurity team and ensures compliance with security policies and regulations, including reviewing third-party relationship cyber practices and procedures. Together with CISO, coordinates incident response.
Security Analyst	Monitors information system for threats and vulnerabilities. Analyzes and plans mitigation of cyber incidents.
Security Awareness Trainer	Responsible for developing or acquiring training programs and deploying them for organization insiders and partners.
Security Engineer	Develops and maintains processes and procedures to protect the IT infrastructure. This includes implementing firewalls, vulnerability scanners, intrusion detection systems, and other technologies.

Profile 1-4. Director of Information Security

Profile: John Craft, Director of Information Security, University of Richmond
Education: Master of Science in Computer Science—Information Security
Certifications/Credentials: Multiple specialized certifications, including Cisco Certified Network Associate and Microsoft Certified Solutions Expert
Job History: US Army Officer, VP Information Technology at a community bank, Deputy CISO for Commonwealth of Virginia, Director of Technology Security Operations for Virginia Retirement System
Current Job Description: Manages the cybersecurity program for the university to include security operations, compliance, risk management, and building a culture of cybersecurity awareness though consistent outreach and training.

security threats. For example, in the event of a malware intrusion, it will be "all hands on deck." The structure and titles of these roles can vary depending on the organization's size and specific industry requirements.

About This Book

As noted at the beginning of this chapter, cybersecurity is important for individuals, every kind of organization, and nations. This book focuses on cybersecurity for business and accounting. The book is (relatively) not technical and is not meant for individuals who are planning to work as information security specialists or cybersecurity analysts. It is, however, a primer for business and accounting students, as well as others looking for fundamental cybersecurity knowledge as it relates to business. Reading this book provides a useful first step for a cyber career or a consulting career in risk management.

Cybersecurity is in the news daily, and it has its own language. This book provides fluency in that language and an understanding of the threats, vulnerabilities, potential impacts, risks, and controls that are part of the cybersecurity landscape. Because cybersecurity has its own language, each chapter in the book includes a list of key terms.

The book lays the groundwork for learning the language of cybersecurity with the next chapter (Chapter 2) on IT infrastructure. That chapter provides a basic understanding of computer hardware and software, with a focus on networks and technologies that impact our digital security and can be viewed as "cybersecurity vocabulary." They include the Internet of Things (IoT), cloud computing, artificial intelligence (AI), cryptocurrency, and blockchain.

You might think of Chapters 3 to 7 as the heart of the subject. Figure 1-4 illustrates the security ecosystem and the topical coverage in that section.

Cybersecurity is intended to control vulnerabilities and mitigate threats. Cybersecurity threats are all the potential attacks that can impact a business. Chapter 3 discusses the "what," "who," "why,"

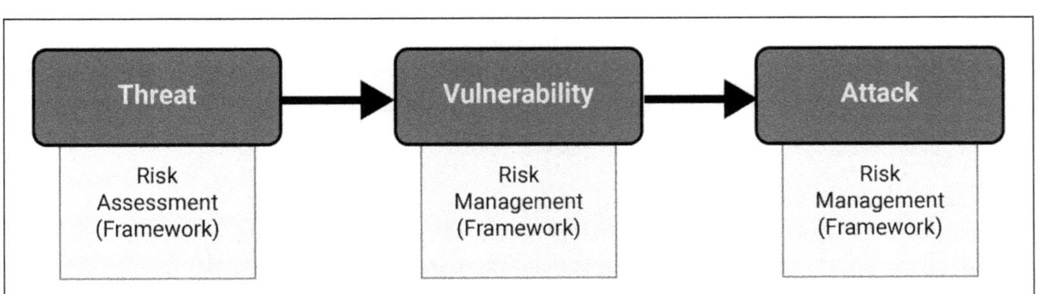

Figure 1-4. The Cybersecurity Ecosystem

and "how" of threats. The "what" are the assets, including data assets, that an adversary is after in an attack. The "who" describes the landscape of adversaries or bad actors who perpetrate cyberattacks, which includes individuals, organizations, and nation-states. "Why" is about these actors' motivations, which can range from "just for the heck of it" to cyberwarfare. The "how" is the set of threats that affect a business with direct or indirect financial and other negative consequences.

A mantra that we will repeat throughout the book is that it takes just one vulnerability for a cyberattack to be successful, but the defensive side must negate each and every vulnerability to reduce the risk of a successful cyber exploit to zero. Chapter 4 describes vulnerabilities or weaknesses that bad actors can exploit. A successful or realized threat is one that has taken advantage of such a vulnerability. This book classifies vulnerabilities into three categories: human, technical, and operational. The chapter provides some guidance on managing vulnerabilities, but you will find most of the coverage of that topic in the chapter on risk management.

Chapters 5 and 6 are about risk—how to assess it and how to manage it. In its broadest context, all of cybersecurity is an effort to manage or mitigate risk. Accountants have a long history in this realm. Risk is a function of the likelihood of a bad outcome, which in the case of cybersecurity is a successful attack. Business entities need to assess risk to determine how much of it they might be willing to accept, and the level of resources warranted to mitigate threats. In Chapter 5, risks are organized into three classifications: operational, reputational, and compliance. Each of these risk categories contains risks that are financial and/or nonfinancial. While most anything can be translated into a financial risk, some risks (e.g., low company morale) are more difficult to quantify than others. Chapter 5 describes both qualitative and quantitative approaches to risk assessment.

As discussed in Chapter 6, managing risk is about controls. Information system controls might be directed at preventing, detecting, or correcting the risk of a realized cybersecurity threat. These are the *functions* of controls. The *types* of controls are technical, administrative, and physical. Technical controls are mostly software controls, such as firewalls and antivirus protection. Administrative controls are all the policies and procedures an organization develops and employs to manage risk. They include software patching policies, administrative structure, and incident response plans. Physical controls ensure that information system resources are appropriately protected with gates, guards, and environmental controls.

Many frameworks address a variety of cybersecurity-related concerns. Some frameworks are specialized and address risk assessment, risk management controls, and specific threats and vulnerabilities. Consulting firms specializing in cybersecurity also may offer proprietary frameworks and methodologies for information security. Various frameworks are described throughout this book. Chapter 7 describes in some detail the two most broadly used general cybersecurity frameworks: the National Institute of Standards and Technology (NIST) Cybersecurity Framework and ISO 27001.

Chapter 8 of this book discusses the line of attestation services that the AICPA offers called System and Organization Controls reports. SOC reports are covered in both the Audit and ISC sections of the CPA examination. The Information Systems and Control discipline in the CPA exam covers these attestation services extensively. These services are built on the AICPA Trust Services Criteria, which also are covered in Chapter 8.

An important part of cybersecurity is compliance with legal and regulatory requirements. Chapter 9 describes laws and regulations governing cybersecurity and how cyber threat actors might be prosecuted for their offenses. Businesses invest in cybersecurity to protect themselves against threats and for compliance reasons. These investments do not generate revenues but instead are an expense that reduces profitability. Justification for this spending is that it reduces risk. Chapter 10 discusses the valuation of cybersecurity damages.

Chapter 11 covers privacy. Accountants have many reasons to be concerned about digital privacy. The practice of accounting involves handling data and information that require various levels of privacy and confidentiality. Whether it is an accountant's personal data, client records, or communications, accountants have a need to safeguard data. Chapter 11 describes why privacy is important, privacy policies, privacy regulations, and digital natives' attitudes toward privacy.

The final two chapters in this book extend beyond accounting and business to consider global issues related to cybersecurity and how cybersecurity impacts individuals. Cybersecurity threats affect nations and individuals just as they pose a threat to business.

Each chapter includes Cases in Point to illustrate concepts. These are real-world current examples of threats, vulnerabilities, attacks, controls, and other cybersecurity-related topics. Cybersecurity is a topic that appears in the news every day. This book provides resources for staying current with related news. End-of-chapter materials consist of multiple-choice questions, review questions, cases, and exercises.

The end of each chapter in this book contains a list of key terms used in the chapter. The book concludes with a glossary of the key terms identified in each chapter.

Summary

Cybersecurity should be a topic of interest to everyone because it impacts every individual, organization, and nation. This chapter defined the concept and provided an introduction. Two sections of the chapter explained the relevance of cybersecurity to business and to accounting. This book is intended for business and accounting students who want to study cybersecurity, professional accountants and business leaders who recognize that they need to know about all the threats and risks their organizations face, and the simply curious.

The book is not a technical primer for those who wish to manage networks or write code to patch vulnerabilities. Rather it is a relatively nontechnical introduction to the language of cybersecurity and cybersecurity concepts. Most professionals will need to interact with cybersecurity professionals at some point in their career. The topics in this book will allow them to talk with those professionals in an intelligent manner rather than from a position of ignorance.

This introductory chapter describes the niche that cybersecurity occupies specifically for business and outlines the organization of the book, with brief chapter descriptions. The chapter also describes cybersecurity education, certifications, and career opportunities.

Chapter Review

Key Terms

Certified Ethical Hacker (CEH)
Certified Information Systems Auditor (CISA)
Certified Information Systems Manager (CISM)
Certified Information Systems Security Professional (CISSP)
Certified Risk in Information Systems and Control (CRISC)
CIA Triad
CompTIA Security+
CPA Evolution
Cybersecurity
Data
External Auditor
Information
Information Technology (IT) Auditor
Internal Auditor
IT Application Controls
IT General Controls
Penetration (PEN) Testing
Service Organization Controls (SOC) Reports

Review Questions

1. What is cybersecurity, and what are the main areas related to cybersecurity?
2. Why do nontechnical business professionals need to understand the fundamentals of cybersecurity?
3. How is cybersecurity related to the accounting profession?
4. How does cybersecurity impact education?
5. What educational areas in accounting could benefit from cybersecurity exposure?
6. List the major cybersecurity credentials and how they apply to individual business careers.
7. Compare and contrast the different career paths discussed in this chapter.

Multiple Choice

Identify the choice that best completes the statement or answers the question.

1. What skill set of accountants makes them well suited for providing consultancy services in cyber risk management?
   a. Social media proficiency
   b. Skepticism, internal control, measurement, and materiality judgment expertise
   c. Data entry proficiency
   d. Technological innovation knowledge
   e. Physical security expertise
2. Which of the following is **not** mentioned as a credential that cybersecurity experts may choose to obtain?
   a. Certified Ethical Hacker (CEH)
   b. CompTIA Security+
   c. Certified Information Systems Manager (CISM)
   d. Certified Public Accountant (CPA)
   e. Certified Information Systems Auditor (CISA)
3. According to the chapter, what might be considered the first line of defense in protecting assets from cyber threats?
   a. Increased connectivity
   b. Digital geography
   c. Metal cabinets
   d. Cyber-literate individuals
   e. Physical security measures
4. Which report, assessing third-party controls over information systems, can be issued only by certified public accountants?
   a. AI Impact Assessment
   b. Privacy Impact Assessment
   c. SOC 2 Report
   d. Cybersecurity Benchmark Report
   e. SEC Compliance Report
5. What is mentioned as a significant human element in cyberattacks in the chapter?
   a. Metal cabinets
   b. Cloud storage
   c. Paper files
   d. Physical security
   e. Phishing attacks
6. What is the primary goal of cybersecurity?
   a. Increasing physical security
   b. Safeguarding computer systems, networks, and data
   c. Creating a new geography
   d. Protecting metal cabinets
   e. Reducing connectivity

Case Studies

Natalie Magee, Cybersecurity Career

Natalie Magee is a recent college graduate from a large Midwestern US university. Natalie graduated with a BS degree. She majored in accounting and earned a minor in business consulting. Although she enjoyed her accounting courses, she has determined that she does not wish to pursue a career in financial audit or tax. She has also determined that she would prefer a smaller, regional company rather than an international public accounting or consulting firm.

During her course of study, Natalie took a course in information security. She really enjoyed the content and believes she would enjoy working for a consulting firm that has an information risk management or cybersecurity practice. She knows that data analytics tools are important. She also understands that she may need to have some credentials before she begins applying for jobs in this field.

Case Requirements

1. Research online to find three examples of dedicated cybersecurity consulting firms. What types of services do these companies offer to their clients?
2. What advice do you have for Natalie for obtaining additional credentials or certifications in cybersecurity?
3. In which data analytics tools do you think it would be most valuable for Natalie to acquire expertise?
4. Research online to find three or more cybersecurity internship opportunities. What are the required qualifications for these positions?

Yusuf Khan, Cybersecurity Training

Yusuf Khan graduated from a large state university four years ago with a business degree in information systems. After graduation, he went to work for the Heron Corporation, a national automobile parts distributor with annual revenues of $150 million and about 1,000 employees. The IT function has 18 employees, and there are four individuals working in IT security: the Director of Information Security, two security analysts, and a security engineer. Yusuf has been working as a security analyst, and he is now being assigned responsibility for designing a cybersecurity awareness training program for all of Heron's employees.

Case Requirements

1. Why is it important for a cybersecurity training program to have different requirements for different job roles within a company?
2. Conduct online research to identify two companies that offer cybersecurity awareness training. What types of training do those companies offer and in what formats?
3. Create a model for a cybersecurity awareness training program for Heron that includes the training duration and format, the frequency of training, and major content areas that would be covered (at least four).
4. Some companies conduct in-house phishing tests as part of their cybersecurity awareness training. Is this a good practice? Cite at least three sources to support your answer to this question.

Exercises

1-1. Many educational resources are available for students interested in developing cybersecurity skills and expertise. Tryhackme.com offers both free content and subscription-based training. Required:
   1. Visit www.tryhackme.com. What educational resources do they offer free of charge?
   2. Depending on your interest and skill level, complete one of the learning paths.
   3. What certifications can you obtain from TryHackMe?

1-2. The CPA examination now includes an Information Security and Controls discipline that covers many topics related to cybersecurity. Required:
   1. What are the three sections in the ISC discipline, and what are the percentages covered in each section?
   2. For each section you have described, what courses within your college or university are most likely to cover the associated topics within it?

1-3. This chapter described many of the reasons why accountants and other businesspeople need to have knowledge about cybersecurity. Another group that increasingly needs to have some cybersecurity expertise is a Board of Directors. Required:
   1. Why should a member of a public company Board of Directors need to know about cybersecurity?
   2. Explore the website for the National Association of Corporate Directors (nacdonline.org) and find an article about cybersecurity. Read the article and summarize it in 50 words or less.

1-4. This chapter provided profiles of four individuals who have a career in cybersecurity. Required:
   1. Which of these individuals has a job that appeals most to you? Why?
   2. Beginning where you are now in your education and experience, what steps would you take going forward to secure a position such as the one held by the individual profile you selected?

References

Cybersecurity Risk Management, Strategy, Governance, and Incident Disclosure, 17 C.F.R. § 229 (2023). https://www.federalregister.gov/documents/2023/08/04/2023-16194/cybersecurity-risk-management-strategy-governance-and-incident-disclosure

Deloitte. (2025). *Our services*. Retrieved October 31, 2025, from https://www.deloitte.com/dk/en/services.html

Eaton, C., & Volz, D. (2021, May 19). Colonial Pipeline CEO tells why he paid hackers a $4.4 million ransom. *Wall Street Journal*. https://www.wsj.com/tech/cybersecurity/colonial-pipeline-ceo-tells-why-he-paid-hackers-a-4-4-million-ransom-11621435636

Fortune Business Insights. (2025, September 15). *Cybersecurity market size, share, and industry analysis*. Retrieved October 31, 2025, from https://www.fortunebusinessinsights.com/industry-reports/cyber-security-market-101165

Hiscox Inc. (2023). *Hiscox cyber readiness report 2023*. https://www.hiscox.com/documents/Hiscox-Cyber-Readiness-Report-2023.pdf

Insikt Group. (2025, January 28). *Cyber threat analysis: 2024 annual report*. https://www.recordedfuture.com/research/2024-annual-report

Makortoff, K. (2025, June 15). "We're being attacked all the time": How UK banks stop hackers. *The Guardian*. https://www.theguardian.com/business/2025/jun/15/uk-banks-hackers-attacks-cybersecurity

Mogg, T. (2025, June 25). *A hospital hack has been linked to a patient's death*. Digital Trends Media Group. https://www.digitaltrends.com/computing/hospital-hack-linked-to-patients-death/

Securities and Exchange Commission. (2023). *Cybersecurity risk management, strategy, governance, and incident disclosure* (Release Nos. 33-11216; 34-97989; File No. S7-09-22). Federal Register, 88(149), 51896–51989. https://www.federalregister.gov/documents/2023/08/04/2023-16194/cybersecurity-risk-management-strategy-governance-and-incident-disclosure

Segal, E. (2023, December 03). The 9 biggest risks and threats that companies will face in 2024. *Forbes*. https://www.forbes.com/sites/edwardsegal/2023/12/03/the-8-biggest-risks-and-threats-that-companies-will-face-in-2024/

US Bureau of Labor Statistics. (2025, August 28). *Occupational outlook handbook: Information security analysts*. Retrieved October 31, 2025, from https://www.bls.gov/ooh/computer-and-information-technology/information-security-analysts.htm

Chapter 2

IT Infrastructure

Chapter Outline

Data
Data Protection
Data Storage Options
Data Backup
Components of IT Infrastructure
The Internet
Artificial Intelligence
Quantum Computing
Cryptocurrency and Blockchain
Auditing IT Infrastructure

Learning Objectives (LO)

After reading this chapter, you should be able to:

- LO1. Understand the basic terminology of information systems.
- LO2. Describe the different data storage options available to organizations.
- LO3. Describe backup configurations and business continuity planning
- LO4. Understand the major components of IT infrastructure.
- LO5. Describe the major sections of the Internet and how they are used.
- LO6. Understand how artificial intelligence (AI), cryptocurrency, and the Internet affect cybersecurity.

Introduction

Cybersecurity is all about protecting data stored via software programs on computing devices and communicated through networks. Most of this chapter describes how data are stored and sent, and the information architecture that lets us access those data. Although IT infrastructure is traditionally viewed as the domain of computer scientists, accounting and business professionals need to have a basic understanding of it to understand the issues that arise in cybersecurity. The IT infrastructure fundamentals covered in this chapter set the stage for the upcoming chapters on threats, vulnerabilities, and risk.

Several recent technological advances have had a significant impact on cybersecurity, none more so than the Internet, including the Internet of Things (IoT), though other emerging technologies all affect information security in important ways. Other important technological advances that impact cybersecurity are artificial intelligence (AI), cloud computing, and cryptocurrency. This chapter discusses each of these emerging technologies in some depth.

Data Raw facts, figures, or characters that are collected, stored, and processed by computer systems, often in digital format.

Information Data that have been processed or organized in a way that provides meaning or value to the user.

Knowledge Information prepared, sorted, and organized so that it can be used in high-level decision-making processes to set policy or strategy.

Big Data Extremely large and complex data sets that are difficult to process, store, or analyze using traditional data processing tools.

ETL Software Software that can extract, transform, and load big data for analysis.

Data Life Cycle The series of stages through which data passes, including creation, storage, usage, sharing, archiving, and deletion, to ensure proper management, security, and value extraction.

Data at Rest Digital information that is stored on a physical medium and is not actively moving across networks or being processed.

Data in Motion Digital information that is actively moving from one location to another, whether across the Internet or a private network or between devices and applications.

Database An organized collection of structured information, or data, typically stored electronically in a computer system.

Database Schema The logical structure that defines how data is organized in a database. It describes the tables, fields (columns), data types, relationships, indexes, views, procedures, and constraints that make up the database.

Data

Information technology is all about **data.** Computing devices were created to process and transfer data. Many people confuse data with information, but they are not the same. Data are the raw material and may be unintelligible without any additional perspective. For example, "1234" is a string of data, but it has no meaning without context. "Employee identification number 1234" provides context and turns this small set of data into **information**, which may be useful for decision making, again depending on the context of the decision that needs to be made. The aggregation and interpretation of different pieces of information become **knowledge** that management can act on to set policy and strategy.

Big data references exceptionally large data sets that can be difficult to process and analyze with traditional software. For example, Excel software is often used to analyze large sets of data, but as of 2025 it can handle only just over a million rows of data. Processing and analyzing big data sets requires **ETL software**, which can extract (E), transform (T), and load (L) big data for analysis.

Data are created, collected, stored, used, transmitted, and discarded. The **data life cycle** describes the period from creation to final disposition. Data are created or gathered using a data collection method, which is often online input. The data may be input by their owner, by a data request from a user, or by being automatically collected. Many businesses try to collect as much data as they can because most organizations in the information age view data as an asset. It is not unusual for customers to be asked to provide their phone number or email address when making a purchase or signing up for a rewards program. Questionnaires are another method of collecting data. Social media platforms may offer "quizzes." By answering the questions in these online quizzes, individuals are providing data about their preferences and creating a profile that can be used, for example, for targeted advertising. (Chapter 11 addresses the importance of data privacy and security.) Finally, websites or programs can automatically record dates, times, locations, IP addresses, and so on, from just visiting a page.

Data are generally in one of two states: **at rest** or **in motion**. Data at rest are in storage and are not currently being accessed, while data being accessed, communicated, or transferred are called data in motion. How data are stored and how they are transmitted have enormous implications with respect to cybersecurity. Data are typically collected in groups, which can range from simple files to complex databases. A data file or table may be structured, with the ability to easily access a particular piece of data, such as a customer record, based on some unique character. Unstructured data files are not organized in any particular way and are more difficult to access.

EXAMPLE

A user can access the customer table shown in Table 2-1 by several unique characteristics. However, assigning a unique customer number as a primary key (PK) allows all other data to be tied distinctly to that customer. That includes inventory items purchased and billing history.

Table 2-1. Example of a Customer Table

Customer ID (PK)	Customer Name	Customer Address	Customer Email	Customer Phone
12345	Ace Company	1 Broad Ave.	aceco@gmail.com	7376162400
12346	Ace Corp	2 Main St.	acecorp@gmail.com	8062314000
12347	Ames Company	3 High Blvd.	ames@aol.com	1234567000

A **database** is a structured set of data consisting of rows and columns, as in the data file in Table 2-1. The data are logically connected and may be accessed by a unique identifier (e.g., the customer identification number). A **database schema** describes how data within a database are organized and stored. The three most common logical database schemas are **star, snowflake,** and **relational**. Their physical representations are as their names imply, and all three are useful for organizing large data sets and making them available for business intelligence and data analytics.

The star configuration will have a central fact table that links to multiple dimension tables. For example, an order table would be the central fact table and include the order number, customer number, inventory number, and so on. It would link to other files, including the customer and inventory tables. The snowflake configuration is more complex, with multiple layers of dimension tables. While this schema can reduce data redundancy and provide faster data retrieval, it is also more difficult to create and maintain. A relational schema uses the relationships between different data sets to form a collective whole. For example, Customers would be related to a Salesperson through a third table called SalesTransaction. In this example, some of the relationships themselves have attributes directly related to each of the data sets. Relational databases are best designed through the process of normalization.

Data normalization is a process that organizes data so that each piece of data (e.g., customer name or customer address) is stored only once. The objective is to improve data integrity by reducing data redundancy. Data are represented in tables such as the customer table in the example above and linked to other tables with primary and foreign keys, which reference primary keys in other tables. In our example, the Customer ID is the primary key in the customer table, and it can be a foreign key in an inventory table.

Structured Query Language (SQL) is a programming language used to access and manage data in a structured database. The basic SQL data-retrieval statement is made up of three commands: SELECT, FROM, and WHERE, as shown in Table 2-2. The table also describes other commonly used SQL statements or commands. We discuss SQL injections, where hackers are able to use SQL commands to access target databases, in Chapters 3 and 4.

Table 2-2. Common SQL Commands

SQL Command	Description
SELECT	Retrieve data from one or more tables.
FROM	Identifies the table(s) from which you are selecting data.
WHERE	Gives one or more criteria for selecting data.
DELETE	Remove rows from a table.
INSERT	Add rows to a table.
JOIN	Combine rows from two or more tables, using related columns.
UPDATE	Modify data within a table.

Data Protection

Regardless of database configuration, cybersecurity is concerned with **data loss prevention (DLP)**, which involves using tools and processes that prevent sensitive data from unauthorized access and loss or misuse. A **data dictionary** is valuable in DLP for identifying and classifying data within a database. The data dictionary is the repository for **metadata**, which are data about the data. For example, Customer ID metadata might include the following: data type (e.g., integer or string), number of characters, data owner, data storage location, date created, and date of last update.

The **CIA triad** emerged several decades ago as an information security model for data protection. The term has nothing to do with the Central Intelligence Agency but rather the "C" represents confidentiality, the "I" stands for data integrity, and the "A" is for availability. **Confidentiality** means that the data can only be accessed or changed by authorized parties. People often confuse confidentiality with privacy, but the two are a bit different. *Private* generally means

Star Schema A type of database schema that organizes data into a central fact table connected to multiple dimension tables.

Snowflake Schema A type of database schema that organizes data into multiple related tables to reduce redundancy.

Relational Schema The formal structure of a relational database, describing how data is organized into tables (relations) and how those tables relate to each other.

Data Normalization The process of organizing data in a relational database to reduce redundancy and improve data integrity by dividing large tables into smaller, related ones and defining relationships among them.

Structured Query Language (SQL) A standard programming language used to create, read, update, and delete data in relational databases.

Data Loss Prevention (DLP) A set of tools, policies, and processes designed to detect, prevent, and respond to the unauthorized access, transmission, or leakage of sensitive data.

Data Dictionary A centralized repository that contains definitions, descriptions, and metadata about the data elements in a database or information system.

Metadata Structured information that describes the characteristics, origin, structure, and usage of data.

CIA Triad A model designed to guide policies for information security within an organization, consisting of three key principles: confidentiality, integrity, and availability.

Confidentiality The assurance that sensitive data is not disclosed to unauthorized parties, whether intentionally or accidentally.

Integrity The assurance of accuracy, consistency, and trustworthiness of data and systems throughout their life cycle. It ensures that information is not altered, deleted, or corrupted—whether accidentally or maliciously—without proper authorization or detection.

Availability The assurance that systems and information will be operational and accessible within the required time frame, even during disruptions or high demand.

that the data belong to an individual or a group of individuals, whereas *confidential* means that the data are to be kept secret or inaccessible to unauthorized users. This distinction is discussed in detail in Chapter 11.

Data **integrity** means that data are complete and accurate. Data has integrity when it has not been compromised/changed either accidentally or purposefully. Physical integrity requires protecting the data while stored and in motion. Logical data integrity ensures that the data structure and relationships are intact.

Data **availability** implies that the data are available on demand and as needed. Many actions and environmental factors can impact availability, including power outages, system maintenance, and ransomware attacks. Data and information systems need to be continuously available to users. System availability is the period during which users can access information systems services and data. The period when they are available is called uptime; the period when they are not accessible is downtime.

The CIA triad is fundamental to any information security infrastructure and is the basis for most security architectures. The objective to keep data confidential, protect its integrity, and make it available is a core principle of an information security program.

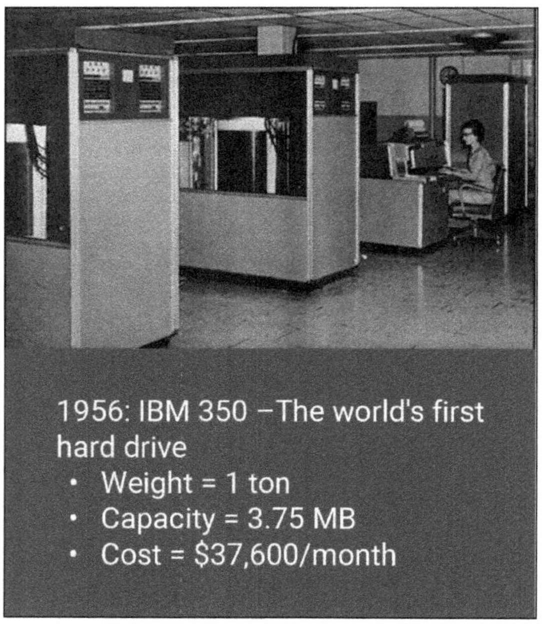

Figure 2-1. The IBM 350 Hard Drive Wikimedia Commons

Data Storage Options

Data may be stored on-premise using different storage media or in the "cloud." On-site storage can be on a computer, the hard drive of a computer (an HDD), a peripheral device such as a **Universal Serial Bus (USB)** device, or a drive on a local area network (LAN). When data are stored on-site with no connection to a wide network, such as the Internet, it is easier to protect them. Highly sensitive or critical data may be **air gapped.** This approach typically means that the data are not available to any online computer network. The data might be physically stored on a hard drive, tape, USB, or CD, but they are physically isolated from any public network connection. These data may be transferred via a portable physical device, from one computer to another, or through a private network that is not connected to any public network. Apart from a physical air gap, data may also be logically air gapped by using encryption or hashing security techniques.

Today, rather than storing data on-premise, many organizations look to the cloud. The "cloud" is simply space that you are renting on someone else's server. If you were to compare current organizations' data centers with data centers from a decade ago, you would see a tremendous shift from many servers storing software and data to a single or very few servers. Cloud providers now host the servers, as the following case illustrates.

Universal Serial Bus (USB) A standard for connecting peripherals such as keyboards, mice, and external drives to computers.

Air Gap A security measure in which a computer or network is physically isolated from other networks, especially the Internet.

CASE IN POINT

Loudon County, Virginia, has the nickname "Data Center Alley" because it hosts the world's largest concentration of data centers. It was estimated in 2025 that about 70% of the world's Internet data flows through this northern Virginia region just outside the US capital. Driving through the area, you would find Amazon, Google, Microsoft, and Oracle, among others, having a presence. While many factors have contributed over time to this concentration of data centers, the area not being prone to natural disasters and the evolution of IT infrastructure in the area (e.g., fiber optic connectivity) are two important ones.

The cloud has a significant impact on cybersecurity, both positive and negative. A loss of control over security may or may not be outweighed by the security that a cloud vendor offers. Different cloud deployment models each include trade-offs in multiple dimensions.

Cloud Deployment Models

Cloud deployment is another name for cloud storage. There are several different types of cloud storage, each with advantages and disadvantages in terms of capacity, cost, and security. Data *in the cloud* are stored in a data lake, data warehouse, or data mart. A **data lake** contains raw, unstructured data, making it easy to store. Data stored in this way may be accessed and used for a variety of purposes. The data lake may consist of an entire entity's store of data. **Data warehouses** maintain the data in a predefined structured format, which allows for more limited data analysis. The data within a data warehouse is easily accessed for business analysis. A **data mart** is a type of data warehouse that has been created for a specific purpose or business unit.

An entity may also choose a public, private, or hybrid cloud storage model. Figure 2-2 summarizes the alternatives.

Cloud Deployment Model How a cloud computing solution is delivered to the end user and how that end user can access the resources and tools provided.

Data Lake A storage repository that holds a vast amount of raw data in its native format until it is needed.

Data Warehouse A system used for reporting and data analysis, storing current and historical data from multiple sources.

Data Mart A subset of a data warehouse, often oriented to a specific business line or team.

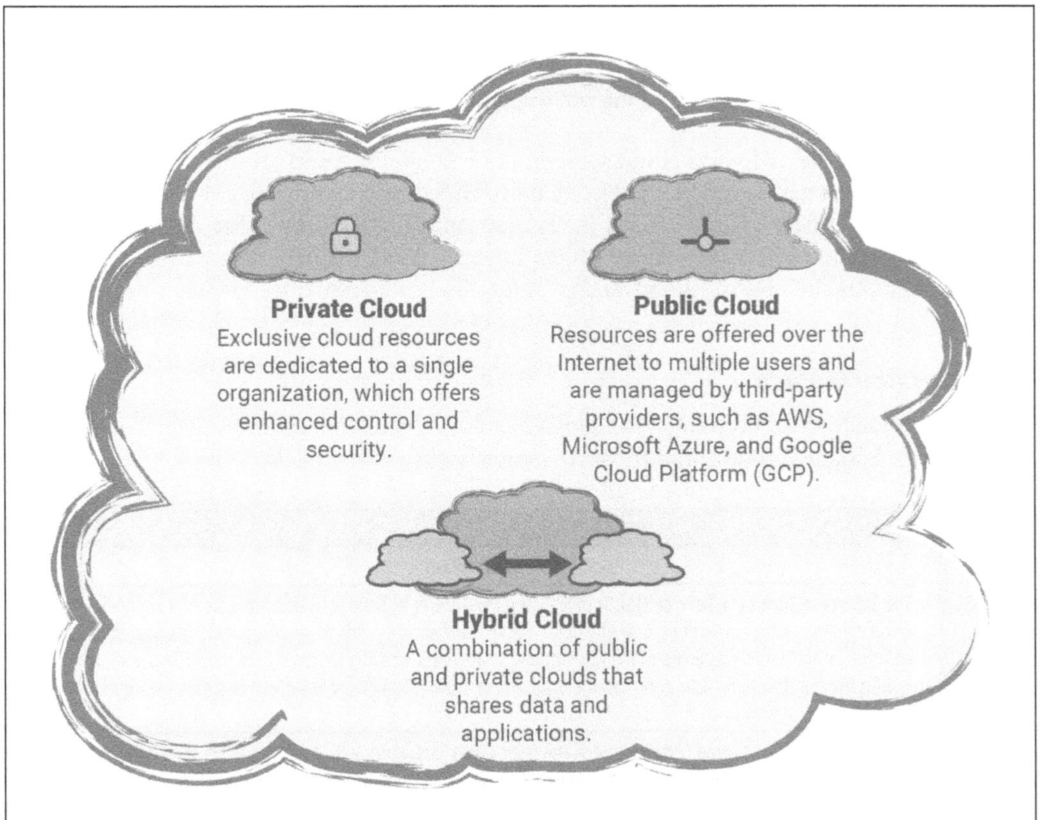

Figure 2-2. Cloud Deployment Options

Public cloud services are off-site storage maintained by a third-party provider. Proprietary data are stored on remote servers owned by a service entity. In a public cloud, data owners can retrieve and manipulate their data just as they would if the data were stored locally but without a need to physically own the computers storing the data. Some of the advantages of public cloud storage are that costs are lower, the cloud company provides data maintenance, and capacity is virtually

Public Cloud Cloud services offered over the public Internet and available to anyone.

limitless. Public clouds can offer more security than possible with local data storage. While cloud storage is not perfect with respect to security, well-known public cloud service providers make use of highly sophisticated security tools to ensure the safety of the data they maintain, as their business model depends on it. As necessary with any third-party service provider, users should conduct due diligence when selecting a public cloud service vendor.

Private Cloud Cloud infrastructure operated solely for a single organization.

A **private cloud** is more expensive but has the potential to be more secure than a public cloud. It costs more because cloud services are proprietary rather than shared. The information technology infrastructure is contained within the data owner's data center, and only that entity's users have access to the data stored on it. The data owner maintains control over the data and their security, rather than outsourcing that responsibility to a service provider.

CASE IN POINT

It is difficult to imagine any organizational entity that needs more security over its data than the US Central Intelligence Agency (CIA). In 2014, the CIA contracted with AWS for a private cloud that would host and manage intelligence data from multiple government agencies. The cloud is hosted on private intelligence community premises, and security agencies including the CIA and the National Security Agency can order on-demand computing and analytic services at lower cost. As Frank Konkel described the approach in *The Atlantic,* "Think of it as the intelligence community sharing information behind a walled castle apart from the rest of the world operating on the Internet" (Konkel, 2014).

Hybrid Cloud A combination of private and public clouds.

As Figure 2-2 notes, a **hybrid cloud** solution is a mix of public and private cloud services. The hybrid environment makes use of a public cloud provided by a commercial cloud services entity and one or more private clouds housed in the user entity's data center. The public and private clouds share data and software applications. In a hybrid cloud, the data owner maintains control over security but is also able to take advantage of the computing power that public cloud providers offer.

Cloud Service Models

In addition to multiple options for cloud deployment or storage, choices are available for the cloud services offered. Table 2-3 describes the three basic cloud service models.

Table 2-3. Cloud Service Model Options

Cloud Service Model	Description
Infrastructure as a Service (IaaS)	In the **Infrastructure as a Service (IaaS)** model, the third-party service provider purchases the servers, software, networking equipment, and physical space. User entities rent the service for a fee and access the resources on demand. There are many advantages to the user entities in this model, including scalability and forgoing the need to purchase and manage computing infrastructure themselves. The user entity pays only for the services needed.
Platform as a Service (PaaS)	In the **Platform as a Service (PaaS)** model, the user entity contracts with the third party for a software development platform. The user entity writes the code and is responsible for managing applications and data but has access to tools, code libraries, and development environments. The provider maintains and manages the platform for developing the applications.
Software as a Service (SaaS)	In the **Software as a Service (SaaS)** model, the third-party service company provides the software application and the infrastructure that hosts it. The user entity simply accesses the software as needed. Both institutions and individuals make use of SaaS, often daily. Shopify, Dropbox, and Zoom are all examples of SaaS providers.

CASE IN POINT

The Coca-Cola Company used the PaaS AWS Lambda to quickly develop a consumer touchless fountain service. This new Coca-Cola product was created during the pandemic so that consumers could dispense their drinks from their phones, thus avoiding having to

> **touch a dispenser. Coca-Cola already had its infrastructure hosted by AWS, and Lambda allowed the company to build the application in real time. With AWS Lambda, users run their own code for all types of applications and backend services rather than acquiring and managing their own servers (Lewis, 2022).**

Cloud service models are popular because they offer significant cost savings, scalability, and reliability. All cloud service models require users to rely on a third party's cybersecurity to some extent, which can be a benefit. However, the following case demonstrates that while SaaS companies invest heavily in cybersecurity, they can be the target of hackers and are not invulnerable to attacks.

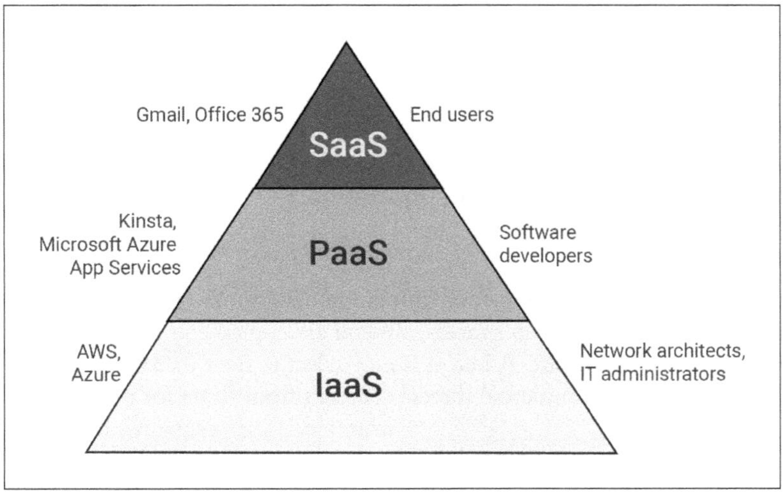

Figure 2-3. Cloud Service Models

> **CASE IN POINT**
>
> **Dropbox is a popular SaaS file-sharing company. In spring 2024, Dropbox became aware of unauthorized access to Dropbox Sign, its e-signature service offering. Hackers were able to access the production platform and customer database, including emails, usernames, hashed passwords, and Multifactor Authentication (MFA) information. Unfortunately, individuals who had not created a Dropbox Sign account, but had used the platform to sign a document, appeared to have had their data compromised along with account holders.**

Many **Enterprise Resource Planning (ERP)** programs offer applications as SaaS. ERP programs contain modules that cover areas such as accounting, manufacturing, and human resources and allow businesses to automate their core activities. SAP is perhaps the largest ERP vendor offering SaaS applications over the Internet.

Enterprise Resource Planning (ERP) Integrated software systems that help manage core business processes like finance, HR, and supply chain operations.

ERP systems offer many benefits to businesses because of the integration of functions. They are efficient in integrating data across various activity areas, as described in the following example.

> **EXAMPLE**
>
> **Upon receipt of a sales order from a customer, the ERP system automatically calculates the total price, including discounts and taxes, and prepares an invoice. At the same time, inventory levels are updated. The marketing function will be able to track the sale and aggregate sales for each product line. The accounting function will receive the sales data and use it in budgeting and financial reporting.**

Accounting Information System (AIS) A structure that an organization uses to collect, store, manage, process, retrieve, and report its financial data.

While an ERP system includes financial/accounting modules, a specialized **accounting information system (AIS)**, such as QuickBooks or NetSuite, is designed specifically for providing financial management and reporting. These software applications are often offered in a SaaS model. Oracle NetSuite is an AIS SaaS platform where customers may access the software for a subscription fee.

Data Backup

One of the most important information security measures is data backup. Simply maintaining data in a file either on-site or in the cloud is not sufficient for the file to be secure. To protect data, files should be stored in more than one place. Consider your own data. Do you keep an important research paper or a gradebook just on your computer hard drive, or do you store it on a network as well? The latter would protect you if your computer drive were corrupted, and the former would protect you if the network data were encrypted or exfiltrated in a data breach. Individuals and organizations need to establish policies and procedures regarding data copying and copies. Data that have no backup are at risk from environmental factors, accidents, and even malicious actions. Most of us know someone who lost important information because they failed to back up a laptop or the information resources on a smartphone. Use of the cloud for data storage provides more opportunities for backup because it expands capacity.

Full Backup The most basic and comprehensive backup method, where a duplicate of all data is sent to another location for storage.

Incremental Backup A process that backs up all files that have changed since the last backup occurred.

Differential Backup A process that backs up only the files that have changed since the last full backup.

Backup Types

A data backup may be full, incremental, or differential (Figure 2-4). A **full backup** is the most comprehensive. During a full backup process, all data are copied and stored in another location. That location may be on- or off-site. A business may select to have its data backed up at a remote location that is subject to environmental threats different from those for the on-site location. For example, if the company's main data center is in a flood zone or area threatened by hurricanes, they may choose a location where those threats are not present. Full backups are certainly thorough, but they can be time consuming and use a lot of storage capacity.

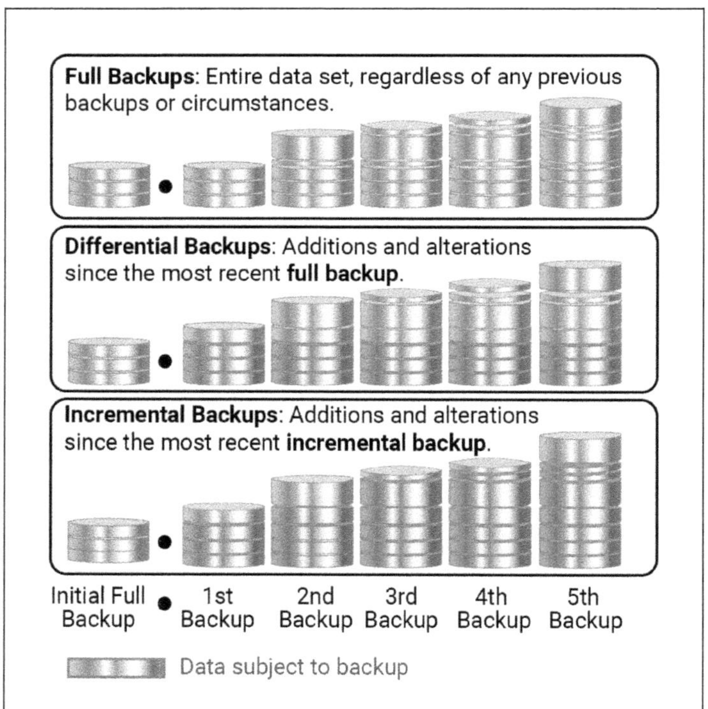

Figure 2-4. The Three Backup Types (Wallen, 2020)

Incremental backups are backups of data that have changed since the last backup, regardless of whether that process was full, incremental, or differential. The advantage of this type of backup is that it is faster than a full backup and requires less storage capacity. However, in the event that data must be restored, that process will require retrieving the last full backup along with all subsequent incremental backups.

A **differential backup** consists of all data that were changed since the last full backup. You may have a series of incremental backups, but the differential backup would ignore them. For example, if you ran a full backup on Monday and then performed an incremental backup on Tuesday and Wednesday, the differential backup on Thursday would include all of the Tuesday and Wednesday data changes, whereas an incremental backup on Thursday would only store the data changed since Wednesday. An advantage of a differential backup is that if data need to be restored, you only need a full backup and one differential backup rather than a full backup with multiple incremental data restorations.

There are many policy decisions to be made about the backup of data. They concern timing or frequency, type of backup, and location of backup data. They might also cover testing the backup system to ensure that data in backup storage are recoverable. In addition, there should be clear policies about the disposal of data that are no longer needed. Such policies should include data retention and data destruction. There are legal and regulatory requirements for some data to be kept, and in other cases it may be important to dispose of unnecessary data when it is no longer used or when the data owner is no longer employed. For example, when an employee leaves an organization, whether voluntarily or involuntarily, some information about that individual may need to be kept, and some may not. An example would be retention of the employee's email messages.

Backup Media, Location, and Method

With respect to the location of backup data, a number of different configurations are possible for an organization. Storing data backups on-site is convenient, but if there is a problem at the site, the backups are useless. Storing data in physical media off-site can be cumbersome if you need to restore your backups, as you would need to have the backups delivered to the main data center. Utilizing cloud services for backups is convenient, but the organization has less control over the architecture and incurs an ongoing expense to maintain the backups.

When you choose the media for storing your backups, you will have a number of pros and cons to consider. Table 2-4 shows the benefits and drawbacks of different **storage media** types for backup storage.

Storage Media The physical materials or devices used to store digital data in a computer system. It allows data to be saved, retrieved, and transferred.

Table 2-4. Backup Storage Media (Garry, 2023)

Media Type	Benefits	Drawbacks
Tape Cartridges	• Mature tech • High capacity • Long life span • Portable	• Sequential access • Slow transfer speeds • Data integrity risks
External HDDs	• Fast access • Simple USB interface	• Individual drive failure risk • Capacity limits
Removable Disks	• High capacity • Random access	• Higher cost • Bulkier than tapes
Cloud Storage	• Highly scalable • Managed service	• Monthly costs • Limited control

Storage of backups, regardless of the media, needs to be both convenient and secure. While storing the data at the main facility is convenient, if the data center has a catastrophic failure (fire, flood, etc.), the backups will be useless/destroyed with the data center. The **data replication method**/configuration, assuming off-site storage of backups, comes with its own benefits and drawbacks, as shown in Table 2-5.

Data Replication Method The technique used to copy and maintain data in multiple locations to ensure consistency, availability, fault tolerance, or performance across systems or networks.

Table 2-5. Off-Site Data Replication Options (Garry, 2023)

Replication Method	Benefits	Drawbacks
Mirrored Data Center	• Continuous real-time updating • Fast Recovery Time Objective (RTO)/ Recovery Point Objective (RPO)	• Higher cost • Dependent on Internet/Wide Area Network (WAN) link
Storage Area Network (SAN)/Network Attached Storage (NAS) Disk Replication	• Scheduled incremental updates • Provides versioning	• Storage space requirements
Secure File Transfer Protocol (FTP) Transfer	• Simple to implement	• Only batch updating • Bandwidth impact
External Drive/ Media Rotation	• Air gapped protection • Unlimited capacity	• Manual process • Slow recovery time

Business Continuity Planning

Business Continuity Planning (BCP) The creation of a comprehensive framework to maintain or quickly resume critical business functions in the event of a crisis, such as a natural disaster, cyberattack, pandemic, or system failure.

Hot Site An alternate business location where the target environment is already up and running and can be immediately activated by a failover.

Cold Site An alternate business location where the contracted facility has no active equipment, requiring extensive setup before use.

Warm Site An alternate business location where the target environment has some active equipment, requiring minimal setup before use.

Failover The amount of time it takes to switch from production systems to backup systems.

Regardless of the backup method or storage option, organizations also need to plan for various contingencies, some of which deal with the possibility that their data center will not be usable. **Business continuity planning** (BCP) efforts should plan for contingencies in which data, processing, and/or work will need to be done at a separate site (hopefully) beyond the range of any possible disaster that could impact the organization. To facilitate this, organizations make use of hot, cold, or warm sites depending on their needs. A **hot site** is a fully operational data center that can provide immediate recovery from a cyber incident or a natural disaster. When an entity has most of its data in the cloud, the hot site provides an immediate physical backup location that can deploy critical applications. The site will include preconfigured servers and network equipment that can connect to the cloud data. A **cold site** is less extensive and less expensive to maintain. This is a data center with power and basic connectivity, but it needs to be set up prior to use. If a disaster disables the primary data center, an organization can restore data from its cloud storage, although at a slower pace than would be the case with a hot site. A **warm site** is a hybrid model, a partially equipped backup facility that includes basic infrastructure and some data resources. Figure 2-5 summarizes the features of these backup facility options. Note that **failover** is the time it takes to switch to backup systems.

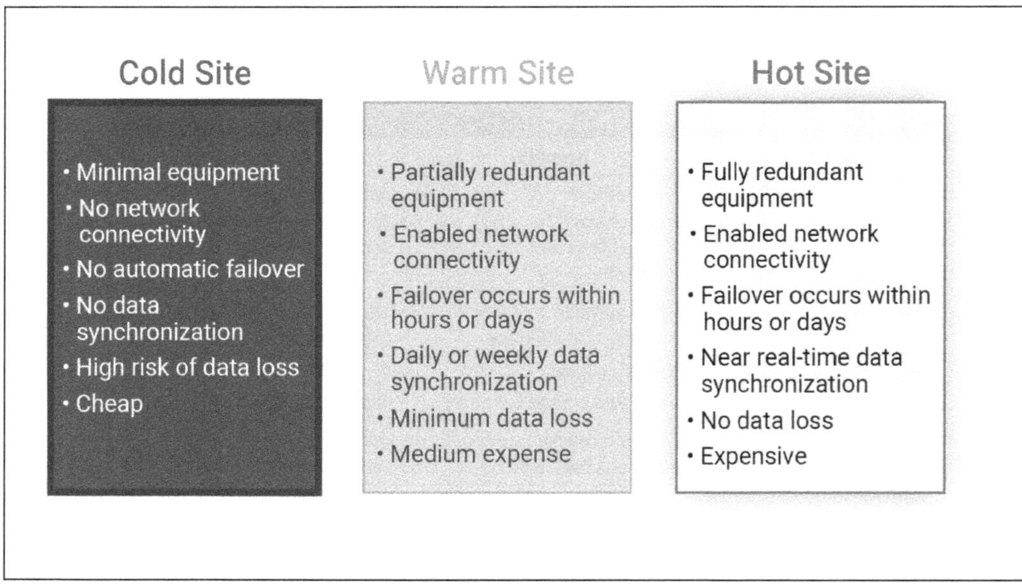

Figure 2-5. Backup Facility Options for Business Continuity

Data Replication The process of having data in multiple places that are continually synced to enhance access.

Data Mirroring The process of copying data from one location to another in real time to ensure data redundancy and high availability.

Also note that there is a difference between data replication and data backup. **Data replication** means that the user data is in several places at the same time and is synchronized using some type of software method. An example of this is Dropbox or Microsoft OneDrive, where copies of the same file are stored on (possibly) multiple devices as well as on the main Dropbox or OneDrive servers. Data (or system) backup means that the data or system is replicated to a storage device and retained in that format until deleted in accordance with company policy. A backup is a fixed point in time that you can use in case the existing systems are corrupted. **Data mirroring** is a type of data replication where the user creates an exact duplicate of data from one device on another device in real time. Both data replication and mirroring create data redundancy and ensure that data have integrity and are available as needed. These are important enhancements to system reliability and are useful in disaster recovery situations.

The link of data backup and BCP to cybersecurity is that when the cybersecurity attack occurs, backups and BCP provide options to continue business while dealing with the cyberattack. Without current backups, a ransomware attack paralyzes the organization. Nothing can be done until the ransom is paid. Without alternative sites, a site that is compromised will have to go dark until the vulnerability is addressed and it is safe to resume business. Researchers at the National Archives estimate that 93% of businesses that have lost their data center operations for more than 10 days file for bankruptcy within one year (Hartman, 2012). With the goal of maintaining the organization, cybersecurity efforts must include a data backup aspect to be successful.

Components of IT Infrastructure

Data in use are in motion over a network of computing devices. A network has four basic elements:

- Hardware
- Software
- Protocols
- Transmission media

Network Hardware

Computer networks connect multiple computers with one another. These computers consist of **clients** and **servers.** The servers are powerful computers located inside or outside a network that act as hubs to connect client computers and provide them with services. The servers may be generally purposed or dedicated to specific functions.

> **EXAMPLE**
> **Google search is a classic example of client-server architecture. The client (any user device) sends a request to the server (in this case a Google server running the search configuration) requesting cat videos. The server uses the Google search algorithm to locate cat videos and returns them to the client for their review. This architecture is the core of Google's services, as evidenced by the estimated 1 million servers exclusively running the search engine.**

File servers, web servers, database servers, and print servers are all examples of dedicated servers. Client computing devices, such as laptops, desktop computers, and printers, interact with the server to run applications and access data.

> **EXAMPLE**
> **A web server may consist of both hardware and software. The hardware is the computer that stores the web server software and website files. The server accepts requests from clients that access it via hypertext transfer protocol (HTTP), the set of communication rules that a web browser (e.g., Google Chrome or Apple Safari) uses to retrieve data and files hosted at the server website. Web servers also support email and file transfers.**

Networks may be local (**local area network [LAN]**), which means they are confined to one physical site, such as a building. A **wide area network (WAN)** is one that spans multiple physical spaces. A WAN may connect multiple LANs. The Internet is an extreme or ultimate example of a WAN.

The way computers are positioned in networks influences the network cost and the data flow. There are many possible **network topologies**, or the way that links and nodes are arranged. Three common configurations are **star**, **bus**, and **ring**. The most common network topology is the star configuration, where each network device is connected to a central hub. The advantage of the star topology from a security standpoint is that any connection failure will impact only one computer or endpoint rather than the entire network. A bus topology has one main line or cable, and each

Network Client A software or hardware component that accesses a service made available by a server.

Server A system that responds to requests from clients by delivering specific functions, such as file storage, websites, applications, or databases.

Local Area Network (LAN) A network that connects computers within a limited area, such as a home, school, or office building.

Wide Area Network (WAN) A network that spans a large geographical area, connecting multiple local area networks (LANs) across cities, regions, or countries.

Network Topology The arrangement of network elements, including nodes and connecting lines, that determines the structure and performance of a network.

Star Network Topology All devices connect to a central hub or switch.

Bus Network Topology All devices share a single communication line (backbone).

Ring Network Topology Devices form a circular chain; data travels in one or both directions.

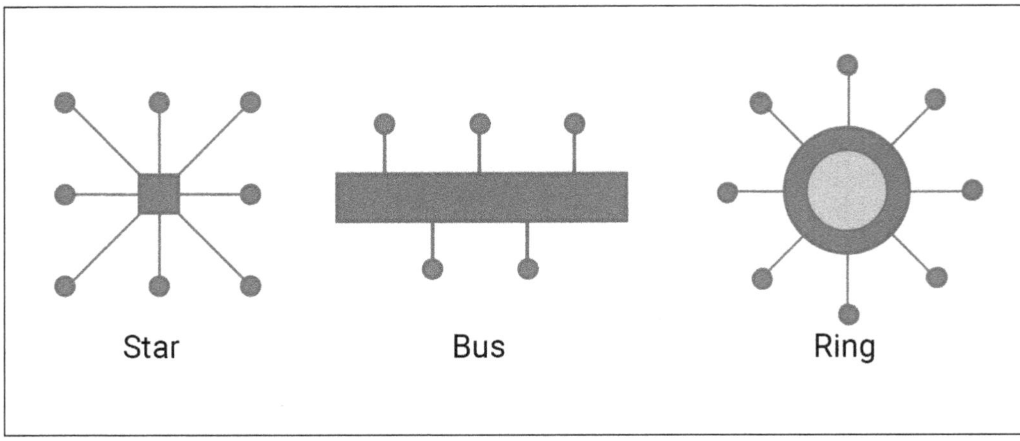

Figure 2-6. Network Topologies

device on the network has a separate connection to it. You might think of a bus line with many stations. This configuration can be cost-effective for small networks, but it does not work well for a network with many devices because data moving along the line will stop at each station until it reaches its destination. In a ring network, each device connects to another on either side of it, forming a circular data path. As data moves along the ring, it goes in one direction along the pathway. A ring topology provides more efficient data transfer than a bus configuration, but if any one path fails, the entire network goes down.

In addition to the computers themselves, many other pieces of network hardware allow clients and servers to communicate with each other. Table 2-6 describes each of them.

Table 2-6. Network Hardware Components

Network Hardware	Description
Bridge	A **bridge** is a hardware device that links multiple networks into a single aggregated network. The bridge may, for example, connect two or more LANs. This allows the separate networks to exchange data and messages and work as though they are one interconnected network. Bridged networks are closed loops in the sense that data can only be exchanged among the networks connected by the bridges.
Endpoints or End-User Devices	An **endpoint**, or end-user device, is a remote hardware device that communicates across the network that connects it with other hardware, software, and data repositories. It is so named because it is physically an endpoint on a network. Examples of endpoint hardware devices are laptops, tablets, desktop computers, smartphones, and servers. Endpoints are important in cybersecurity, as they are difficult to secure and therefore frequently subject to attacks. Hackers may attack an endpoint device to access a protected network.
Hub	A network **hub** is a simple connector that works best within small, relatively simple Ethernet LANs. Ethernet refers to a common type of wired network. Hubs are less used today than previously because of their limitations. They are less intelligent and have fewer capabilities than switches and routers.
Modem	A modulator-demodulator, or **modem**, is another piece of hardware needed for network connectivity. The primary purpose of a modem is to serve as a bridge between an Internet Service Provider (ISP) and the user. It is typically a small device with several lights and open ports. Modern cable or fiber optic digital modems convert electromagnetic waves into a digital format. This translation is necessary for an internal network to connect to the Internet.
Network Interface Card (NIC)	For computers to connect to a network and receive information packets, they need a **network interface card** (also called a network adapter). These circuit boards provide the client or server with a dedicated network connection for wired or wireless communications. These communications may be within a LAN or over remote networks using Internet protocols. A network interface card may be internally installed on the computer motherboard, which contains a slot for its insertion. If a laptop or other computing device does not have an internal card, an external wireless network card can be inserted into a USB port. These external cards can provide a quick connection and are portable.

Table 2-6. Network Hardware Components (continued)

Network Hardware	Description
Router	A **router** is an intelligent device that can forward data outside its connected network clients and servers. Routers usually have a WAN interface in addition to the LAN interface(s) and are more useful than bridges when joining remote networks. Routers may be used for both wired and wireless connections, with the connections made via IP addresses (discussed later in this chapter). Individuals with home networks are familiar with routers, as they are the devices used to connect computing devices within a home to the Internet. A wireless router in your home usually has a cable connection to a modem (some routers are combined with a modem in a single unit), allowing it to exchange information with the Internet. The router lets multiple devices and users connect to the network, and the modem provides Internet access. Antennas let the router communicate with your home's Wi-Fi network to give all the endpoint devices in your home (e.g., television, security system, laptop computers) Internet access via the modem.
Switch	Unlike bridges and routers, **switches** connect devices within a network, and they rely only on wired connections. The primary role of a switch is to connect the endpoint devices (computers, printers, workstations) with each other to form a network. These devices make use of packet switching to exchange data among devices. Switches are more sophisticated than hubs, as they can exchange data with designated devices, whereas hubs will direct data to all the connected devices.
Workstation	A **workstation** is a computer that is generally owned and/or operated by a single user, perhaps to perform a specialized task. Workstations have greater processing capabilities than client computers because of their interaction with the network, storage capacity, and graphics processing capability. Workstations are useful for performing graphic design or complex computations needed for engineering or science functions.

Figure 2-7 illustrates the main network hardware and how the components work together through an ISP to deliver Internet resources to endpoint devices.

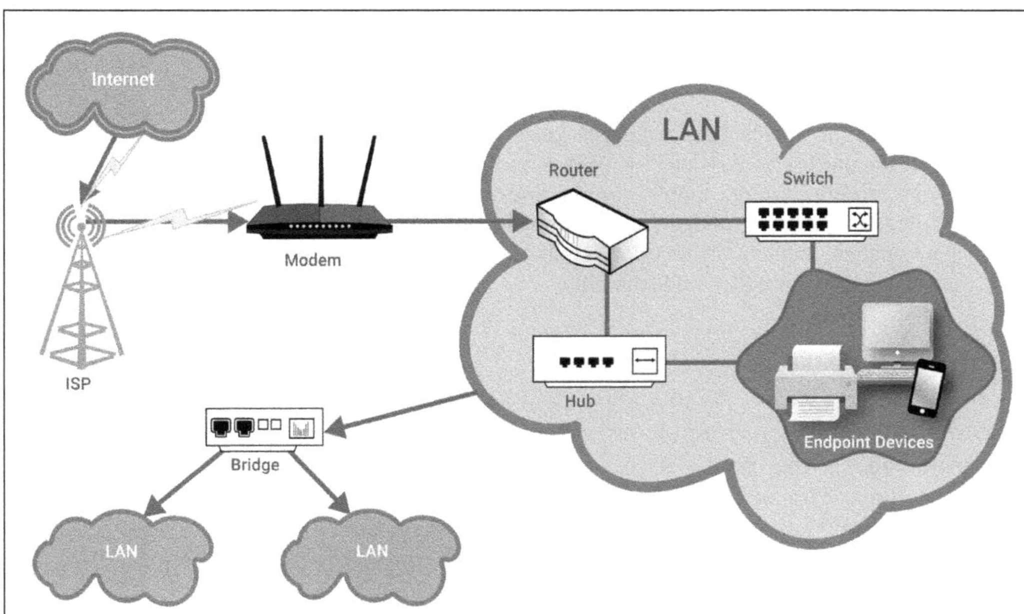

Figure 2-7. Network Hardware

Network Software

Network software is a broad term for all the programs that make network applications work securely. This includes various types of security software, such as asset management and patch management software, and these types of software will be discussed in subsequent chapters. This chapter will focus only on network management software and network operating systems.

Network Management Software

Network Management Software An application or set of applications that lets network engineers manage a network's independent components.

Network management software is a broad array of software tools that do just what the term says—they assist in the management of networks. The software accomplishes this through an array of functions that include administration, maintenance, operation, and provisioning. Network administration allows users to manage the network hardware, software, and transmission media. This administrative function also monitors the performance of network software and makes updates as needed. The network maintenance function maintains, updates, and repairs network resources on an ongoing basis. The network operation application monitors ongoing network activities and resolves performance issues as they occur. Provisioning entails the configuration of network resources to add devices as needed into the network environment to support network services, including adding more users and extending the network. Network management software interacts with network administrators to maintain network performance and security, though many functions may be handled automatically.

Network Operating Systems

Operating System The software that manages a computer's hardware and software resources, providing services to applications and users.

Network Operating System Software that manages network resources and allows communication between computers and devices on a network.

Every computing device has an **operating system**. Operating systems, such as Microsoft Windows, Apple macOS, or UNIX, manage software programs, applications, and the hardware functionality of endpoint devices. Some of the basic functions of an operating system are booting up the system (powering it on), managing memory, executing programs, managing storage, and providing security features. Operating systems have a user interface that interacts with the user as they enter and retrieve information.

Network operating systems are the specialized software that allow clients and servers on the network to run and communicate with other clients to share files, applications, databases, printers, and other network resources. Examples of network operating systems are UNIX, Linux, macOS Server, and Windows.

> **CASE IN POINT**
>
> **Linux is an open source (free!) network operating system developed by Linus Torvalds at the University of Helsinki in 1991. It can be installed on a variety of computing devices, including tablets, mobile devices, and desktop computers. Linux can be used as the operating system for many of the hardware devices in the network, including routers and switches. Some of the functions Linux can provide are routing, bridging, networking, monitoring, and network troubleshooting.**

Protocols

Network Protocols Rules and conventions for communication between network devices.

A protocol is a set of rules. As children we had a bedtime protocol that included brushing our teeth, washing our face, picking up our dirty laundry, and going to bed at a certain time. **Network protocols** are the specific set of rules that determine how computers (or any endpoint devices) communicate with each other within a network. Computing devices must speak a common language to communicate, and network protocols establish that language. This common language lets computing devices talk to each other regardless of their differences in structure, processes, hardware, or design. Protocols can have features that provide both security and efficiency, and network connectivity cannot exist without these common rules. There are many different types of protocols, and Table 2-7 describes the most common protocols that have implications for cybersecurity.

Table 2-7. Network Protocols

Protocol	Description
Ethernet	**Ethernet** was defined earlier as a simple form of wired network, since within an Ethernet LAN all the devices are hardwired or attached via cables. However, Ethernet is also a standard communication protocol within a network.
File Transfer Protocol (FTP)	**File Transfer Protocol (FTP)** defines the rules for transferring files between/among servers in a client-server architecture network. Network users authenticate with a username and password to sign in to download, upload, or transfer files. Due to security concerns and the movement to cloud file storage, FTP is being phased out and is no longer supported by many web browsers, including Google Chrome.
Hypertext Transmission Protocol (HTTP) and Hypertext Transmission Protocol Secure (HTTPS)	Like FTP, **Hypertext Transmission Protocol (HTTP)** relies on a client-server architecture for transferring files over a network, in this case, the World Wide Web. HTTP is the protocol that allows a web browser to retrieve Hypertext Markup Language (HTML) resources, including text, images, and videos. **Hypertext Transmission Protocol Secure (HTTPS)** uses Transport Layer Security (TLS) to encrypt communications, providing more secure network communications. Originally, HTTPS was used primarily for websites that collected particularly sensitive data (e.g., financial institutions), but the trend is toward all websites using HTTPS to prevent bad actors from accessing network communications.
Internet Protocol (IP)	Each computing device that connects to the Internet has an IP address associated with it. **Internet Protocol (IP)** provides the rules for formatting network communication sender and recipient addresses. IP addresses have an important role in cybersecurity for two reasons. One is that if hackers know your IP address, they might use it to obtain information about where you are and who you are. This information could allow an adversary to send you unwanted messages or restrict some access rights. IP address knowledge works two ways, though: the device a hacker uses might also be identified by its IP address, alerting law enforcement about the identity and location of a bad actor. Internet addresses are unique strings of numbers. Since 1981, IP version 4 has been in use. IPv4 uses a 32-bit numerical string for Internet addresses consisting of four sets of numbers ranging from 0–255, as shown in Figure 2-8. An example would be 214.120.6.15. As more and more devices are connected across the Internet, we are running out of these addresses. IPv6, which uses 128-bit addressing and has eight segments, is replacing the older version. Since this system will accommodate about 340 trillion trillion trillion different unique IP addresses, it should last awhile! You can also see an example and explanation of an IPv6 address in Figure 2-8.
Simple Mail Transfer Protocol (SMTP)	Dedicated mail servers use the **Simple Mail Transfer Protocol (SMTP)** application for sending and receiving email messages. The application is an Internet standard communication protocol that defines the rules for electronic mail communications. The rules in SMTP concern delivery and receipt. For example, a protocol might be defined that specifies the frequency with which a delivery service attempts to send undelivered email before signaling the sender that the message cannot be sent to that address. You can easily locate your own SMTP email server address, usually by accessing the settings for your email account.
Transmission Control Protocol (TCP)	**Transmission Control Protocol (TCP)** is a primary protocol within the suite of Internet protocols. While it is a separate network protocol from the Internet Protocol (IP), it is complementary, and the suite is commonly called TCP/IP. As described previously, the IP portion of the protocol is responsible for addressing. The TCP portion of the suite controls the delivery of data once the IP address has been identified. TCP rules enable all devices connected to the Internet to communicate with each other. TCP contrasts with HTTP in that HTTP manages the data resources within the communication and TCP manages the communication itself.

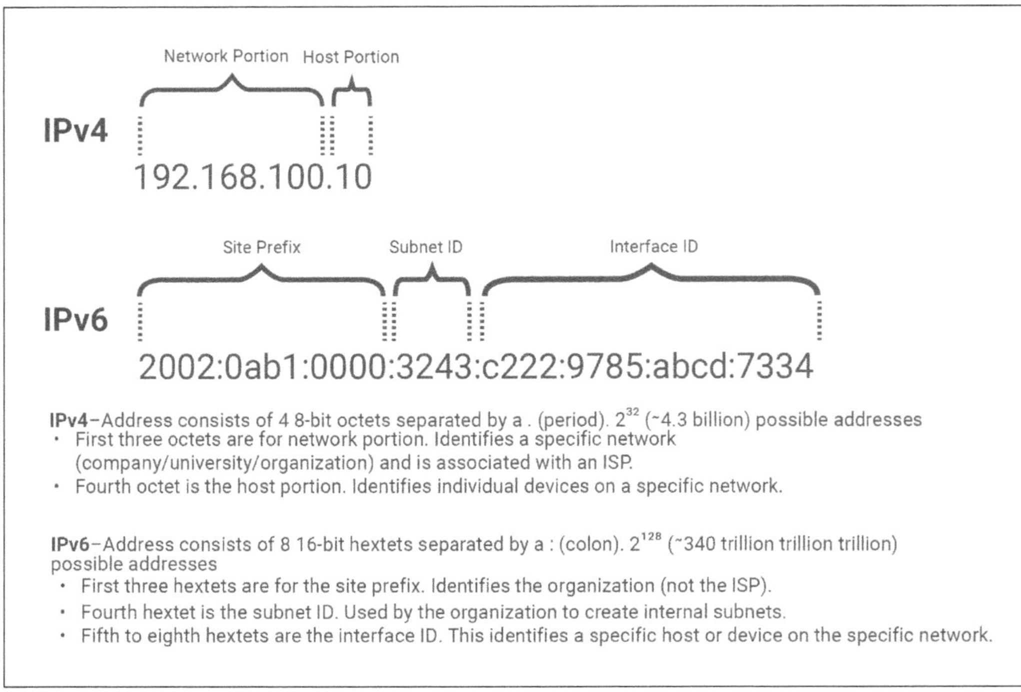

Figure 2-8. IPv4 vs. IPv6 Address Comparison

The following two cases highlight the importance of these protocols in cybersecurity.

CASE IN POINT

In 2018, Google Chrome began assigning an insecure symbol to sites that did not use HTTPS. When you use Chrome to access a website, you will see a symbol directly to the left of the site name in the address window. You will often see a padlock or lock symbol. This means that the website is secure. Clicking on that symbol will give you information about cookies and site details and permissions. An icon with an "!" in the center or the text "Not Secure" or "Unsecure" indicates insecure communication. Selecting the icon will bring up a warning that the connection is not secure and you should not provide any sensitive information to the site.

CASE IN POINT

Technically, a hacker might be traced through their IP address. Hackers know this, so if they are at all sophisticated, they will disguise their address. There are several ways to do so. One would be to simply borrow someone else's computing device—and put the blame on the owner of that device (a good way to lose friends). Bad actors might also be bad neighbors by hacking into a neighbor's Wi-Fi network router and changing an IP address. An experienced adversary can use several more technical approaches as well, including routing their traffic through a VPN.

Transmission Media
The physical or wireless medium through which data is transmitted, such as fiber optics, twisted pair cables, or radio waves.

Transmission Media

The fourth component of any network is the **transmission media**, or the channel through which network communications pass among connected devices. The media provides the connection, which may be wired or wireless. Data move through the channel between senders and receivers as electromagnetic signals. Table 2-8 describes transmission media alternatives, which are basically wired, using some type of physical cable, or wireless, and Figure 2-9 shows a representation of the different cable types.

Table 2-8. Network Transmission Media

Transmission Media	Description
Coaxial Cables	**Coaxial cables** consist of a thick copper electrical cable with two layers of conducting wire. Signals travel along one conductor and then move back through the other for two-way communication. The cable has an outer plastic sheath that protects the wires from environmental hazards.
Twisted Pair Cables	**Twisted pair cables** use intertwined cables. They consist of two separate insulated conducting wires that are twisted together and placed parallel to two other twisted together wires. Both coaxial cable and twisted pair cables carry analog and digital signals and may be used in networks and for voice communications, though coaxial cables are not in much use today except in legacy systems.
Fiber Optic Cables	**Fiber optic cables** transmit data by converting electromagnetic signals into light. The cables consist of very long, thin strands of glass that have a radius that is just a fraction of the size of a human hair. Their appearance can be likened to fishing line. Advances in glass technology allowed fiber optic cables that could transmit information much more quickly, much more efficiently, and over much longer distances than cables that were based on electromagnetic signal transmission. Optical fiber cables are lighter than coaxial or twisted pair cables, and the fibers are not flammable or impacted by electromagnetic pulses. From a cybersecurity standpoint, fiber optic cables are more secure because they do not allow for electrical signal interception, making it difficult for bad actors to eavesdrop on communications.
Radio Waves	Data transfer in wireless communications occurs via **radio waves**. The radio signals are electric currents moving at light speed. The wavelength is the distance from one point on a wave to the same spot on the next one, and the frequency is the speed, or the number of waves passing through a point during one second of time. In wireless communications, a transmitter sends the electronic field forward from an antenna to the receiver on the same wavelength, which then picks up the signal and translates it back into its original form. This form could be sound, image, or data. Wireless communications are by nature less secure than wired networks because data are transmitted through the air over radio waves that can be intercepted. This makes them more vulnerable to cyberattacks than a network connected by physical cables. Secure configurations and encryption can help mitigate this vulnerability, and hacking is rare but possible. For example, a Bluetooth connection can be accessed by a bad actor, allowing insertion of malware.

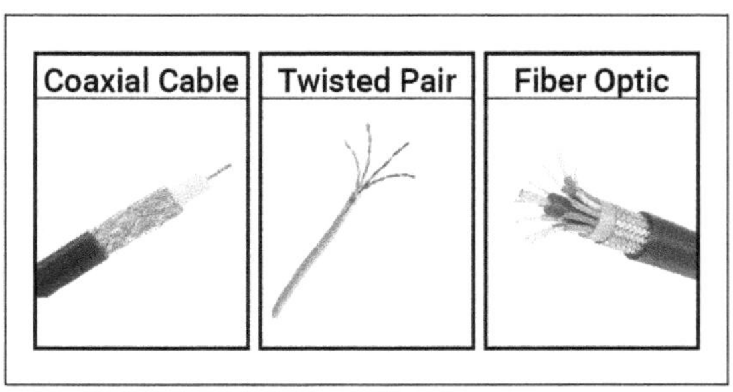

Figure 2-9. Cable Types

The following case illustrates the importance of making a choice between cables and radio wave transmissions in cybersecurity.

> **CASE IN POINT**
>
> **In December 2021, Vice President Kamala Harris came under fire from some segments of the news media for her refusal to use Bluetooth earbuds. When talking to the media, the vice president relied on a wired headset. This practice is consistent with the advice of the FBI, which suggests that people should use headsets with a cord to connect to computing devices when possible, as it is more secure.**

The Internet

Internet A global network of interconnected computers that communicate using standard protocols to share information and services.

World Wide Web (WWW) A system of interlinked hypertext documents and resources that can be accessed via the Internet using a web browser.

Uniform Resource Locator (URL) The web address used to access resources on the Internet, typically in the format https://www.example.com, for example.

Domain Name A human-readable address, such as www.example.com, that is mapped to an IP address and used to access a website on the Internet.

The **Internet** is a network that connects computer networks globally using TCP/IP for network and computing device communications. This network is so transformative that it is akin to the Industrial Revolution in its impact on the world. Thanks to the Internet, more than half of the global population can access a world of knowledge with the click of a button on a personal communication device.

The Internet itself is the hardware that connects networks. The **World Wide Web (www)** is the information and resources in web pages that users access using browser software. These pages are connected via **uniform resource locators (URLs)** and hyperlinks. A URL is a specific web address.

EXAMPLE

https://www.richmond.edu is a web address. The domain name is richmond and the .edu indicates that the web address is related to an educational organization. A URL is a specific address, whereas a domain name is an area that can consist of different top-level domains, such as .org or .com (Figure 2-10).

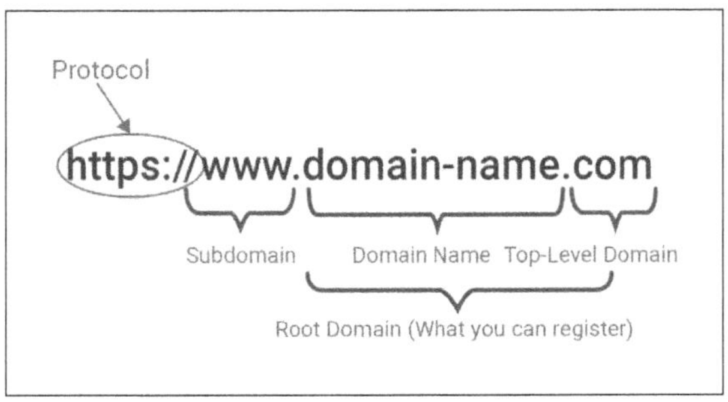

Figure 2-10. Domain Name Explanation

Domain Name System (DNS) The system that translates domain names (e.g., www.example.com) into IP addresses that computers use to communicate over the Internet.

Web Browser Software used to access and navigate the World Wide Web on the Internet.

Domain names are registered through a valid registry authorized by the Internet Corporation for Assigned Names and Numbers (ICANN). This nonprofit organization oversees the assignment of domain names and IP addresses. **DNS** stands for the **domain name system** that acts like a directory (similar to the contact list you maintain on your phone), identifying all the possible Internet domain names in existence. A server running the DNS translates domain names into their corresponding IP addresses, allowing the user's **web browser** to load Internet resources or web pages. Unfortunately, DNS servers provide an opportunity for bad actors, as described in the following example.

EXAMPLE

Hackers can introduce malware by compromising a router, and the DNS is an attractive target. A threat actor can use a brute force attack or other technique that provides them with logon credentials. Once in, the hacker makes changes to the DNS default settings. Since the DNS translates a domain name into an IP address, the user inputting a domain name may be accessing a site where the hacker is controlling the web page. The web page the user accesses might look like it is the legitimate page they sought, but it is actually the hacker's server, and the hacker captures any data entered.

Users access Internet resources via their web browser, such as Google Chrome, Firefox, Apple Safari, or Microsoft Edge. The browser is the software that retrieves the Internet resources and displays them.

The Internet has three main parts (Figure 2-11). The **surface web** is the section of the Internet that contains publicly available web resources. This part of the Internet contains company websites, news sites, information sites, and basically any Internet resource that does not require a credential to access it. The **deep web** is the part of the Internet that is not visible to users without a credential. The deep web accounts for about 90% of web resources and is by far the largest segment of the Internet. Your bank account, email accounts, subscription services, social media pages, company databases, and any other resources that require you to log on for access are all examples of the deep web. The **dark web** is the portion of the Internet that ordinary web browsers cannot access because they identify your IP address as you access websites. For example, when you use Google Chrome, your browser is identifiable to the site you are accessing. This allows the web server to monitor and filter web traffic. Your IP address is known and may be either blocked or allowed access. The dark web, accounting for just about 6% of all Internet resources, requires a special browser that anonymizes the IP address of the user accessing the resources.

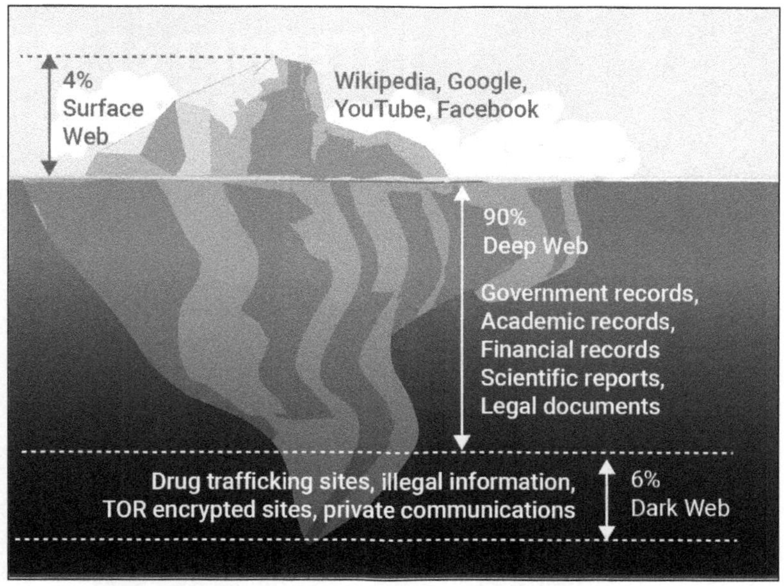

Figure 2-11. The Internet as an Iceberg

The Onion Router (TOR) is free, open-source software that provides browser access to the dark web. TOR was developed by the US Navy to protect intelligence communications. It provides privacy through encrypting all communications traffic as it moves requests and data from IP address to IP address to retrieve the desired information. User anonymity can be desirable for multiple reasons, including efforts to evade laws and regulations. It is not illegal to access the dark web, but many illegal marketplaces operate within this domain. Drugs, pornography, and weapons are all accessible on the dark web, as well as trading in personal identification. For example, if your credit card information has been stolen, it is possible that it will be sold in this portion of the Internet. While the dark web is the place where many illegal transactions take place, it can also be a haven for individuals who need the protection afforded by anonymity. Dissidents who fear political reprisal can find a haven on the dark web, as can individuals suffering from abuse.

CASE IN POINT

Between 2011 and 2013, the dark web was home to the Silk Road, an Internet marketplace where customers could purchase drugs, stolen credit cards, and the services of a hit man, but no weapons or child pornography. Ross Ulbricht founded the site in 2011 and gave himself the moniker Dread Pirate Roberts, after a character in *The Princess Bride*. During its two years of operations, the Silk Road processed about $9 billion in transactions. The anonymity afforded by browser encryption and Bitcoin allowed the site to thrive.

The Internet is one of the greatest inventions of humankind, but its limitless connectivity is not without peril for users. To be safe online, as you access the Internet through wireless connections, requires security protocols. **Wired Equivalent Privacy (WEP)** was introduced in

Surface Web The portion of the Internet that is indexed by standard search engines and accessible to the general public.

Deep Web The part of the Internet that is not visible to users without a credential, meaning it requires a login.

Dark Web A part of the Internet that is not indexed by traditional search engines and requires specific software, configurations, or authorization to access. It is often associated with illegal activities.

TOR (The Onion Router) A technology that uses layered encryption to route Internet traffic through multiple volunteer-run servers (nodes), hiding the user's identity, location, and activity.

Wired Equivalent Privacy (WEP) An outdated and vulnerable security protocol for wireless networks, replaced by WPA, WPA2, and WPA3.

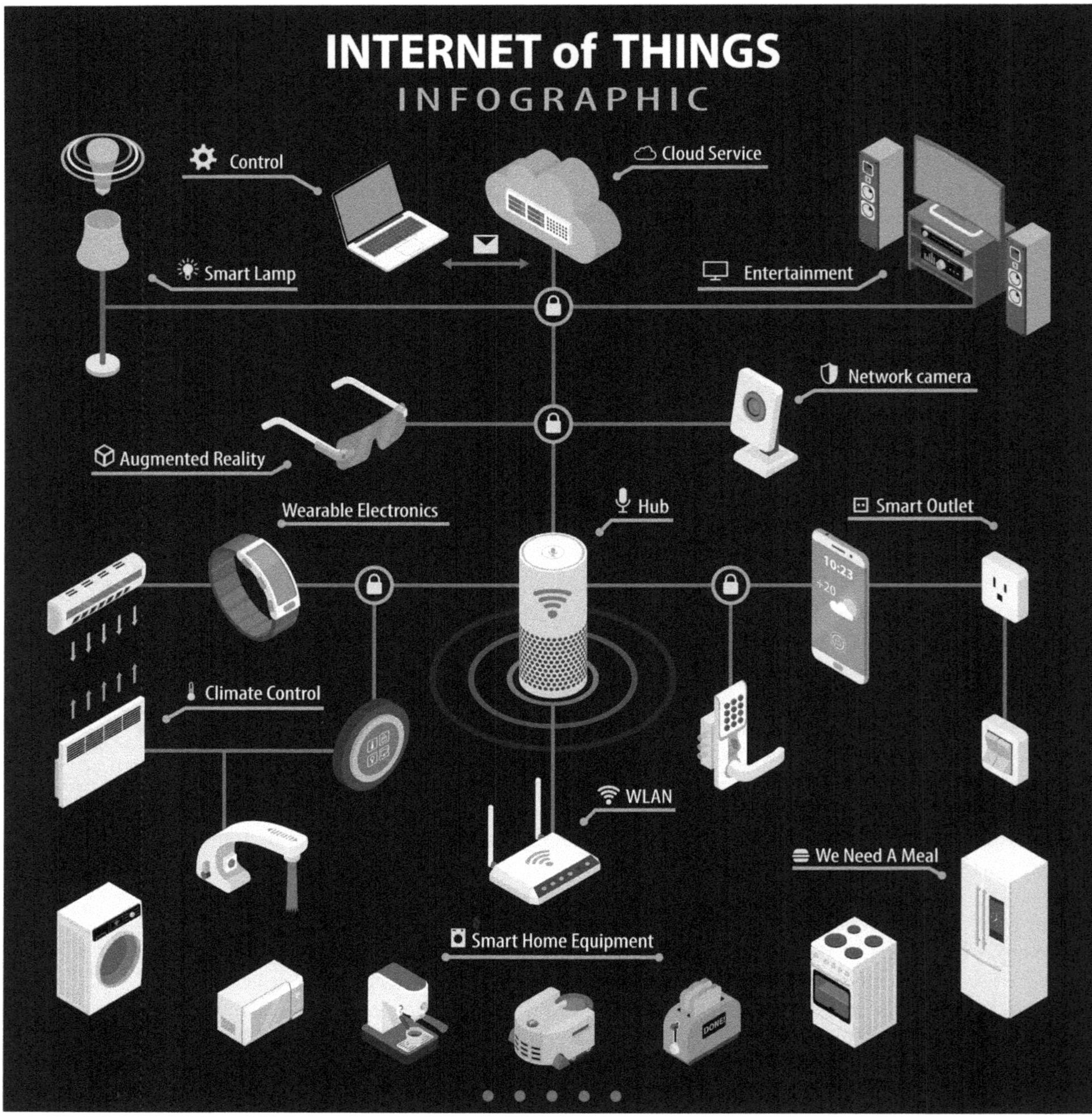

Figure 2-12. The Internet of Things

Wireless Protected Access (WPA, WPA2, and WPA3) Security protocols used to secure wireless networks, with WPA2 offering stronger encryption than WPA and WPA3 stronger than WPA2.

1999 to provide data privacy on wireless networks. This protocol has been superseded first by **Wi-Fi Protected Access (WPA)**, then **WPA2** and **WPA3**. Turning on the WPA feature on a router provides encryption for data being sent and received across the network. WEP is still present on legacy systems and older devices that have not been updated. While it provides some basic level of security through encryption, it can be hacked relatively easily.

The **Internet of Things (IoT)** refers to all the objects that are connected to the Internet. They include endpoint devices, which can include everyday items that people use. An interesting exercise is to take an inventory of all the IoT devices you might have in your home. This list might include your watch, television, phone, thermostat, security system, intelligent personal assistant, baby monitor, kitchen appliances, automotive equipment, and so on. The IoT allows for seamless communication between these objects and the Internet. For business enterprises, the list of IoT devices is huge and includes many that are used in security and production. Although this technological advancement is a boon for human productivity and quality of life in many ways, it also expands the attack surface, providing increasing opportunities for cyber adversaries. Figure 2-12 depicts some of the objects that make up the IoT.

Internet of Things (IoT) A network of physical objects embedded with sensors, software, and other technologies to connect and exchange data with other devices and systems over the Internet.

Artificial Intelligence

Many emerging technologies have implications for cybersecurity. **Artificial intelligence (AI)** can impact information security in negative ways as adversaries use it to perpetrate attacks, but it also has many benefits for protecting the CIA triad. For example, the AI ability to create deep fakes or write well-crafted phishing messages is beneficial to hackers. But cybersecurity professionals can use AI to find patterns in huge data sets that previously required extensive statistical methods and knowledge of how to interpret the results.

Artificial Intelligence (AI) The simulation of human intelligence in machines that are programmed to think and learn like humans.

Artificial intelligence is machine intelligence, as opposed to human knowledge, behaviors, and aptitudes. AI is not new; it has been a field of study since the mid-1950s and has been featured in countless movies, books, and television shows. The field gained traction in 1997 when IBM built a computer that was able to beat a champion chess player and, more recently in 2011, when IBM's Watson beat human trivia champions on the game show *Jeopardy*. In recent years, AI has been deployed for many purposes, and it is particularly useful in data science, where it can be deployed to mine big data.

Machine learning, a subfield of AI, is particularly useful in cybersecurity. It is the process of computers gaining knowledge over time. Automated information security tools that deploy this technology can learn from attempts to penetrate security defenses and adapt to defend against changing attack techniques. Adversaries are quite resourceful in creating new forms of malware and other threats. AI that incorporates intelligence that has the ability to predict patterns or behaviors gives information security professionals a chance to detect such threats before they become realized. Not only can AI scan Internet databases and information resources to detect threats, but it also can conduct behavioral analysis of potential adversaries and help business entities assess and manage risks. Increasingly, security software products incorporate AI technology.

Machine Learning A subset of artificial intelligence that involves training algorithms to learn from and make predictions based on data.

Intrusion detection and prevention system (IDPS) software is being replaced with newer network security software that incorporates AI tools. These programs, **network detection and response (NDR)** software, go beyond the monitoring of network perimeters and analyze all network traffic to detect anomalies and threats. NDR uses machine learning not only to detect malicious activity but also to respond to it. These tools can protect against a variety of threats at an early stage of attack.

Intrusion Detection and Prevention System (IDPS) Hardware and/or software tools that detect and respond to suspicious activities on a network or system, combining both detection and prevention functionalities.

Network Detection and Response (NDR) Security solutions that use machine learning and behavioral analytics to detect and respond to threats on a network.

Generative AI is artificial intelligence that produces new content based on patterns in other content. This type of artificial intelligence started to gain broad adoption in 2022 when OpenAI introduced ChatGPT. Since that time, many types of generative AI have become publicly available. These tools, as with other AI applications, may be used for both good and bad purposes. As an example of a positive use in cybersecurity, Norton, an identity protection vendor, created a tool that allows clients to upload a suspicious message and have the company's AI tool, Genie, determine if it is a scam. Figure 2-13 shows how AI is a two-edged sword that can be used for good and bad purposes in cybersecurity. As AI continues to evolve and improve, both threat actors and cybersecurity experts will be working to find ways to use it for their purposes.

Generative AI Artificial intelligence that can create new content, such as text, images, or music, based on patterns learned from existing data.

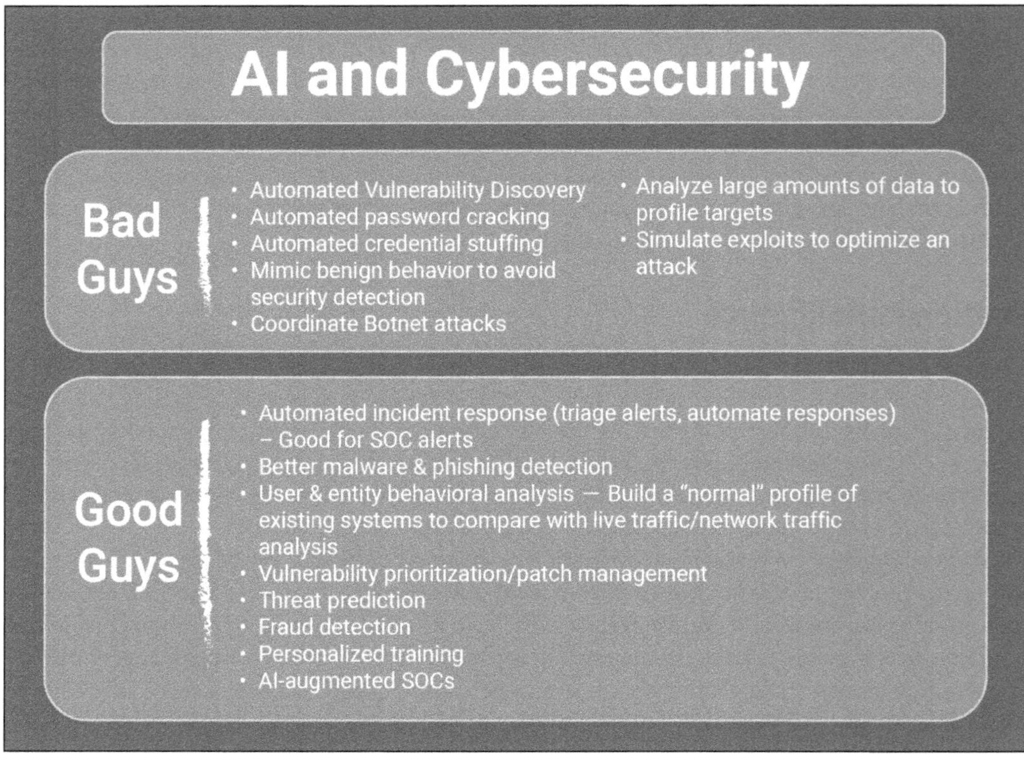

Figure 2-13. Two Sides of AI and Cybersecurity

Quantum Computing

Quantum Computing A theoretical field of computing that uses the principles of quantum mechanics—the physics governing atoms and subatomic particles—to process information to solve certain classes of problems (e.g., factoring large numbers) much faster than classic computers.

Quantum computing, which is currently available only in research labs and specialized highly technical environments, will be a great leap forward from classic computing. Unlike today's computing technology, it uses quantum bits, called qubits. Such bits are different from those in traditional computing, which can only be 0 or 1 interchangeably rather than simultaneously. Qubits allow for superposition, which means that many possibilities can be explored simultaneously, for example, considering multiple solutions at once.

While quantum computing is still a few years away for commercial use, it is expected to impact cybersecurity in some significant ways. The primary expected impact is on cryptography. Quantum computers will be able to break most, if not all, encryption used today. Some threat actors are stealing encrypted data today with the hope that they will be able to crack the code later as quantum computing becomes available. Of course, as with many advanced technologies, quantum computing is a two-edged sword, as it could be used to create more complex encryption that could thwart hackers, particularly those without the resources to use this new technology. It may also help in creating other, more secure, access methodologies that could help to keep data safe from threat actors.

Cryptocurrency and Blockchain

Cryptocurrency A digital or virtual currency that uses cryptography for security and operates independently of a central authority.

Cryptocurrency is a particularly relevant technological advancement, as it enables "pseudonymous" transactions that can allow bad actors to escape consequences for attacks. Note that this is not quite the same as anonymous. Transactions are not linked to a real name, but they are linked to a public address.

In 2008, Satoshi Nakamoto published a paper describing a new digital cash system. This paper introduced **Bitcoin** as a peer-to-peer electronic transaction system that did not need to rely on a third party, such as a financial institution or government entity, for cash exchanges. US Generally Accepted Accounting Principles (GAAP) and International Financial Reporting Standards (IFRS) treat cryptocurrencies as long-term intangible assets rather than cash, cash equivalents, inventories, or investments. These assets can be extremely volatile, with the markets for some of them having collapsed entirely.

Bitcoin A decentralized digital currency that operates without a central bank, based on blockchain technology.

Blockchain A distributed ledger technology used to record transactions in a secure, transparent, and immutable manner.

Cryptocurrency transactions are recorded and secured on **blockchains**. Cryptocurrency itself is digital, and ownership is proved via cryptographic keys allowing access to blockchain-stored transaction data. The blockchain is a ledger (shared and decentralized database) that exists on a network of computers. The ledger is immutable, meaning that it cannot be altered. Individual transactions are recorded in a block, which then links to other blocks to form a chain. This blockchain becomes a permanent record of transactions. To access cryptocurrency resources on the blockchain, the owner needs to use a private key. This private key is part of a key pair and is known only to the cryptocurrency owner. It typically consists of a large, unique alphanumeric code, and it allows the owner to access and transact in the cryptocurrency. Private key owners need to remember their code and keep the key safely stored to ensure that they do not lose access to their crypto assets or allow unauthorized access. Access to the crypto assets is encrypted with the other half of the key, the public key, which can only be decrypted by the private key. Cryptocurrency presents significant cybersecurity issues because private keys can be stolen or compromised, and hackers can steal huge amounts from cryptocurrency exchanges.

> **CASE IN POINT**
>
> **There have been several cases of breached cryptocurrency exchanges, the platforms that allow cryptocurrency owners to purchase and sell their assets. In 2024, the UK exchange Lykke suffered an apparent hack and halted trading. The company was prompt in containing the breach, and users were assured that their investments would be protected. In 2025, the cryptocurrency platform Coinbase was subject to a ransomware attack that stemmed from insiders. The threat actor appears to have obtained proprietary documentation from contractors or employees that had access to Coinbase systems in their roles. Coinbase reported that the incident compromised some usernames, contact details, and partial financial information. Rather than pay the ransom, the CEO offered a commensurate $20 million bounty to catch the hackers.**

Auditing IT Infrastructure

Auditors want to be sure that IT infrastructure supports business objectives and keeps data and other assets secure.

The internal auditor (IA) is concerned with ensuring that proper internal controls are in place and operating effectively. Their role is to provide continuous assurance on risk management by evaluating controls and IT governance. Tasks they perform include assessing access controls, change management of software applications, and monitoring backup and recovery plans and procedures. The IA will also make sure that data privacy and confidentiality are protected. In assessing risk, they will examine controls over hardware and software and policies and procedures. For example, the IA will review third-party contracts, such as those with cloud service providers. The IA will want to be sure that the data center is secure and will check that access is controlled and environmental controls are in place (e.g., cooling systems).

The external auditor (EA) is, as always, most concerned with internal controls that ensure financial statement accuracy and reliability. With respect to IT infrastructure, this means that the EA should evaluate the IT systems impacting financial reporting and ensure compliance with laws and regulations, including the EU's General Data Protection Regulation (GDPR, discussed in Chapter 11). Whereas the IA's work is ongoing, the EA will typically conduct assessments annually, evaluating the effectiveness of internal controls to determine the scope of control testing. The EA may use the work done by the IA as part of their IT infrastructure assurance, provided that the IA is deemed competent and objective and the work meets external auditing requirements.

When auditing IT infrastructure, many of the controls that need to be tested relate to the physical security of the data center as well as the other hardware around the company. A physical security audit can also include overall physical security at the organization, as well as the IT infrastructure, and can be done by either the IA or EA. When auditing a data center, the auditor should evaluate the following controls for adequacy and effectiveness:

1. Perimeter security
2. Entry and access measures
3. Internal security zones
4. Data hall/server room setup
5. Fire safety/emergency response
6. Power/environmental measures
7. Inventory management
8. Documentation/compliance
9. Personnel and training
10. Testing and review

Although this list is a mixture of logical and physical controls, the physical controls should be explicitly evaluated when examining a data center. This is a very important topic that is typically covered in an advanced auditing or IT auditing class.

Summary

This chapter provides a technical foundation for understanding cybersecurity. It begins with a description of the most important element in IT architecture: data. Data are what cybersecurity protection is all about, and this chapter provides an understanding of how data are stored, transmitted, and backed up. The chapter introduces the CIA triad, the information security model for data protection. Data on a stand-alone computing device, with no network connection, can be protected similarly to paper files. Most data, however, flow among computing devices over networks, and this chapter describes all the elements that make up networks, including network hardware, software, protocols, and transmission media. No discussion of networks is complete without talking about the Internet, and this chapter describes the elements of the Internet that influence cybersecurity, including the dark web. The final sections of the chapter discuss important emerging technologies impacting information security—artificial intelligence (AI), cryptocurrency/blockchain—as well as describe the physical security measures necessary to safeguard IT infrastructure.

Chapter Review

Key Terms

Accounting Information System (AIS)
Air Gap
Artificial Intelligence (AI)
Availability
Big Data
Bitcoin
Blockchain
Bridge
Bus Network Topology
Business Continuity Planning (BCP)
CIA Triad
Cloud Deployment Model
Coaxial Cable
Cold Site
Confidentiality
Cryptocurrency
Dark Web
Data
Data at Rest
Data Dictionary
Data in Motion

Data Lake
Data Life Cycle
Data Loss Prevention (DLP)
Data Mart
Data Mirroring
Data Normalization
Data Replication
Data Replication Method
Data Warehouse
Database
Database Schema
Deep Web
Differential Backup
Domain Name
Domain Name System (DNS)
Endpoint
Enterprise Resource Planning (ERP)
Ethernet
ETL Software
Failover
Fiber Optic Cables
File Transfer Protocol (FTP)
Full Backup
Generative AI
Hot Site
Hub
Hybrid Cloud
Hypertext Transmission Protocol (HTTP)
Hypertext Transmission Protocol Secure (HTTPS)
Incremental Backup
Information
Infrastructure as a Service (IaaS)
Integrity
Internet
Internet of Things (IoT)
Internet Protocol (IP)
Intrusion Detection and Prevention System (IDPS)
Knowledge
Local Area Network (LAN)
Machine Learning
Metadata
Modem
Network Client
Network Detection and Response (NDR)
Network Interface Card (NIC)
Network Management Software
Network Operating System.
Network Protocols
Network Topology
Operating System
Platform as a Service (PaaS)
Private Cloud
Public Cloud
Quantum Computing
Radio Waves
Relational Schema
Ring Network Topology
Router
Server
Simple Mail Transfer Protocol (SMTP)
Snowflake Schema
Software as a Service (SaaS).
Star Network Topology
Star Schema
Storage Media
Structured Query Language (SQL)
Surface Web
Switch
TOR (The Onion Router)
Transmission Control Protocol (TCP)
Transmission Media
Twisted Pair Cables
Uniform Resource Locator (URL)
Universal Serial Bus (USB)
Warm Site
Web Browser
Wide Area Network (WAN)
Wired Equivalent Privacy (WEP)
Wireless Protected Access (WPA, WPA2, and WPA3)
Workstation
World Wide Web (WWW)

Review Questions

1. Why is it important for nontechnical people to understand IT infrastructure?
2. How does a knowledge of IT infrastructure help in understanding cybersecurity?
3. What are data and why are they important in the context of cybersecurity?
4. Explain the differences among the cloud deployment models.
5. What are the different cloud computing service models, and what are their implications for cybersecurity decision making?
6. List the three network topologies and explain the differences among them. How does the selection of a topology impact data vulnerability?
7. List the main components of network hardware and explain how they work together to facilitate data transfer and data storage.
8. Describe each of the primary networking protocols.
9. List the main transmission media types and identify their strengths and weaknesses in relation to cybersecurity.
10. Discuss the three main parts of the Internet and the security function of each part.
11. What is the purpose of the TOR browser, and how is it used to securely transmit data?
12. How does artificial intelligence (AI) help and hinder cybersecurity practices? Do the benefits outweigh the costs?
13. Explain how cryptocurrencies are related to the blockchain.

Multiple Choice

Identify the choice that best completes the statement or answers the question.

1. In the 1960s, what network did the US Defense Department develop to link Pentagon research computers over telephone lines?
   a. ARPANET
   b. World Wide Web
   c. Internet Protocol
   d. TCP/IP
   e. Packet Switching Network
2. What distinguishes data from information?
   a. Data is stored, while information is transmitted.

b. Data has context, while information is raw material.
c. Data is unintelligible, while information is processed.
d. Information is collected, while data is created.
e. Information is electronic, while data is physical.

3. How is confidentiality different from privacy in the context of data?
   a. Confidentiality refers to private data, while privacy involves unauthorized access.
   b. Privacy means data belonging to individuals, while confidentiality means keeping data secret.
   c. Confidentiality involves data availability, while privacy focuses on data integrity.
   d. Confidentiality and privacy are synonymous.
   e. Privacy refers to data completeness, while confidentiality involves authorized access.
4. What is the role of Hypertext Transfer Protocol (HTTP) in a web server?
   a. A hardware element of the server
   b. Communication rules for file transfers
   c. Database server functionality
   d. A set of rules for connecting clients
   e. Software used for client interactions
5. What is the advantage of the star topology from a security standpoint?
   a. Lower network cost
   b. Efficient data transfer
   c. Circular data path
   d. Impact on one computer in case of connection failure
   e. Connection to a computer on either side
6. What is the role of network operating systems in allowing clients and servers to communicate?
   a. Transferring files among servers
   b. Managing memory
   c. Connecting devices within the network
   d. Managing storage
   e. Managing operating systems
7. What is the primary benefit of artificial intelligence (AI) in the context of cybersecurity?
   a. Generating new forms of malware
   b. Perpetrating attacks on information security
   c. Learning from attempts to penetrate security defenses
   d. Analyzing human trivia champions on game shows
   e. Building computers that can play chess

Case Studies

Amazing Enterprises—Cloud HR Management Systems

Natalie Magee has just been hired by Herron Consulting LLP. Herron is a midsize regional consulting firm, with offices in four US states. The office in Natalie's hometown has a staff of 40 professionals, including three partners, along with a full contingent of support staff who facilitate professional client service. Natalie reports to Chris Mason, and her title is Risk Management Consulting Associate.

Natlie has just been assigned to her first consulting engagement. The client, Amazing Enterprises, Inc., is a large Midwestern arcade game manufacturer. The company has an ERP system and about 40 other application software programs. Amazing has not found the human resource (HR) management module in the ERP software program to be as robust as it would like. It is considering using a cloud-based HR management program.

Case Requirements

1. Write a recommendation for two potential HR management software vendors. Why did you choose those two vendors?
2. What are the cybersecurity concerns associated with the decision to use a cloud-based HR management application? What can Herron LLP do to provide its client with assurance regarding the cybersecurity of its vendor recommendations?

Herron Consulting—AI

Natalie's boss, Chris Mason, wants to take advantage of her research skills. He recognizes that, as a new college graduate, Natalie has experience in researching various topics and understands some of the newest technological developments. His boss, Amanda McDermott, has been advocating for the potential of AI to increase their workplace productivity. She also has an interest in helping clients use AI technologies to enhance their cybersecurity.

Case Requirements

1. What are the advantages and disadvantages of ChatGPT versus Microsoft Copilot for enhancing workplace productivity in a consulting firm?
2. Research how AI can be used to improve cybersecurity. In your research, describe three different information security applications (e.g., firewalls, antivirus software) and how they have incorporated AI.
3. AI can make cybersecurity better, but bad actors are also using this technology for their own purposes. One way hackers might use AI is to create phishing messages. Ask your favorite AI platform to create such a message. What is its response? How might a hacker circumvent controls that an AI platform might incorporate to protect against misuse?

Ohio Southern University—Data Center Internal Audit

Nancy Tran is an internal auditor at Ohio Southern University. She is developing an audit plan to review the university's data center. The center has changed quite a bit over the past few years, as many applications are now offered and managed in public clouds. However, the data center does still maintain a few servers, for example one that holds proprietary faculty research.

Case Requirements

1. Part of Nancy's review will be to assess the physical security of the data center, including access and environmental controls. Identify five specific physical controls Nancy should find (e.g., access logs).
2. With respect to the IT infrastructure in the data center, what steps should Nancy take in reviewing hardware, software, and network security?
3. Nancy wants to ensure that the data center has a backup plan that can ensure business continuity in the event of a cyber incident or natural disaster. What would be a reasonable schedule for on-premises backup?

Exercises

2-1. This chapter discussed some of the advantages and disadvantages of public versus private cloud deployments and cloud service models. Required:
   1. For a medium-sized company in the retail sector, which software applications/functions are likely to be maintained and accessed in a public versus a private cloud? What criteria would contribute to making these decisions?
   2. For a medium-sized company in the retail sector, how might it use SaaS for its point-of-sale (POS) system? In your answer, describe the data that these types of systems handle, two specific vendor choices, and the efficiencies that this model creates.

2-2. Networked systems may be wired using physical cables or they may be wireless. Required:
   1. What are the implications for the cybersecurity of these two transmission media options (assume optical fiber cables)?
   2. Air gapping is a security measure that segments a server and data from networks, including the Internet. What are three advantages and two limitations of air gapping with respect to cybersecurity?
   3. In the case of the Stuxnet malware, an air gapped computer was breached. How did that happen?

2-3. The Internet has three parts: the surface web, the deep web, and the dark web. Required:
   1. Identify three sites you commonly access on the surface web and three sites you commonly access on the deep web.
   2. The US Naval Research Laboratory invented the TOR browser. What was the rationale for its early development?
   3. The dark web is a part of the Internet where many illegal activities can take place. However, there are also reasons why people may seek anonymity for more positive purposes. Describe two instances where this has occurred.
   4. The Silk Road was an Amazon-like marketplace where illegal drugs and other unlawful goods were bought and sold. What happened to it and its founder Ross Ulbricht?

2-4. Blockchain is a technology that has enabled cryptocurrency. Required:
   1. How does blockchain secure transactions?
   2. What is the difference between pseudonymity and anonymity, and how does blockchain provide the former but not the latter?
   3. A case within this chapter briefly described the 2025 cyber breach at the cryptocurrency trading platform Coinbase. Research the case and describe in some detail how the breach occurred. The CEO of Coinbase made the decision not to pay the ransom but instead to offer the same amount to bounty hunters. Why do you think he made this decision, and do you agree with it? Provide a rationale for your decision.

References

Garry. (2023, October 8). Why backups of data should be kept in a different location to the original? *Darwin's Data.* https://darwinsdata.com/why-backups-of-data-should-be-kept-in-a-different-location-to-the-original/

Hartman, S. (2012, October 12). Understanding data center reliability, availability, and the cost of downtime. *Schneider Electric Blog.* https://blog.se.com/datacenter/2012/10/12/understanding-data-center-reliability-availability-and-the-cost-of-downtime/

Konkel, F. (2014, July 17). The details about the CIA's deal with Amazon. *The Atlantic.* https://www.theatlantic.com/technology/archive/2014/07/the-details-about-the-cias-deal-with-amazon/374632/

Lewis, P. E. (2022, September 27). *See how Coca-Cola uses the compute power of AWS to innovate consumer beverage experiences at scale.* AWS. https://aws.amazon.com/blogs/industries/see-how-coca-cola-uses-the-compute-power-of-aws-to-innovate-consumer-beverage-experiences-at-scale/

Wallen, Dave. (2020, March 31). Types of Backup: Understanding Full, Differential, and Incremental Backup. Spanning. https://www.spanning.com/blog/types-of-backup-understanding-full-differential-incremental-backup/

Chapter 3

Threats to Information Security

Chapter Outline

Source of Threats—Who Is Responsible?
Assets—What Do Threat Actors Want?
Threat Motivation—Why Do Threat Actors Do It?
How Do Threats Lead to a Cyberattack?
Types of Threats
Threat Modeling
Threat Intelligence

Learning Objectives (LO)

After reading this chapter, you should be able to:

- LO1. Define threats and the threat landscape.
- LO2. Know the different types of threat actors.
- LO3. Describe the assets threat actors seek and why they conduct cyberattacks.
- LO4. Understand how threats exploit vulnerabilities to launch an exploit.
- LO5. Know the different types of threats and describe examples of each type.
- LO6. Be familiar with tools organizations can use to analyze and predict threats.

Introduction

Threats are basically bad things that can happen. We all face a variety of threats daily. You might feel the threat of becoming ill with the flu or having a traffic accident. A cyber **threat** is the potential for a threat actor to cause harm through a digital attack. A realized threat is an attack that occurs because the threat actor has been able to exploit a vulnerability. We will discuss vulnerabilities in the next chapter. Cyber threats are escalating. They rose during the COVID pandemic because of remote work, but AI is now causing the numbers to skyrocket.

Threat Any circumstance or event with the potential to cause harm to an information system.

> **EXAMPLE**
> **C. J. Moses, Chief Information Security Officer at Amazon, blamed AI for an exponential increase in cyber threats in 2024. He noted that, on average, there are 750 million cyberattack attempts on the company daily. One reason is that AI allows unskilled would-be hackers to use computers to launch attacks (Rundle, 2024).**

This chapter describes threats to information security in terms of who, why, where, what, and how. We then describe three types of threats—malware, social engineering, and web-based

Threat Landscape The overall view of the threats facing an organization, including the types and sources of threats.

Hacker Someone or some group that uses their technical skills to gain authorized or unauthorized access to systems, networks, or data.

threats—and provide examples for each type. Some threats do not easily fall into one of these types, and we discuss them as other common threats. Business leaders and accounting professionals need to analyze and predict the entire **threat landscape** that might impact their organization. This landscape is the entire ecosystem of potential and existing threats to them. The last sections of the chapter describe threat modeling and threat intelligence, two tools available for analysis and prediction.

Source of Threats—Who Is Responsible?

We often use the term **hacker** to describe a threat actor. This is not entirely accurate, at least in terms of motivation. Hackers can be good guys, bad guys, or somewhere in between. You might have heard of a "black hat" hacker as a bad threat actor, who may have criminal intent. Cybersecurity professionals use hat colors to describe different kinds of hackers, as shown in Figure 3-1.

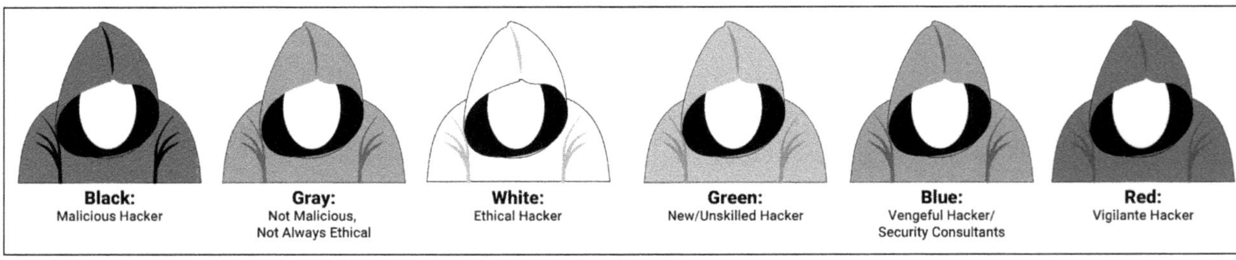

Figure 3-1. Types of Hackers

White hat hackers are ethical hackers who help an organization evaluate its cyber risk by exposing vulnerabilities. Red hat hackers are similar, though they may be acting more as vigilantes and can be more aggressive than white hats. Hackers wearing green hats are unskilled novices (i.e., green), and blue hat hackers are ethical hackers trying to spot vulnerabilities and are usually researchers or consultants. Those wearing a gray hat occupy a space somewhere between the black and white hacker. They might be mostly good or bad, but somewhere in between the extremes, depending on their motives. Threat actors may wear one of these hats, but they also can be described in other ways, such as insiders versus external parties. These adversaries are often characterized by their motivation or intent, just as is the case with the hat color system. For example, an external hacker may wear a black hat or a white hat, depending on whether the intent is to cause harm or to be of help.

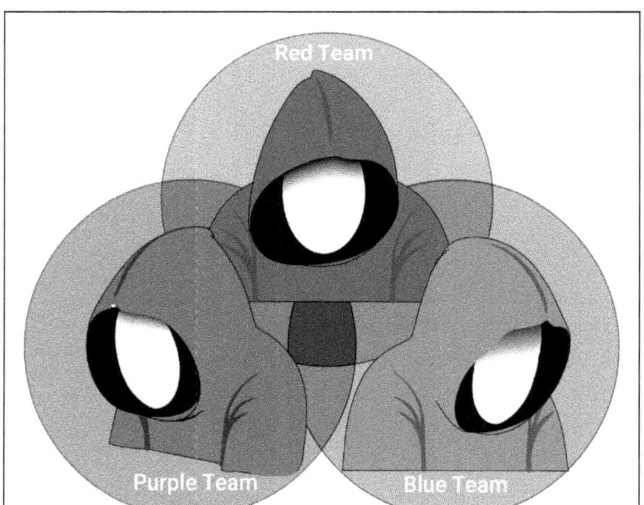

Figure 3-2. Hacker Team Types

In addition to the colored hats, there are colored hacker teams (red, blue, purple—see Figure 3-2) that are employed by or contracted with organizations. Blue teams deal with the day-to-day maintenance and operations of an organization's security. Red teams simulate real-world attacks on the company to test the existing security. Purple teams are a combination of red and blue teams working together to enhance, test, and implement the organization's security.

The following section describes different types of hacker groups. They vary primarily based on their motivation. Common types of hacker groups are cybercriminals, hacktivists, insiders, nation-states, and social hackers.

Cybercriminals

Some threat actors are just plain criminals. **Cybercriminals** commit crimes by using computers. They may be sophisticated and part of organized cybersecurity gangs, such as Darkside or REvil, or petty criminals in search of small stakes financial gain. Many are just groups that have realized they can make more money as cybercriminals than they can through traditional crime vectors, such as physically robbing a bank.

Sophisticated cybercriminals access markets and websites within the dark web, where they can trade hacking tools, malicious code, and stolen data. Some cybercriminal gangs specialize in **Ransomware as a Service (RaaS)**. This criminal business model works similarly to **Software as a Service (SaaS)**, where clients obtain a subscription to lease the software or outright purchase a turnkey ransomware attack. With RaaS, a hacker can obtain developed ransomware tools that they can use to launch incursions. This allows the hacker to operate without having sophisticated technical knowledge and skills. The purchased ransomware kit might include many of the types of products and services provided in SaaS, such as user reviews and client support.

Cybercriminal An individual, group, or organization that uses computers, networks, or digital technologies to commit illegal activities for personal, financial, political, or malicious gain.

Ransomware as a Service (Raas) A hacker business model in which cybercriminals lease ransomware tools and infrastructure to other attackers for a share of the ransom profits.

Software as a Service (SaaS) A cloud computing service model that delivers software applications over the Internet on a subscription basis.

CASE IN POINT

REvil, short for Ransomware Evil, is a cybercriminal gang thought to be based in Russia that was behind many of the largest ransomware attacks in 2020 and 2021, including the attack on the meat supplier JBS and the software company Kaseya. REvil operated a successful RaaS until mid-2021, when its online presence briefly disappeared. The reason why this happened is not known for sure, but the group reemerged a few months later.

Petty cybercriminals can launch unsophisticated attacks themselves or purchase RaaS for financial gain. Apart from ransomware, these criminals engage in a variety of scams, ranging from identity theft to wire fraud.

Petty Cybercriminals Individuals who engage in minor cybercrimes, often for personal gain or mischief.

Hacktivists

A **hacktivist** is an individual who uses computer technology to nonviolently advance a social cause or political agenda. Hacktivists are not necessarily motivated to seek personal rewards, though they may steal financial resources or data to further their cause. They might see themselves as martyrs or vigilantes who use computer hacking for good. These hacker-activists use the same techniques as other hackers. However, because their motives are not the same, they may be inclined toward different types of attacks. For example, to harm an organization that they believe to be representative of an agenda at odds with their own social cause, the hacktivist might launch a distributed denial of service (DDoS) attack and temporarily block customers from accessing the company website. Another tactic of hacktivists would be to hack into a database and then publicly disclose inflammatory or proprietary information that they may believe the public at large should know.

Hacktivist An individual or group that uses hacking techniques to promote political or social agendas.

Anonymous An organized group of political activists who often engage in hacktivism. Anonymous hackers first came together on 4chan, a private internet chat room. "Anons," as Anonymous hackers are known, frequently conceal their faces in public with Guy Fawkes masks.

CASE IN POINT

Anonymous is one of the largest and most famous hacktivist groups in the world. This group consists of hackers loosely affiliated by a common cause to attack individuals, companies, and government organizations to create social change. Group members may be unknown to each other and to those outside the group. Fueled by outrage over the death of George Floyd, the group appeared to have been behind a DDoS attack of the Minneapolis Police Department. To hide their identity, members of Anonymous wear a Guy Fawkes mask, which some see as a symbol of rebellion against repression, mirroring its depiction in the film and graphic novel *V for Vendetta*. Figure 3-3 shows a physical protest by members of Anonymous, though their attacks are usually digital.

Figure 3-3. Anonymous Protest

Insiders

Insiders may be employees, contractors, vendors, consultants, or those in other third-party relationships. Employees can pose a cybersecurity threat either inadvertently or intentionally. Inadvertent threats are simple human error. A common example is an employee who clicks on a link in a phishing email. Employees also can launch intentional cyberattacks, whether because they are disgruntled or because they seek to gain resources they may turn into profits. It seems cold and unfeeling to require a security guard to escort a terminated employee out of an office building, but eliminating access to computers and computer networks protects against retaliatory cyberattacks.

Third parties are another source of insider attacks. Third parties are individuals or entities with which an organization maintains a relationship. Even small companies maintain multiple third-party relationships, such as with builders, maintenance workers, suppliers, customers, and consultants. These third parties may have physical access to the organization's premises or virtual access to parts of its information system. Doing due diligence on third parties, including obtaining an understanding of their cyber hygiene and level of information security, is critical in protecting against breaches from this source, which may be unintentionally compromised (creating a vulnerability) or may be acting as an intentional threat. The 2013 Target data breach, described in the following case, illustrates the harm that a third party can cause.

> **CASE IN POINT**
>
> **The 2013 Target data breach began with a phishing email to an employee of their HVAC vendor. The attackers were able to use that breach to obtain login credentials that let them access Target's network. Once inside, they captured debit and credit card data from about 40 million accounts. The attack showed that even though a target (no pun intended!) organization may itself have good cybersecurity defenses, threat actors can exploit less protected third parties to launch attacks.**

Nation-States

Cyberwarfare The use of digital attacks by a nation-state or organization to disrupt the activities or infrastructure of another entity.

Intellectual Property (IP) Creations of the mind, such as inventions, literary works, designs, or symbols, that are legally protected from unauthorized use.

There are many quotes, dating back several years, about how WWIII will be fought in cyberspace. **Cyberwarfare** is the use of cyberattacks versus a nation-state with the goal of causing harm, stealing **intellectual property (IP)** or proprietary intelligence, or disrupting the economy and/or government operations. An advantage of cyberwarfare over other modes of war is that the barrier to engage is much lower than traditional warfare, which relies on expensive weaponry. Engaging in cyberwarfare requires only some computing power and expertise in cybersecurity. While not inexpensive, the cost is much less than investments in aircraft carriers, for example. Chapter 12 of this book discusses geopolitical cybersecurity threats in some depth.

Nation-states can be threat actors who commit cyberterrorism or hack for financial gain. These adversaries may try to obtain IP that will gain them a national advantage, or they may harbor cybercriminals without penalty. Most industrialized nations have done at least one of the above, with many countries engaged in cyberwarfare against countries whom they perceive as threats or enemies.

Most Internet searches identify China, Russia, Iran, and North Korea as the countries that have particularly targeted the United States in cyberwarfare or by giving safe harbor to cybercriminals. Note that this is just a sample group—many countries engage in cyberwarfare against the United States, and the United States uses cyberwarfare as well. Each country may have different motivations, including the defense of their national interests. For example, one country might be most interested in obtaining IP, while another may wish to cause distractions and introduce chaos by interfering in elections. Still other countries may not enforce laws against cybercriminals at all for a variety of political reasons.

> **CASE IN POINT**
> In 2024, it was discovered that threat actor Salt Typhoon had infiltrated many of the largest US telecom companies. China's Ministry of State Security is believed to be behind Salt Typhoon and many incursions into US company networks. The hack against the telecom companies compromised both text and voice data, including date stamps of messages, the parties engaged in the communications, and, in some cases, audio calls themselves. Upon discovery of the attack, Virginia Senator Mark Warner, who chaired the Senate Intelligence Committee, called it the "worst telecom hack in our nation's history—by far" (Nakashima, 2024).

Social Hackers

Social hackers are typically motivated by ego, boredom, or because they view hacking as a game. **Script kiddies** are a type of social hacker with minimal technical skills and knowledge. They use hacks (scripts) developed by others to engage in hacking. These hackers will usually not be able to breach a well-defended information system, but they can be dangerous to organizations that do not invest in basic cyber hygiene. For example, they can exploit network hardware where the user has not changed default passwords. Script kiddies and other social hackers can take advantage of publicly announced vulnerabilities when an organization has failed to apply a software patch. Hackers sometimes launch attacks just to see if it is possible to do so successfully or even to demonstrate security weaknesses. More sophisticated social hackers might also launch a cyberattack just to see if they can be successful.

Script Kiddies
Inexperienced hackers who use existing scripts or tools to launch attacks without understanding the underlying technology.

> **CASE IN POINT**
> Kevin Mitnick is perhaps the most famous hacker of all time. In the 1990s, he used social engineering techniques, such as impersonating employees, to hack into about 40 large corporations and was the FBI's most wanted cybercriminal. For his crimes, he served a five-year prison sentence, after which he became a successful information security consultant. Why did he do it? He has said that it was not for financial gain but rather for curiosity, adventure, and intellectual challenge.

Assets—What Do Threat Actors Want?

Adversaries launching cyberattacks may not be after any specific assets, but if they are, usually they are seeking either monetary or data assets. The following case describes how creative these hackers can be in seizing an opportunity for monetary gain, usually in the form of **cryptocurrency**. Cryptocurrency, such as Bitcoin, is a popular reward for hackers because it provides them with anonymity. This example also demonstrates how threat actors can exploit people as vulnerabilities, in this case stressed-out parents.

Cryptocurrency A digital or virtual currency that uses cryptography for security and operates independently of a central authority.

> **CASE IN POINT**
> During a baby formula shortage in 2022, many parents turned to social media to create support networks. Desperate parents of infants looked to other parents who might have access to formula and exchanged information and even shipped found supplies to each other. Unfortunately, scammers saw this as an opportunity and posed as parents or suppliers. The FTC issued a warning that scammers were setting up profiles on social media and fake websites that mimicked known brands of formulas and their logos to get parents to send payments.

When it comes to data, many types of data assets have value to cybercriminals and others. You can easily find out the value of data assets such as credit card numbers or a driver's license by searching for the current rate online. For example, at the time of this writing, credit card data was

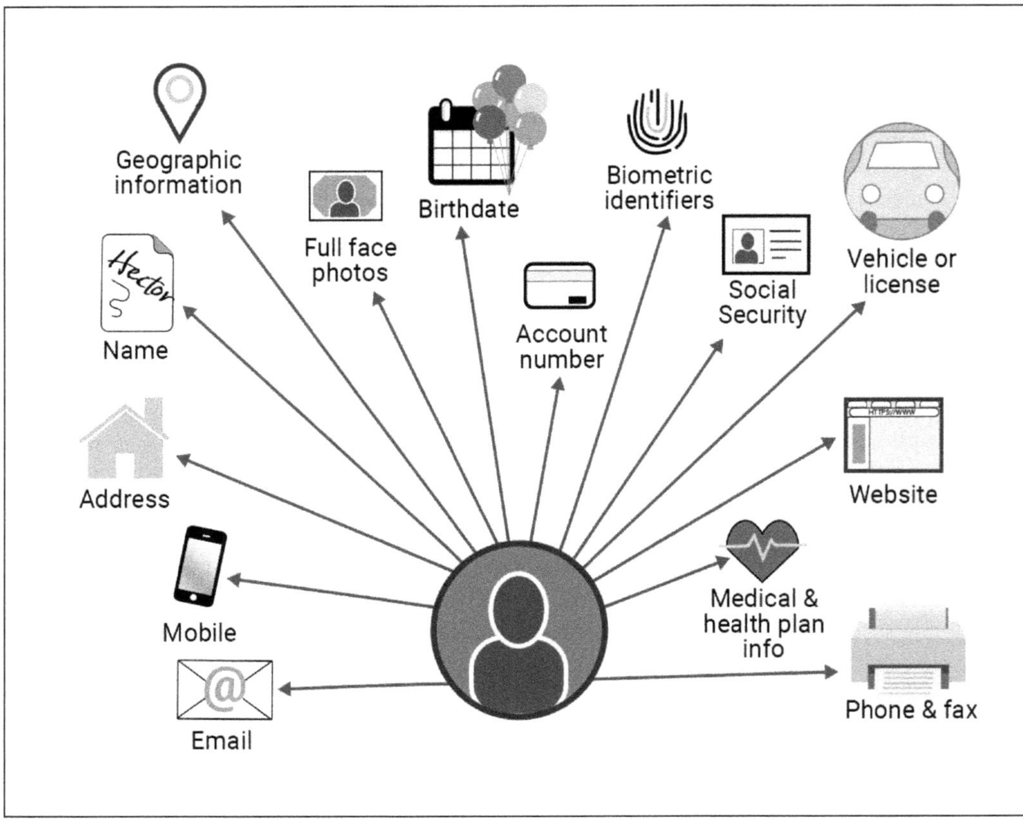

Figure 3-4. Examples of PII

worth about US$12–20. The value of stolen credentials varies from country to country. As data breaches in the United States have become more common, the value of an individual credential has decreased because the countermeasures have become more efficient.

Personally Identifiable Information (PII) Information that, when used alone or with other relevant data, can identify an individual.

Protected Health Information (PHI) Any information about health status, provision of healthcare, or payment for healthcare that can be linked to an individual.

Cardholder Data (CHD) Any personally identifiable information associated with a credit card holder, such as the card number, expiration date, and service code.

Adversaries may launch cyberattacks to obtain other types of corporate data. Such data might include employee information, customer or supplier data, and financial data. Two types of data that hackers often seek are **personally identifiable information (PII)** and **protected health information (PHI)**. PII is data that can be associated with a specific individual, and it includes name, Social Security number (or other unique national identifier), passport number, street address, and bank account numbers. Unauthorized access to these personal data is a privacy violation, and a criminal who obtains these data may engage in identity theft. PHI is particularly sensitive with respect to privacy, which is why there are specialized standards of protection for health information. Another type of data that has special protection is **cardholder data (CHD)**. CHD includes a credit card customer's primary account number (PAN), name, expiration data for the card, and a service code on magnetic stripe credit cards that contains information about magnetic stripe transactions. Figure 3-4 shows some of the PII cybercriminals value most.

Data can be sold and translated directly into a financial reward, or it might be used to obtain other types of assets or services. Theft of PHI may allow a threat actor to use the victim's prescriptions to obtain drugs or to access medical treatments. By obtaining an individual's proprietary access credentials, a threat actor might steal other assets, such as computer hardware or weapons. The following example illustrates how hackers can even steal your car!

EXAMPLE

Keyless entry to automotive equipment provides hackers the opportunity to steal cars in various ways. One method is to trick the vehicle into sensing that the key fob is located

nearby. This technique, known as relay theft, has been particularly effective in gaining access to specific car brands.

Threat Motivation—Why Do Threat Actors Do It?

We have described who commits cyberattacks and what they might steal, and in many cases, the "what" explains the "why." Motives do vary, however. The first motive that comes to mind for most people is financial, but money is not the only reason for cyberattacks. Hackers might attack an organization to obtain IP, as an act of terrorism or activism, for patriotic reasons, as a bullying tactic, to sabotage or engage in warfare, or just to see if it can be done.

Obtaining IP is a possible motivation for competitors, nation-states, and cybercriminals. Valuable IP includes:

- Computer programs
- Copyrights
- Proprietary designs
- Manuscripts (written works)
- Patents
- Recipes
- Research
- Strategic plans
- Trade secrets

Competitors might be motivated to steal IP to replicate products. Food companies protect their recipe IP, which may be their most valuable asset. Universities are a frequent target of cyberattacks, and one reason is that they have proprietary research. Several years ago, a cyberattack at the University of Virginia was thought to have been perpetrated by two adversaries in search of proprietary research.

Other industries that might be targets because of their IP are biotechnology, defense, government, medical and pharmaceutical, and transportation (e.g., automotive). When government or defense industry companies are targeted, the motive may extend beyond IP into espionage, where hackers are spying as well as collecting data. IP may be one of the most valuable assets a business has, as in the following example.

EXAMPLE

The 11-ingredient herb and spice recipe behind Colonel Sanders' Kentucky fried chicken was a closely guarded secret for many years and may have been the main contributor to the fast-food chain's global success.

Another threat motivation is activism, which occurs when hackers are interested in promoting a political or social agenda or perhaps stopping an action as a protest. Sometimes these motives lead hacktivists (described earlier in this chapter as a type of threat actor) to shut down a website or slow an organization's progress toward a goal. Just causing some downtime by disabling servers is a type of threat that might be attributed to activism.

Emotions, including jealousy, love, hate, envy, and vengefulness, can all be motives for a threat actor to launch a cyberattack (Figure 3-5).

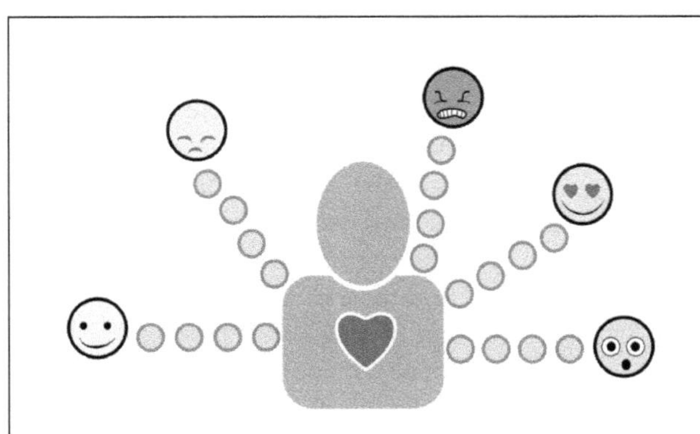

Figure 3-5. Individual Emotions That Can Be Motives (Which one does each symbol represent?)

A threat actor may not care about any financial reward or satisfaction other than to get even. This can lead to many kinds of threats, such as trolling, cyberbullying,

Trolling The deliberate act of posting provocative, offensive, misleading, or disruptive messages online with the intent to provoke emotional reactions, cause conflict, or derail conversations.

Cyberbullying The use of digital platforms to harass, threaten, or humiliate someone.

extortion, or harm. Social media, including Meta's platforms, TikTok, and others, amplify the ability of threat actors to act. **Trolling** occurs when an adversary makes inflammatory remarks online, perhaps on social media, hoping for an angry reaction. **Cyberbullying** is more serious and can be a form of abuse. The abuse could include making fun of someone publicly, sending malicious messages, or engaging in continuous online harassment.

> **CASE IN POINT**
>
> **In 2006, Megan Meier tragically committed suicide because of cyberbullying. Megan had suffered from depression and low self-esteem. A few weeks prior to her death, she was asked to be friends on a social networking website with a 16-year-old boy named Josh. Josh communicated frequently with Megan, and after some time, the messaging became mean and cruel. Other classmates and friends also began posting bullying messages on the social media platform, and this led to Megan's suicide. It was later discovered that "Josh" was not a real person but rather an identity that a neighbor created. Following her tragic death, Megan's mother created a foundation in her name and worked to have legislation against cyberbullying enacted.**

Sometimes a motivation for a threat is "just because," or ego, as we described with social hackers such as Kevin Mitnick. Hacking can involve solving a puzzle or conquering a challenge. Ideally, intellectual curiosity finds nonthreatening outlets, but occasionally a threat actor engages in activating a threat in search of knowledge and learning. Often these types of hacks are posted on underground forums on the dark web to establish bragging rights and to publicize the vulnerability.

How Do Threats Lead to a Cyberattack?

This chapter is about threats, and the next chapter covers vulnerabilities. The combination of these two creates a cyberattack, exploit, or cyber incident.

Exploit Code or techniques used to take advantage of a vulnerability to perform unauthorized actions.

Cyber Kill Chain A framework developed by Lockheed Martin to understand and defend against cyberattacks.

Many cyber threats remain unrealized. They simply don't happen. A company like Amazon may have hundreds of millions of phishing attempts each day, but they will not evolve into an actual **exploit** or cyberattack unless the threat actors are able to find a vulnerability. Vulnerabilities are described in detail in Chapter 4, but they include employees practicing poor cyber hygiene or failing to update software.

Cyberattacks have multiple stages. One version of the cyberattack life cycle is the **cyber kill chain**, a framework originally developed by the aerospace and defense company Lockheed Martin. The process, known as reconnaissance, begins with a threat actor identifying and researching a prospective target. Next is weaponization, where the adversary creates a weapon, such as malware. The attacker then delivers the weapon, perhaps through phishing emails. Next, the hacker releases the weapon and then installs it in the victim's system. Now the threat actor can take control of parts of the victim's network and move laterally throughout their information system. In the final stage, the attacker takes action to achieve their goals, which might be stealing or destroying data. Figure 3-6 depicts the seven stages in the kill chain. Threat actors may or may not proceed through all of the steps, or they may take additional actions, such as demanding ransom. The time from reconnaissance to command and control may be short, such as a few days, or it could last for years as the exploit lies dormant and goes undetected.

Supply Chain Attack A type of cyberattack where attackers target less-secure elements within an organization's supply chain to compromise the larger network. These attacks can occur in both software and hardware supply chains.

A form of attack that can cause harm to multiple organizations is a **supply chain attack**. This type of attack enables the cyber kill chain on every victim information system in the supply chain. Supply chain attacks work by compromising a third party to gain access to many companies with one malware infiltration. This type of attack provides the potential for nation-state cyberwarfare to be even more harmful.

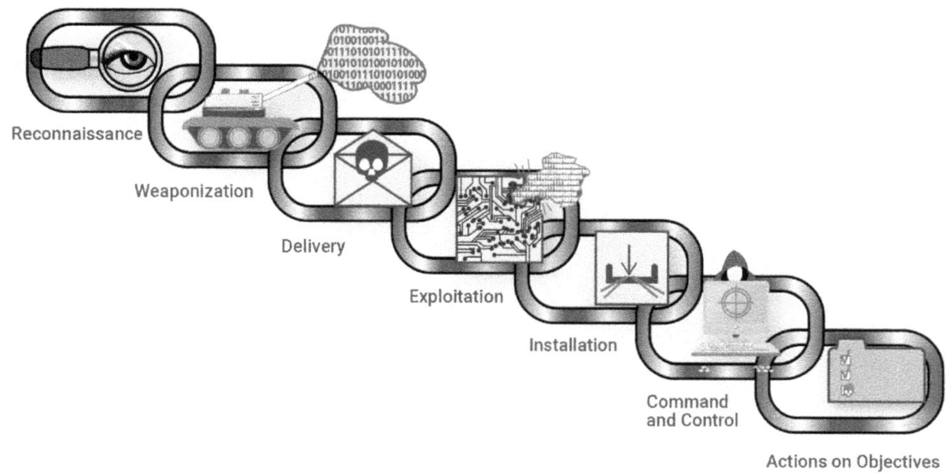

Figure 3-6. The Cyber Kill Chain Illustrated

> **CASE IN POINT**
> **SolarWinds is a US software company that offers IT products and services such as network and IT security software. In 2020, hackers launched a supply chain compromise by inserting malicious code into Orion, SolarWinds' IT performance monitoring system. When SolarWinds customers updated the Orion software, the update inserted the malware into the customer's own networks. These customers included the US Department of Homeland Security, US Treasury, FireEye, Microsoft, and Deloitte. Once these organizations were compromised, the malware could also breach third parties associated with them. A similar type of supply chain attack occurred in 2024, when a software update from cybersecurity company CrowdStrike, created a blue screen of death (BSOD) on client computers. This was not a cybersecurity attack per se but rather a coding error, but the result was a devastating impact across the supply chain.**

Types of Threats

Every day, individuals, companies, governments, and entities of all types face literally hundreds and sometimes millions of cyber threats. Some of these threats are common enough that they are categorized into types, such as malware. Others are unique threats. New threats and threat types appear constantly, and the incidence of threats and threat types varies. Some threats are cyclical and may occur at certain times of the year, such as at the peak of an organization's revenue cycle. In this section, we describe some of the more common types of threats, grouping them by these types: malware, social engineering, web-based, and other common threats.

Malware

Malware Malicious software designed to damage, disrupt, or gain unauthorized access to a computer system.

One of the most insidious threats to cybersecurity is **malware.** Malware is short for "malicious software" and is any kind of software program intended to steal data and/or cause damage to computers or computer systems. A malware attack typically follows all the stages in the cyber kill chain. Adversaries will first conduct research and due diligence to choose a target, determine specific vulnerabilities, and select a mode of attack. Next, the threat actor exploits a vulnerability to insert the malware into a computer system. Malware finds its way into your system via hacked websites, through Internet ads, from infected file downloads, through program and app installations, from emails and text messages, or from just about anything you access electronically. After insertion, the malware establishes a foothold and avoids detection as it explores the target. At this point, the adversary has obtained remote access to the target. Next, the malware works to compromise more computers on the network and infiltrate additional systems and accounts. In the last stage, the attack delivers its **payload**, which may be data exfiltration or system corruption and disruption.

Payload The part of malicious software (malware) or exploit code that actually delivers the intended harmful action once the attack successfully compromises a system.

Malware can take many different forms, from adware to worms, as we will describe here.

Adware

Adware Software that automatically displays or downloads advertising material when a user is online.

Adware is advertising-support software that displays unwanted or unrequested advertising on your computer screen when you are connected to the Internet. The malware automatically delivers these ads to your user interface. The biggest danger adware poses is that it degrades the user experience. It may also slow your computer's performance. Adware is not intended to harm the computer itself, but rather it generates revenue for the owner simply by displaying the ad.

Bots

Bots A bot is a software program that operates on the Internet and performs repetitive tasks. While some bot traffic is from good bots, bad bots can have a huge negative impact on a website or application.

Botnet A network of compromised computers that are remotely controlled by a hacker, often used for large-scale cyberattacks.

Bots (short for robots) can be good or bad. Good bots can execute useful functions such as repetitive, predefined tasks, and since they are automated, they accomplish these tasks more efficiently than humans. Malicious bots or zombies are malware on an individual workstation programmed to take over your computer and carry out harmful actions, such as stealing data or passwords, sending spam, or launching DDoS attacks. A **botnet** is a network of malware-infected systems that can consist of literally millions of individual infected workstations. Figure 3-7 depicts how malicious bots attack a network.

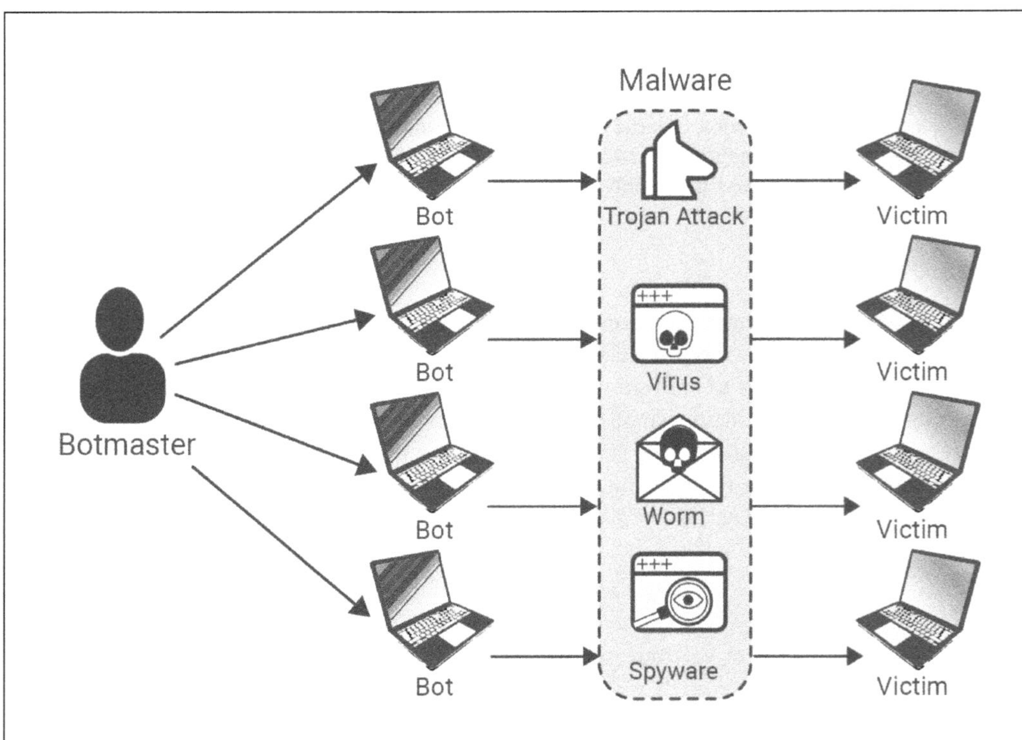

Figure 3-7. Botnet Attack

Ransomware

Ransomware remains one of the most common threats facing organizations of all sizes and industries. Ransomware is a favorite mode of attack among cybercriminals because it can easily exploit vulnerabilities and result in anonymous cryptocurrency payoffs.

Ransomware Malicious software that encrypts a victim's data and demands payment for the decryption key.

> **CASE IN POINT**
>
> **Institutes of higher education are a frequent target for ransomware. These organizations may not have the resources to maintain state-of-the-art information security, they have highly valuable data, and they take in a new class of users each academic term. In September 2021, a ransomware attack shut down online and hybrid classes at Howard University, a prestigious historically black university in Washington, DC. The school shut down its network to investigate the attack. Howard now joins Michigan State University and the University of California San Francisco, among many other colleges and universities, as victims of ransomware attacks.**

In a ransomware attack, the victim might open their workstation to find a message from the threat actor communicating that their data is now held hostage. The data has been encrypted, and unless the victim enterprise pays a ransom, using a form of cryptocurrency, the adversary will not provide a decryption key. The attacker is holding the data hostage. Without adequate available backup data and processes, the organization may not be able to recover. Many entities now protect their data from this ransomware threat through frequent backup procedures and by maintaining off-site recovery capabilities. Unfortunately, ransomware attackers have a secondary threat available to them. If hackers have **exfiltrated** the data, they can threaten to expose PII and other confidential data publicly. These attackers often sell their stolen data on the dark web regardless of whether the ransom is paid.

Exfiltrate To steal and transfer data out of a system or network without authorization. It's the digital equivalent of secretly smuggling information out of a secure facility.

As in the case of physical hostage-taking situations, law enforcement and government discourage paying any ransom in these malware attacks. Unfortunately, even when the ransom is paid, the attacker may not provide the encryption key, or some or all of the data may not be recoverable. Still, some chief executives feel that they have no choice but to pay the ransom. This was the case with the Colonial Pipeline attack in 2021. The CEO apologized for paying the ransom but felt that he had no other option because of the criticality of maintaining gasoline supply. The good news is that there may be a downward trend in the percentage of companies willing to pay ransom, as noted in the following case.

> **CASE IN POINT**
>
> **Coveware, a consulting firm that helps organizations respond to and recover from ransomware attacks, reported some encouraging statistics in 2024. The company revealed that ransom payment rates fell from more than 80% in 2019 to 32% in the third quarter of 2024. During that same quarter, they identified healthcare as the number one targeted industry for ransomware attacks (Coveware, 2024).**

Spyware

Malware that obtains information about an individual or an organization on behalf of another entity with intent to harm is called **spyware**. Spyware captures personal data, such as login IDs, passwords, browsing activity, and banking information, by monitoring activity. **Keyloggers** are a type of spyware installed on end-user computing devices that records every keystroke typed. The record is then communicated via the Internet to the remote hacker responsible for installing the malware.

Spyware Malicious software that secretly monitors and collects information about a user's activities.

Keylogger Software or hardware devices that track and record a user's keystrokes.

Trojans

Trojan A type of malware that disguises itself as legitimate software or code to trick users into installing it on their systems.

During the Trojan War, the Greeks created a wooden horse and presented it as a gift to the city of Troy. Unfortunately for the Trojans, Greek warriors were hiding inside the hollow horse. Once the horse was inside the city, the warriors emerged and opened the city gates to attack. **Trojans** have come to symbolize a ruse to gain unauthorized entry, and Trojan malware does just that. Unwitting users click on a seemingly innocuous file link that appears legitimate, only to find that they have introduced malware into their computer network.

> **EXAMPLE**
> **Users are tricked into entering their banking user login and password at a fake financial institution website. This is known as a "banking or banker Trojan," and hackers subsequently use the stolen credentials to extract financial resources from the victim's account.**

Rootkit Trojan A Trojan malware type that operates at a deep level within the system, often evading detection by standard security measures.

Remote Access Trojan (RAT) A Trojan malware type that allows the attacker remote access to and control of the infected system.

There are several different types of Trojans. A **rootkit Trojan** prevents malware that has already infected a system from being discovered. Another type of Trojan malware is a **remote access Trojan (RAT)**. In this type of attack, the adversary can obtain complete remote control over a computer system. In a RAT attack, the hacker can maneuver the user's mouse and keyboard, access files, and access any other network resource for which the target user has access.

Viruses

Virus A computer program that spreads by first infecting files or the system areas of a computer or network router's hard drive and then making copies of itself.

A computer **virus** derives its name from human viruses because it infects and spreads in much the same way. A software virus infects by modifying computer programs with its own malicious code. The virus spreads from computer to computer in a network when the host software or document it is attached to is transferred. This can be done through infected attachments or other file sharing methods. For example, an infected spreadsheet program may be shared, and each new user will be spreading the virus. Antivirus software can protect against viruses with known signatures, but it is not 100% effective, as it cannot guard against zero-day viruses, which have yet to be identified. By incorporating AI, antivirus software can analyze patterns and behavior to protect against new viruses. Still, hackers are becoming increasingly sophisticated in developing virus code that can escape detection.

Worms

Worm Malicious software that replicates itself and spreads to other systems without user intervention.

A malware **worm** is very similar to a computer virus, with the primary difference being that a worm does not require a host to copy itself and spread. This self-replication feature is a distinguishing characteristic, though viruses and worms are often confused with one another.

> **CASE IN POINT**
> **The first computer worm, often misidentified as a virus, was named Creeper. Creeper was created in about 1970 by Robert Thomas, an engineer at BBN (forerunner to Raytheon). Creeper was a self-replicating test program that had no malicious intent. As the worm moved throughout the system, infected computers displayed the message "I'm the Creeper. Catch me if you can."**

Social Engineering

Social Engineering The manipulation of individuals into divulging confidential information or performing actions that compromise security.

As will be described in the next chapter, the most significant vulnerability in information systems is humans. People are fallible, and in addition to making mistakes, they can be manipulated. **Social engineering** is a hacking technique used to trick people, through deception and persuasion, into revealing confidential information or performing actions that are adversarial to information security. Social engineering can be accomplished through any type of human interaction, including email, text messaging, phone calls, or physical behavior.

Phishing (and various other forms)

Phishing is the most prevalent form of social engineering today. In these attacks, a bad actor sends mass emails to potential targets, hoping to entice one or more individuals to reveal personal information or to download malicious software. Victims receive emails that ask them to either click on a malicious link or open a compromised attachment. The email will be disguised to look like it is from a legitimate source, such as a financial institution or vendor. In the past, these emails were fairly easy to detect as fraudulent because of spelling errors, poor grammar, or a subtle change in the domain name. As hackers have become more sophisticated, phishing emails have become harder to spot. Hackers can use AI to craft sophisticated messages and overcome any language challenges. Phishing emails frequently cite urgency, trying to get the victim to act quickly so that they won't have time to double-check the email source. Phishing emails also tend not to address the recipient by name; they may say there is a problem, such as suspicious activity in an account; or they might lure a victim to open a link or attachment by saying that they have won a prize or reward.

Increasingly, phishing attacks are more targeted to the individual recipient. These are called **spear phishing** or **whaling** threats. Spear phishing is an online attack against one or more specific victims. The threat actor will try to glean some personal information about the target, such as email address, location, employer, or purchasing habits. They will then use any insight they gain to craft an email that appears legitimate because of the insider knowledge included in the message. The goal, as with other phishing techniques, may be to get the victim to click on a link or open an attachment, or to reveal PII that will allow the perpetrator to access financial accounts or engage in identity theft. Whaling is a variation of spear phishing in which the target is a senior executive or high-profile individual.

Email is the most prevalent avenue for phishing attacks, but hackers are also luring victims into revealing PII through text messaging and sometimes phone calls. Phishing through texts is called **smishing**. Attempts by threat actors to obtain a victim's data through phone conversations is called **vishing**, which is short for "voice phishing." Spam phone calls are ubiquitous, and most are ignored or dropped. Vishing soared in 2024 (Crowdstrike, 2025), becoming a popular way for hackers to gain initial access.

> **EXAMPLE**
>
> **A hacker calls a university Help Desk and says that they are a professor who has forgotten the password associated with their login ID. Using information found on a social media site for career networking, the threat actor is able to convince the Help Desk staff member to share the password.**

Threat actors can also compromise QR codes in a **quishing** attempt. Victims click on an infected QR code that directs them to a fake website or downloads malware.

Physical Social Engineering

Physical social engineering attacks may take the form of **tailgating**, also sometimes called **piggybacking**. In this type of attack, a hacker follows an individual who has authorized access into a physical structure such as an office building or, more specifically, a restricted data center. **Dumpster diving** gets its name from the image of someone picking through trash in a dumpster to find information that can be used to attack a computer network. Many people shred their sensitive paperwork for privacy reasons, and companies have disposal policies that make use of shredders and shredder services to both protect privacy and avert cyberattacks that use this technique. **Shoulder surfing** is yet another type of physical engineering, where the threat actor may wander through an office, stopping at a desk as someone is signing on to a network.

Phishing A social engineering technique that attempts to acquire sensitive data through a fraudulent solicitation in email or on a website, in which the perpetrator masquerades as a legitimate business or reputable person.

Spear Phishing A targeted phishing attack aimed at a specific individual or organization, often using personalized information to make the attack more convincing.

Whaling A type of phishing attack that targets high-level executives or other important individuals in an organization, often involving personalized and sophisticated tactics.

Smishing Derived from "SMS" and "phishing," is a type of cybercrime that uses deceptive text messages to manipulate victims into divulging sensitive information.

Vishing Short for voice phishing, refers to fraudulent phone calls or voice messages designed to trick victims into providing sensitive information or access.

Quishing Short for QR code phishing. A type of phishing attack where cybercriminals use malicious QR codes to trick victims into visiting fraudulent websites, downloading malware, or revealing sensitive information.

Tailgating Same as Piggybacking. A physical security breach where an unauthorized person gains access to a restricted area by following an authorized individual.

Piggybacking Same as Tailgating. A physical security breach where an unauthorized person gains access to a restricted area by following an authorized individual.

Dumpster Diving The practice of searching through trash to find sensitive information that has been discarded by employees/private citizens.

Shoulder Surfing A social engineering technique where an attacker observes someone's private information by either watching them directly or through visual aids (cameras, mirrors, etc.) without the target realizing it.

Pretexting

Pretexting The use of a fabricated story, or pretext, to gain a victim's trust and trick or manipulate them into sharing sensitive information, downloading malware, sending money, or otherwise harming themselves or the organization they work for.

One more type of physical social engineering attack is **pretexting**. This technique involves a hacker posing as a trusted individual or as someone investigating a victim. For example, a threat actor may pose as someone from the FBI or IRS, or as a vendor or contractor, to gain trust and acquire PII or data that may facilitate a cyberattack.

CASE IN POINT

In 2023, a social engineering ploy sent a company valued at more than $30 billion back into the predigital era of paper and pencil. A hacking group studied LinkedIn profiles of MGM Resort employees and obtained sufficient information to convince a Help Desk employee at MGM to part with login information. That 10-minute phone call to the Help Desk led to a ransomware attack that caused a disabled website, inoperable digital room keys, and nonfunctioning reservation systems and slot machines.

Hackers can research victims to learn about them by studying their online profiles in social media or networking websites or apps. Many individuals do not realize how much information about themselves appears on the Internet that could allow a hacker to "know" them.

EXAMPLE

Have you ever taken a quiz offered on social media? Perhaps the quiz asks you the name of your best friend or your pet? These are clues about you that hackers can use in identity theft and other scams.

Business Email Compromise

Business Email Compromise Similar to spear phishing, where a scammer gains access to a trusted person's email account and requests actions in the name of the account owner, usually a boss or manager.

In a **business email compromise (BEC)** scam, a threat actor sends an email message, posing as a trusted business stakeholder (e.g., a boss or a customer). It might appear to be a legitimate request for information or for money. The victim doesn't know the email system has been compromised or infiltrated, so they may be tricked into giving away classified information or company resources. The message does not contain malware or a suspicious link; it is simply a fake communication that has been able to bypass an organization's email security filter.

Web-Based Threats

This chapter describes threats in terms of an exploit itself (e.g., malware) and how it might be delivered (e.g., phishing). The two are intertwined and commonly thought of together as threats. Another example of this type of threat are web application attacks, which can include malware. **Web application threats** stem from infected websites. The sites may be infected with a form of malware, or an adversary may use another approach to compromise a site. There are many examples of web application threats and attacks, with a few of the more common ones described here.

Web Application Threats Cyber threats specifically targeting web-based applications, often through techniques like SQL injection, cross-site scripting (XSS), or session hijacking.

Brute Force or Dictionary Attacks

Brute Force Attack Attempts to gain unauthorized access to a system by systematically trying all possible combinations of letters, numbers, and symbols for passwords or encryption keys.

Dictionary Attack A method of breaking a password or encryption by systematically entering every word in a predefined list of likely passwords.

A **brute force** or **dictionary attack** occurs when a threat actor attempts to penetrate a secure application, such as a web application, using every possible combination of words, special characters, and letters. A hacker can employ an automated script or bot to try every possible grouping of characters to log in or guess a password or PIN and gain access to a site or application. A dictionary attack is more targeted, using a source of common passwords or login IDs to generate character combinations to obtain access.

Injection Threats

Adversaries can gain access to web applications in several ways by injecting malicious code. One approach is **cross-scripting (XSS)**, when an attacker injects malicious script into web pages at trusted websites. Another threat of injection is **SQL injection**. Structured Query Language (SQL) is a powerful computer language that is used to communicate with and manage data in relational databases. The threat of SQL injection is that an attacker can insert malicious commands into the application that allow them to exfiltrate, modify, or delete data.

Cross-Scripting (XSS) A type of security vulnerability typically found in web applications, where attackers inject malicious scripts into content from otherwise trusted websites.

SQL Injection A type of attack where malicious SQL code is inserted into a database query, allowing the attacker to gain unauthorized access to data.

CASE IN POINT

One of the earliest major SQL injection exploits was the 2007 attack on the 7-Eleven convenience store chain. In that breach, the bad actors accessed 7-Eleven's servers through a SQL injection (probably similar to the following example) on their website. Having gained access to the 7-Eleven servers that processed the ATM transactions, they installed malware that allowed them to steal the logins and card data as they were transmitted between the servers and the ATM machines. For about 12 days, customers who logged into ATMs at the stores had their debit and credit card numbers and related credentials stolen by the adversaries.

EXAMPLE

Here is an example of using SQL code to hack a login screen that requires users to input their username and password:

```
SELECT * FROM users WHERE username = 'user_input' AND password = 'password_input';
```

A threat actor can insert ' OR '1' = '1' and any input in the password field to try to obtain a list of all the records in the user data table.

Distributed Denial of Service (DDoS) A cyberattack in which multiple compromised systems are used to flood a targeted system with traffic, overwhelming it and causing it to be unavailable.

Distributed Denial of Service (DDoS) Threats.

A **DDoS threat** (see Figure 3-8) is an exploit in which a hacker overwhelms a server or network with traffic, thus disabling the site or application. First, the threat actor infects large numbers of computers and IoT devices with malware to obtain remote control. These compromised devices become bots, and upon receipt of attack instructions, they will all send requests to a target's IP address to flood it with requests and thus disable its ability to serve legitimate requests.

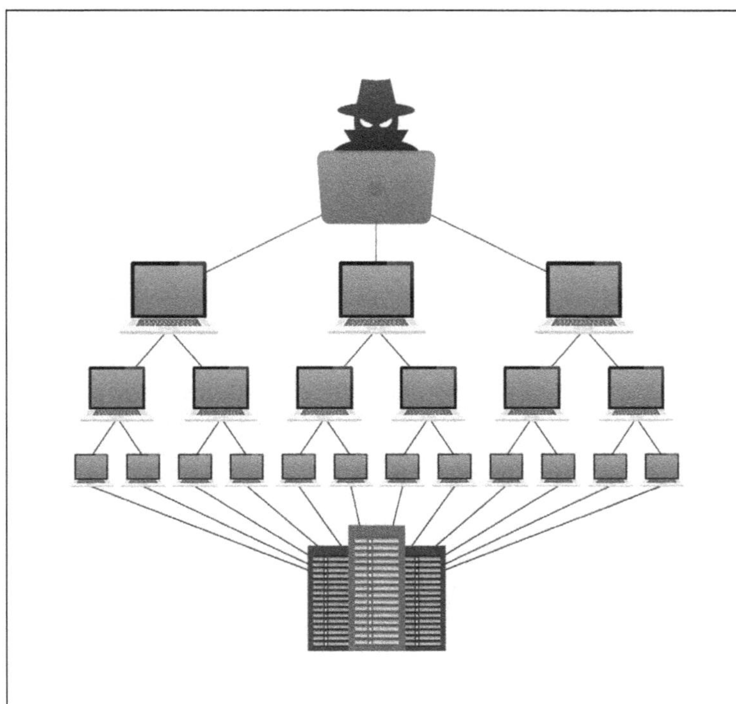

Figure 3-8. A DDoS Attack

Other Common Threats

In addition to the threats already described, many other threats do not fit into those categories. Several are listed here, with the most common being described more fully afterward.

- Advanced persistent threats (APTs)
- Audit log injection
- Cloud jacking
- Deepfakes/synthetic identities

- Man-in-the-middle attacks
- Buffer overflow
- Mobile code
- Race conditions
- Covert channel
- Replay and return-oriented attack

Note that this is not a comprehensive list, and threat types continue to expand at a rapid pace.

Advanced Persistent Threats (APTs) A long-term targeted cyberattack in which the attacker gains unauthorized access to a network and remains undetected for an extended period.

Advanced Persistent Threats (APTs)

As the name implies, **advanced persistent threats (APTs)** are advanced, and they continue over time. This type of threat may take many forms, including malware, but its hallmark is that it is typically highly sophisticated and likely to have been developed by a nation-state, a state-sponsored threat actor, or a highly skilled adversary deploying an attack against one or more governmental or corporate entities. These attacks require a high level of technical skill and significant funding

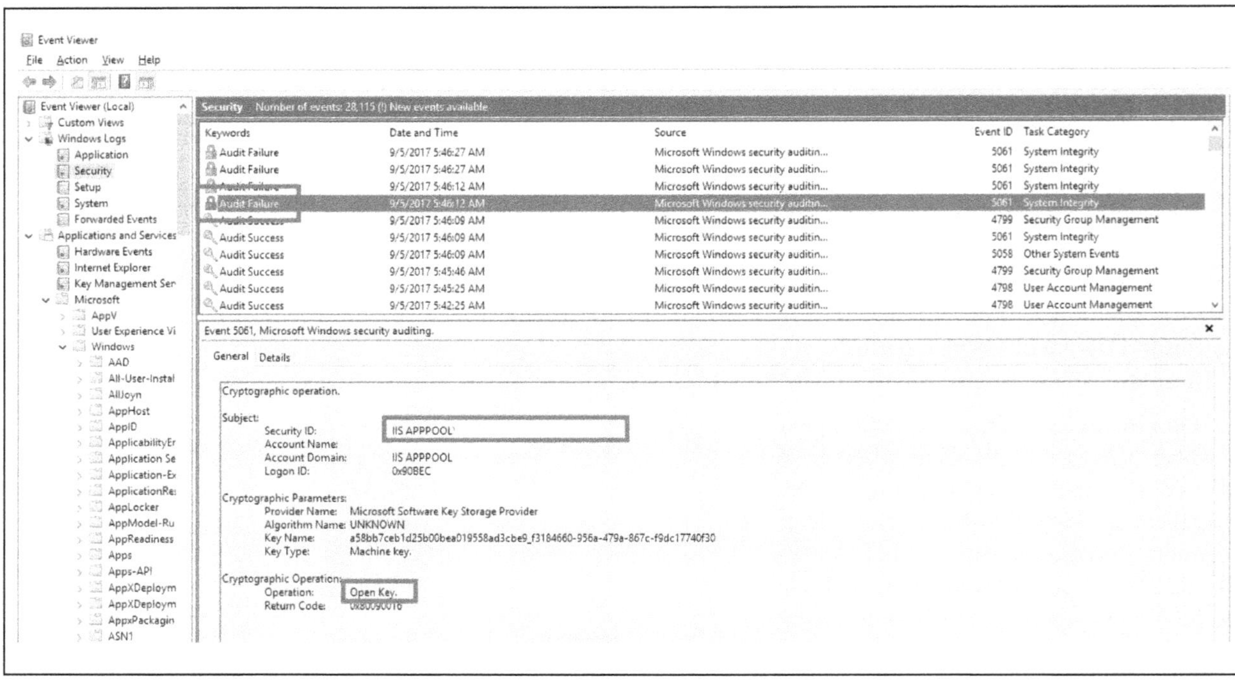

Figure 3-9. Example of an Audit Log

to execute. In an APT attack, the threat actor obtains unauthorized access to a computer network and remains undetected for some time until the payload is triggered.

Audit Log Injection A type of security vulnerability in which an attacker inserts malicious or deceptive content into an application's audit or security logs. The goal is usually to manipulate, hide, or falsify log records.

Audit Log Injection

An audit log (Figure 3-9) records activity within a computer system, network, or application. It is detailed and allows auditors and others to identify suspicious activity, comply with legal data requirements, troubleshoot problems, and provide accountability if needed.

In an **audit log injection**, the threat actor accesses the log and inserts fictitious content. The injection might cause the audit trail to be unreliable in an investigation or cover traces of an attack.

Cloud Jacking

As more individuals and organizational entities move to storing data in public and private clouds, **cloud jacking** or cloud hijacking is likely to increase as a threat. This threat involves hackers accessing and using a victim's cloud resources for their own purposes. Cloud accounts hold PII and PHI for individuals, governments, and businesses, along with every other type of data and information that can be of interest to hackers. As a threat, cloud jacking is like any other threat to data, except that the platform changes from on-perimeter computers, servers, or other digital devices to data storage maintained by third parties.

Cloud Jacking The act of taking over or compromising a cloud computing environment or resource.

Deepfakes/Synthetic Identities

Sometimes it is difficult to know what is real and what is not. **Deepfakes** are a form of synthetic identity where pictures or videos are compromised with fake identities. Your image can appear in situations or contexts that you never experienced. Celebrities, like political figures, may appear in deepfake videos that show them doing something that is contrary to their political stance or values, such as with President Obama in the following Case in Point. Unfortunately, as AI technology continuously improves, synthetic or fake identities are increasingly difficult to spot and have the potential to do harm as they deceive viewers to achieve myriad goals related to behavioral influence. Social media platforms provide a mechanism for distribution of deepfakes, and regulations and social media platform policies have not kept up with the evolution of this technology, which can have an impact on personal behaviors and even elections.

Deepfake A piece of synthetic media in which a person's likeness is replaced with another's or is manipulated in other ways, often used maliciously to create fake videos or audio clips.

CASE IN POINT

Actor Jordan Peele created a fake video as a public service announcement that shows a video clip that is presumably of President Barack Obama talking about the dangers of fake news and misinformation. If you watch the video, you will believe that you are seeing Barack Obama, but it is really Jordan Peele. AI technology allowed the actor to substitute his personal facial motions onto Obama's facial characteristics. The video had millions of views on Facebook, Twitter, and YouTube.

Man-in-the-Middle

A **man-in-the-middle (MitM)** threat is the chance of a cyberattack that intercepts a digital communication. The purpose may be to either simply listen in or to modify the communication. The threat actor may pose as an imposter for one of the two parties participating in the conversation. MitM threats can take many different forms, including session hijacking and spoofed HTTPS. A simple form is Wi-Fi eavesdropping, when the attacker monitors communications on public or unsecured wireless connections. The hacker might also create a fake Wi-Fi network that sounds real. A user in a public space might connect to it and then have their credentials stolen by the hacker.

Man-in-the-Middle (MitM) Attack A cyberattack where an attacker secretly intercepts and relays communication between two parties who believe they are communicating directly with each other.

Threat Modeling A structured process for identifying, evaluating, and addressing potential security threats.

Threat Modeling

With so many different types of threats, an information security professional can be overwhelmed trying to systematically assess their risk. **Threat modeling** is a tool that provides a framework for categorizing and analyzing threats. Threat modeling techniques can help an organization allocate security resources and decide on optimal cyber defense strategies. Many threat modeling tools are available. One of the earliest such tools is the **STRIDE threat model** (Figure 3-10). Two Microsoft engineers developed this model in 1999. STRIDE is an acronym for the particular set of threats it covers: spoofing, tampering, repudiation, information disclosure, denial of service, and elevation of privilege. For each threat, the STRIDE model provides a definition and notes the information security property that it violates.

STRIDE Threat Model A developer-focused model to identify and classify threats under six types of attacks—spoofing, tampering, repudiation, information disclosure, denial of service (DoS), and elevation of privilege.

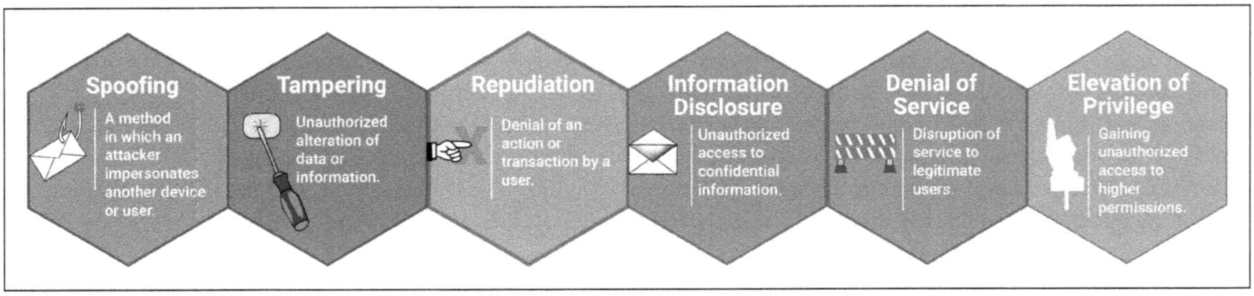

Figure 3-10. The STRIDE Threat Model (Klein, 2025)

Tactics, Techniques, and Procedures (TTPs) The methods and strategies that threat actors use to carry out cyberattacks.

Other threat models include MITRE ATT&CK, OCTAVE, and PASTA. MITRE ATT&CK is a matrix of threats for each step in an attack life cycle (similar to a cyber kill chain). For example, it identifies **tactics, techniques, and procedures (TTPs)** threat actors may use, which allows security professionals to model threats and allocate resources to defend against them. We discuss MITRE ATT&CK in more depth in Chapter 7.

Operationally Critical Threat, Asset, and Vulnerability Evaluation (OCTAVE) is a threat modeling approach that focuses on cybersecurity risks. Rather than emphasizing technology vulnerabilities, this modeling process focuses on organizational risks and how they might affect an entity's operations. Threat modeling teams include both technical and nontechnical members, who work together to identify critical assets, threats, and vulnerabilities, culminating in the development of strategies and procedures to manage cyber risks.

PASTA also takes a risk-centric approach to threat modeling, allying business objectives with cybersecurity initiatives. The process involves seven steps, including defining the objectives and scope and identifying threats and vulnerabilities. The process concludes with a risk and impact analysis.

Threat Intelligence Evidence-based knowledge about past, current, and future threats, including information on threat actors and their motives.

Threat Hunters Security professionals who proactively search for cyber threats both within and outside an organization's network.

Threat Maps Also known as a cyberattack map, a real-time map of computer security attacks that are going on at any given time. Can be at the company, regional, national, or worldwide level.

Threat Index A measure used to quantify the level of threat that various cyber threats pose.

Information Sharing Analysis Centers (ISACs) Groups that share threat intelligence to protect critical infrastructure.

Threat Intelligence

Threat intelligence is evidence-based knowledge about past, current, and future threats, as well as information on threat actors. It can come from many sources, including online resources, professional conferences, professional associations, social media, the dark web, and information security consultants. Often, threat intelligence focuses on adversaries or threat actors, offering insights about their possible motives, the environments in which they operate, types of attacks, and possible vulnerabilities they would exploit.

Different forms of intelligence about threats include threat reports and trackers, indices, graphs, and maps. Many cybersecurity consulting firms employ **threat hunters** and produce proprietary reports and analyses of threats. Threat hunters are IT professionals who seek out and report on advanced threats. They can work for a business and have the responsibility of protecting against such threats by addressing vulnerabilities and risk, or they may work for consultants and help detect threats that can impact clients.

Some of the top cybersecurity firms offer threat reports in different forms, often for free, as they seek information about inquirers who may be prospective clients. **Threat maps** show, in real time, a global view of attacks. These maps may include information about country and industry sources and targets of threats. Reports may take the form of a **threat index**, which measures, analyzes, and ultimately provides a comparative score for threat levels over a period.

CASE IN POINT

ENISA, the European Union Agency for Cybersecurity, created by the European Union in 2004, provides many public reports on cybersecurity threats. They include a variety of threat landscape reports on topics such as supply chain threats and threats related to

> **Internet infrastructure, along with articles, speeches, and other documents. All are available for download at their website:** https://enisa.europa.eu.

An important aspect of threat intelligence is shared knowledge of threats. **Information Sharing Analysis Centers (ISACs)** and **Information Sharing and Analysis Organizations (ISAOs)** are two entities that have this goal. ISACs were created to share knowledge that would protect critical infrastructure. They are industry-specific, and they gather and share threat intelligence, including between private and public sector groups, to protect critical infrastructure from cyber threats. As they have evolved, they have expanded into more industrial sectors. ISAOs originated with a White House Executive Order calling for voluntary threat intelligence sharing within industry groups.

Information Sharing and Analysis Organizations (ISAOs) Groups that voluntarily share threat intelligence within industries.

Summary

Threats are one of the main components of cybersecurity. This chapter described the who, what, why, and how behind threats to information security. The "who" are the bad actors or adversaries presenting the threats. The "what" are the information security assets that the threats target, and the "why" are the motivations that inspire them. The "how" is the approach taken to launch the threat. Many threats may be categorized as malware, social engineering, or web-based, though a few others do not easily fit into these categories. Threats continue to evolve and multiply, and it is impossible to provide a comprehensive database of all threats. However, to provide a good starting point, this chapter described many of the most common ones. Threat modeling and intelligence include frameworks, reports, tracking, indices, mapping, and other tools that provide information on past, present, and potential threats.

Chapter Review

Key Terms

Advanced Persistent Threats (APTs)
Adware
Anonymous (Hacker Group)
Audit Log Injection
Blue Screen of Death (BSOD)
Botnet
Bots
Brute Force Attack
Business Email Compromise
Cardholder Data (CHD)
Cloud Jacking
Cross-Scripting (XSS)
Cryptocurrency
Cyber Kill Chain
Cyberbullying
Cybercriminal
Cyberwarfare
Deepfake
Dictionary Attack
Distributed Denial of Service (DDoS)
Dumpster Diving
Exfiltrate
Exploit
Hacker
Hacktivist
Information Sharing Analysis Centers (ISACs)
Information Sharing and Analysis Organizations (ISAOs)
Intellectual Property (IP)
Keylogger
Malware
Man-in-the-Middle (MitM) Attack
Payload
Personally Identifiable Information (PII)
Petty Cybercriminals
Phishing
Piggybacking
Pretexting
Protected Health Information (PHI)
Quishing
Ransomware
Ransomware as a Service (Raas)
Remote Access Trojan (RAT)
Rootkit Trojan
Script Kiddies
Shoulder Surfing
Smishing
Social Engineering
Software as a Service (SaaS)
Spear Phishing
Spyware
SQL Injection
STRIDE Threat Model
Supply Chain Attack
Tactics, Techniques, and Procedures (TTPs)
Tailgating
Threat
Threat Hunters
Threat Index
Threat Intelligence
Threat Landscape
Threat Maps
Threat Modeling
Trojan
Trolling
Virus
Vishing
Web Application Threats
Whaling
Worm

Review Questions

1. In the context of cybersecurity, define a threat and explain why it is important to understand threats.
2. What are the common motivations for cybersecurity attacks? How are they similar? How do they differ from each other?
3. List the assets targeted in cybersecurity attacks and the motivation behind the attempt to obtain these assets.
4. Who are the main sources of cybersecurity threats? What are their motivations? How do their targets differ?
5. What types of threats are there? How are they similar or different from one another?
6. What is malware? List the common types of malware and how each type operates in an attack.
7. Describe social engineering and its common techniques.
8. What is threat modeling? List and describe some of the common threat modeling frameworks.
9. What is threat intelligence? Why is obtaining threat intelligence important?

Multiple Choice

Identify the choice that best completes the statement or answers the question.

1. What is the advantage of cyberwarfare over traditional warfare?
   a. It involves diplomatic negotiations.
   b. It is less damaging to infrastructure.
   c. It requires expensive weaponry.
   d. It has a lower barrier to engage.
   e. It relies on physical confrontation.
2. What types of data do hackers often seek in cyberattacks?
   a. Social Security numbers and passport numbers
   b. Employee information and customer data
   c. Personally identifiable information and cardholder data
   d. Financial data and protected health information
   e. All of the above
3. Which type of malware spreads from computer to computer by modifying computer programs with its own malicious code?
   a. Ransomware
   b. Worms
   c. Spyware
   d. Viruses
   e. Trojans
4. What is a tactic hacktivists often employ to harm organizations they oppose?
   a. Blocking access to websites
   b. Conducting ransomware attacks
   c. Launching phishing attacks
   d. Providing technical support
   e. Offering financial rewards
5. Why did cybersecurity threats escalate significantly during the global COVID pandemic?
   a. Advancements in threat intelligence tools
   b. Decreased reliance on email communications
   c. Increased remote work
   d. Reduced availability of professional conferences
   e. Decreased vulnerability in information security
6. What is the payload in the context of information security threats?
   a. The motivation for compromise
   b. The vulnerability or weakness in information security
   c. The potential for an adversary to negatively impact an asset
   d. The impact of an exploit
   e. The entire set of potential and existing threats
7. What is the term for phishing attacks conducted through text messages?
   a. Vishing
   b. Dumpster diving
   c. Spear phishing
   d. Tailgating
   e. Smishing
8. What is a potential motive for cyberattacks beyond financial gain?
   a. Solving a puzzle
   b. Promoting a political agenda
   c. Obtaining intellectual property
   d. Seeking revenge
   e. All of the above

Case Studies

Briarcliff College—Information Security

Briarcliff College is a four-year liberal arts college in Westchester County, New York. The school has about 2,000 undergraduate students and is known for its English literature and American history programs.

Briarcliff has five IT staff members, including the Chief Information Officer, two system administrators, a network engineer, and a Help Desk technician. The staff is supported by three part-time student interns. No staff member is dedicated to cybersecurity, but each role bears some responsibility for it. For example, one of the system administrators oversees the firewall, antivirus, and attack detection software. The Help Desk technician is charged with overseeing cybersecurity training for faculty, staff, and students.

Case Requirements

1. Should Briarcliff College hire a dedicated cybersecurity staff member? Why or why not? Support your answer with research about the education industry and arguments for assigning an information security role to one individual versus matrixing the protection across all IT staff.
2. Of the cybersecurity threats identified in the chapter, which threat type is likely to pose the most risk for Briarcliff College?

Equifax—Threat Actors and the Cyber Kill Chain

One of the most significant cyberattacks to date was the 2017 Equifax data breach, where hackers stole the PII of more than 140 million individuals. Equifax is one of three US national credit agencies and holds personal and financial data for individuals, which can be used to assign credit ratings and facilitate financial transactions.

Case Requirements

1. Who were the threat actors responsible for this attack, and what was their motivation?
2. This chapter described the cyber kill chain. Research the Equifax breach online and describe how the attackers carried out each stage of the attack.
3. In 2025, another credit reporting agency also suffered a data breach. Using AI or other online resources, briefly describe what is known about this cyber incident.

Yellow Subs—Cyber Threats

Yellow Subs, a Herron Consulting (Natalie's employer in Chapter 2) client, has 400 sandwich shops in the United States. Some of the stores are company-owned, but the majority are owned by franchisees. The company has expanded rapidly during the past five years and has only recently become concerned about cybersecurity threats. Natalie has been assigned to conduct an analysis of the threat landscape for the client.

Case Requirements

1. What assets would a fast-food restaurant chain have that might attract a cybercriminal?
2. Who are the likely threat actors who pose a cybersecurity threat to Yellow Subs?
3. Using online search tools and/or AI, research the restaurant industry and describe the types of threats that are most common for a company like Yellow Subs.
4. Using online search tools and/or AI, research to find cases where fast-food chain restaurants like Yellow Subs have undergone a cyberattack.

Exercises

3-1. This chapter described various types of PII. Required:

1. Using online research tools, including AI, find the current values for each of the following: US passport, US Social Security card, merchant credit card, UK driver's license. Which of these would fetch the highest price? Why do you think that is the case?
2. PHI is even more valuable than PII. Explain how a threat actor could use a victim's PHI to obtain prescription drugs. Can you find an example of this crime online?

3-2. Phishing is one of the most common threats facing business entities, and it is a relatively easy way for threat actors to launch an attack. A threat actor can send out thousands of messages with embedded malware, knowing that at least a few recipients are likely to take the bait. Required:

1. Identify four characteristics that would alert you to a suspicious email message.
2. Some companies send out phishing emails to employees as a method of cybersecurity training. In some cases, those who click on a link in the message are required to undergo a course in cybersecurity training. Do you think this is an effective approach to help an organization protect against phishing threats? Explain your reasoning.

3-3. SQL injection and cross-scripting are two kinds of web-based threats. Required:

1. What are the similarities and differences between these two types of threats?
2. How common are SQL injections? How can you protect against them?

References

Coveware. (2024, November 1). *Law enforcement doxxing raises risk profile for threat actors.* https://www.coveware.com/blog/2024/11/1/law-enforcement-doxxing-raises-risk-profile-for-threat-actors

CrowdStrike. (2025). *2025 Global threat report.* https://www.crowdstrike.com/en-us/global-threat-report/

Klein, Charlie. (2025, April 28). STRIDE Threat Model: A Complete Guide. Jit. https://www.jit.io/resources/app-security/stride-threat-model-a-complete-guide

Nakashima, E. (2024, November 21). Top senator calls Salt Typhoon "worst telecom hack in our nation's history." *Washington Post.* https://www.washingtonpost.com/national-security/2024/11/21/salt-typhoon-china-hack-telecom/

Rundle, J. (2024, November 21). The AI effect: Amazon sees nearly 1 billion cyber threats a day. *Wall Street Journal.* https://www.wsj.com/articles/the-ai-effect-amazon-sees-nearly-1-billion-cyber-threats-a-day-15434edd

Chapter 4

Cybersecurity Vulnerabilities

Chapter Outline

Identifying Vulnerabilities
The Attack Surface and IoT
Types of Vulnerabilities
Managing Vulnerabilities

Learning Objectives (LO)

After reading this chapter, you should be able to:

- LO1. Define vulnerabilities.
- L02. Understand how the attack surface relates to vulnerabilities.
- LO3. Know how information systems staff can identify vulnerabilities.
- LO4. Understand how threats and vulnerabilities together create successful attacks.
- LO5. Be familiar with human, operational, and technical vulnerabilities.
- LO6. Be able to describe examples of human, operational, and technical vulnerabilities.
- LO7. Be familiar with common approaches to identifying and managing vulnerabilities.

Introduction

Simply put, vulnerabilities are weaknesses. Just as people can be vulnerable, so can information systems. You may walk into the woods without mosquito spray, and insects will take advantage of that vulnerability and bite you. In cybersecurity, a vulnerability creates an opportunity for a successful attack or exploit. This could involve a threat actor taking advantage of a weak firewall or poor password protection, for example. As described in the previous chapter, a cyberattack occurs when a threat is realized through the successful exploitation of a vulnerability. Remember:

Threat + Vulnerability = Cyberattack

> **EXAMPLE**
> **Malware is a threat that can exploit a vulnerability in access management, such as weak passwords. This can lead to a ransomware attack.**

In this chapter, we first note how information security professionals identify system vulnerabilities. Chapter 3 described how attacks happen, and this chapter briefly discusses attacks in the context of vulnerabilities. The heart of the chapter describes specific examples of human, operational, and technical vulnerabilities that may lead to an attack. We conclude with a short discussion of vulnerability management. This topic is explored in more depth in Chapter 6, regarding risk management, which describes internal controls that mitigate risk from vulnerabilities.

Identifying Vulnerabilities

Vulnerability A specific weakness in a system that a threat can exploit.

NIST defines a **vulnerability** as a "weakness in an information system, system security procedures, internal controls, or implementation that could be exploited or triggered by a threat source" (NIST, 2025). Every computer system is vulnerable to threats, and no amount of spend on information security can provide complete protection. Cybersecurity needs to "plug every hole" or control for every vulnerability, whereas a threat actor needs to find only one weakness to launch a successful exploit. Still, identifying vulnerabilities is a necessary part of risk assessment and risk management, which we will describe in the next two chapters. Vulnerabilities can be inherent in purchased software or network configurations, or they may be introduced through the absence of internal controls.

Regardless of best efforts, vulnerabilities in software, hardware, and, of course, people, exist. An organization should assign responsibility for vulnerability detection and management. Since vulnerabilities and threats are so closely connected, this role should have responsibility, accountability, and budget for both. Those responsible need to document every step and role assignment in the plan for identifying and managing vulnerabilities.

Common Vulnerabilities and Exposures (CVE) A database of publicly known cybersecurity vulnerabilities and exposures.

Zero-Day Vulnerability An undiscovered flaw in an application or operating system; that is, a gap in security for which there is no defense or patch because the software maker does not know it exists—they have had "zero days" to prepare an effective response.

Processes to identify vulnerabilities include automated vulnerability scanning, penetration testing, and identifying **common vulnerabilities and exposures (CVEs)**. CVEs, discussed in more depth at the end of this chapter, are known vulnerabilities that have been discovered and publicly communicated. Protecting against vulnerabilities is more complex in the case of **zero-day vulnerabilities**. These are weaknesses that have yet to be detected or protected against. The term signifies that there are zero days to develop and apply a security patch that controls the vulnerability. For example, someone may discover that there is a coding error in a software application for which an update has yet to be released. An adversary who learns of or finds this vulnerability may exploit it during the period between its identification and the release of the patch.

CASE IN POINT

In 2021, two cybersecurity researchers at Computest were awarded a $200,000 "bug bounty" for discovering a zero-day vulnerability in the cloud-based video conferencing software Zoom. During the pandemic, Zoom's usage rate grew exponentially, making it difficult for the software company to stay on top of software vulnerabilities. Some of them allowed nonparticipants to zoom bomb or disrupt virtual meetings, but the zero-day vulnerabilities that the ethical hackers at Computest discovered were much more serious, potentially allowing threat actors to infiltrate and take command of user computing devices (Stahie, 2021).

Automated Vulnerability Scanning A process that automatically scans a system or network for security vulnerabilities.

Penetration (PEN) Testing A security exercise where a cybersecurity expert attempts to find and exploit vulnerabilities in a computer system. The purpose of this simulated attack is to identify any weak spots in a system's defenses.

Automated vulnerability scanners are software tools that allow organizations to monitor networks, information systems, and software applications for security vulnerabilities. The scanners inventory IT assets on the network, such as servers and laptops, and for each asset, the scanner identifies its operating system and applications, along with attributes such as open ports. Scanners can provide information about detected weaknesses and associated risks, as well as recommended mitigation procedures. A variety of scanners are available, depending on the hardware or software to be scanned. Two examples are network or application scanners.

Scanners may not discover every vulnerability, and periodic **penetration (pen) testing** is an important part of vulnerability management. These tests, whether manual or automated, find weaknesses in your system and attempt to gain unauthorized access to assets in the same way that an adversary might. Pen testers are sometimes referred to as white hat hackers, discussed in Chapter 3, because they are a type of ethical security hacker. Pen testing has several phases, much as does an actual exploit. These include planning, reconnaissance, the attack itself, and post-attack analysis/reporting, as shown in Figure 4-1.

In the planning phase, pen testers determine the scope of the testing. For example, which applications are to be tested? Threat actors try to find out as much about their target as they can, and pen testers do the same in the reconnaissance phase. A pen tester could use different kinds of attacks. For example, the ethical hacker might launch a noncompromising phishing attack or some other simulated social engineering scenario. Post-attack analysis provides management with feedback on vulnerabilities discovered in the test.

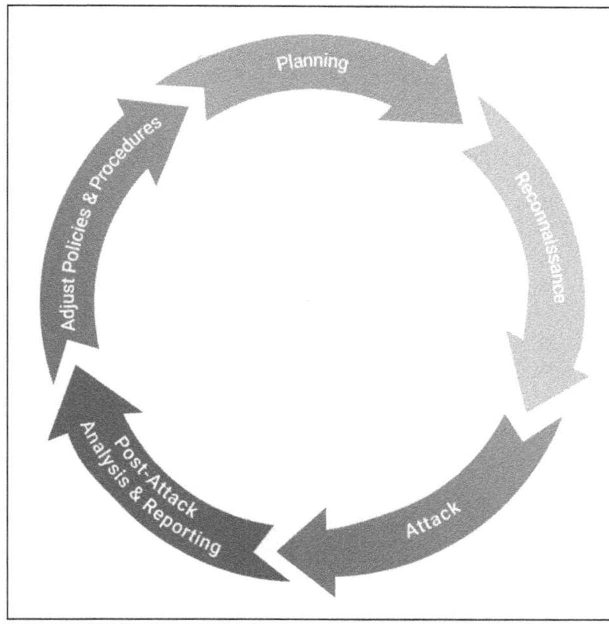

Figure 4-1. Penetration Testing Phases

The Attack Surface and IoT

Attack vectors are the pathways or methods threat actors use to exploit information system security vulnerabilities. These vectors allow attacks to occur. Examples of attack vectors are malware, a malicious third party, and unpatched system vulnerabilities. Adversaries may use one or a combination of attack vectors to launch a data breach or other type of exploit.

An **attack surface** consists of the sum of all attack vectors that the adversary can exploit. It is the total possible ways a hacker can successfully launch an attack. Yet another way to think about the attack surface in an organization is to consider it as all the entity's digital assets that can be compromised. These include servers, switches and routers, and endpoints such as mobile phones and laptops. Reducing the attack surface is a basic control measure in information security. One way to do this is to discover and inventory assets and then eliminate unused or unnecessary ones. The idea is to shrink the surface by providing only necessary services to users who require them and eliminating software programs and code that serve little purpose.

Attack Vectors The methods or pathways that attackers use to breach a system's defenses.

Attack Surface The set of points on the boundary of a system, a system element, or an environment where an attacker can try to enter, cause an effect on, or extract data from that system, system element, or environment.

> **EXAMPLE**
>
> **The attack surface in a university includes software, hardware, the network itself, data, people, and third-party vendors, such as cloud service providers. Unique to universities are student portals, such as learning management systems and registration systems; servers with proprietary research; and a transient student population.**

Unfortunately, the attack surface for just about every business is expanding rather than shrinking. It can be a surprising exercise to catalogue all the items both in your home and that you carry with you that are connected to the Internet. Watches, home security systems, smart TVs, voice-controlled virtual assistants, thermostats, and fitness monitors are all examples of such items. The **Internet of Things (IoT)**, defined in Chapter 2, for business and other organizations has grown exponentially in the past decade, with continuing growth in the next. That means bigger attack surfaces for threat actors to exploit. Figure 4-2 shows the growth in IoT devices.

Internet of Things (IoT) A network of physical objects embedded with sensors, software, and other technologies to connect and exchange data with other devices and systems over the Internet.

IoT devices present several information security vulnerabilities, in part because they often lack the security protocols built into traditional computing devices, such as antivirus software. These devices may not include computational abilities, so they cannot incorporate security that relies on complex functions. IoT users themselves are also a vulnerability because they might misuse the devices and create opportunities for hackers. For example, patching security vulnerabilities can be nearly impossible because of sheer numbers and the ability to catalogue and track equipment.

The cybersecurity vulnerabilities in billions of IoT devices can allow bad actors to obtain remote control to exploit them and obtain network access. Networks can be contained with perimeter defenses that cannot work when IoT endpoint devices are able to obtain access remotely. Adversaries can also crash devices with malware to block Internet and network access.

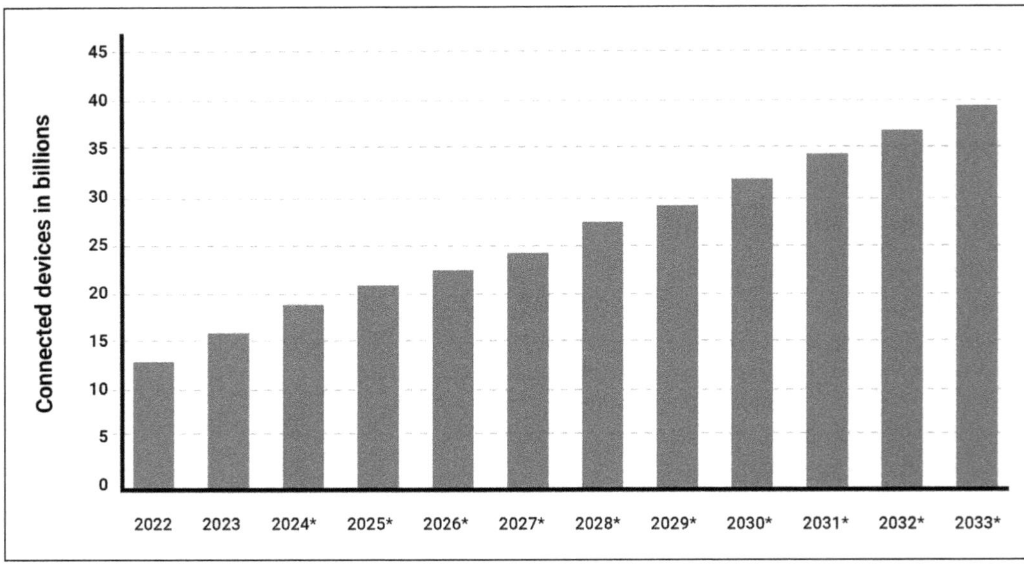

Figure 4-2. Internet of Things (IoT) Connections Worldwide (Statista, 2025)

Another reason behind growing attack surfaces is remote work. Cybercrime went up dramatically in the early days of the COVID pandemic, and one big reason is that people stayed away from their offices and worked remotely. When everyone in an office is working through a local area network, the perimeter of the network can be safely secured. Think about a castle with a moat around it. However, when individuals leave the office, or castle, and access resources within the office/castle remotely, those resources become more vulnerable to threats. The attack surface is simply bigger when employees work outside the office, and that means increased vulnerability to cyberattacks.

In addition to expanding the attack surface, remote work introduces other vulnerabilities. Many remote workers remotely log into their office computers using their own personal computers, which may not be protected, or use their work laptops for personal purposes. In addition to vulnerabilities introduced by performing personal tasks on a business laptop from home or a public venue, remote workers might also loan their work computers to other family members or friends, adding another vulnerability.

Types of Vulnerabilities

In the simplest and broadest terms, the root cause of all vulnerabilities is *people.* Whether it's clicking on a compromised link, coding software with exploitable weaknesses, or creating policies and procedures that are inadequate and not monitored, people are often the problem in cybersecurity. But they are not the only problem. In addition to human vulnerabilities, operational and technical weaknesses (though these are designed by people, too!) provide an opportunity for a threat to become realized. This section of the chapter will describe and provide examples in each of these three classifications of vulnerabilities: human, operational, and technical.

Human Vulnerabilities

Humans are the weakest link in many cybersecurity architectures. User errors can easily expose sensitive data, create weaknesses for attackers to exploit for profit, or disrupt systems. The cartoon in Figure 4-3 is an example of human fallibility. It is followed by descriptions of some common types of human vulnerabilities.

Figure 4-3. Unsecure Emailing

Default Settings

Hardware and software come with built-in **default settings**. Many of these settings concern privacy and security. Your cell phone and other computing devices have many such settings. They control which applications have access to your camera, location, data, and other features. Options for location settings might be:

Default Settings Pre-configured settings or values that are applied to a device or software when it is first installed or reset.

- Never
- Ask Before Allowing
- Allow While Using the App

Depending on the device manufacturer, the default setting might be to share location, so users should consider whether this is the option they prefer. Some devices offer enhanced privacy, such as lockdown mode, which protects against spyware. These settings are not enabled by default, so users must activate them if they wish to add that layer of protection.

Software applications, including Internet browsers, also come with a variety of default settings. You may have the option to place your browser into a "private browsing mode," where it is not tracking your browser history. However, the default setting might be "normal browsing mode," where your searches are being tracked. Failure to change default settings is one way threat actors can access and collect data that they may use to steal an identity or account access credentials. The following example highlights just one privacy setting on a social media platform. These types of applications have many options for privacy and security.

EXAMPLE

The default privacy setting on Instagram is Public. This means that anyone can see your profile and posts, whether or not the person is on the platform. You may change the setting to Private, which will make your profile and posts available only to those you choose to identify as followers.

Employees

Employees are just one type of human that represents a vulnerability for information security. However, as they are the class of insider with the most access and ability to compromise data confidentiality, integrity, and availability, we highlight them here. Employees provide vulnerabilities due to errors and mistakes and sometimes through malicious actions. Creating a culture of cyber awareness; regular training and security reminders; strict access and authentication processes, including assigning least privilege (discussed later in this section); and zero-trust frameworks are all security measures that can mitigate vulnerabilities from employees. However, it takes just one careless response to a phishing email to open the gates to threat actors intent on doing harm. Such a vulnerability poses an outsize threat.

Not all employee vulnerabilities are free from malicious intent. Employees may see opportunities where controls are not in place to keep them from accessing valuable organizational assets. The motive here might be financial gain, but disgruntled and terminated employees are a different vulnerability because they may cause intentional harm just to retaliate, as in the following case.

> **CASE IN POINT**
>
> **Christopher Dobbins lost his job with Stradis Healthcare, a medical packaging company, in 2020. When his job was terminated, he also lost computer system access. However, while still employed, he had created a fake user account. He then used that account to edit and delete thousands of records, creating disruptions and delays in the company's shipments of PPE that were needed by healthcare providers during the height of the pandemic. Mr. Dobbins was sentenced to federal prison for his crime.**

Lack of Endpoint Device Protection

Another people-related vulnerability is failure to add passcode protection and/or security software on endpoint devices. Enterprise network users have mobile phones and laptops that, even without remote access to a company network, are likely to contain work-related email. Many individuals use their work email accounts for personal use, and even if they do not, they may keep all email accounts on their cell phone. The proliferation of IoT and portable computer devices creates many more points of vulnerability for bad actors to exploit. Endpoint devices should require face or fingerprint recognition or a passcode to activate.

Organizational Structure and Culture

Lack of communication between parties and indistinct reporting lines are two examples of vulnerabilities that can lead to information security failures. Auditors often talk about the "tone at the top," which means that culture begins with the organizational head, whether that be a Board of Directors, CEO, or CIO. It is usually easy to spot a culture that has lax information security measures versus one that values security and privacy. An auditor trying to determine this might ask some of the following questions:

- What are the password policies?
- Does everyone in the organization get trained about information security?
- Are office spaces and areas where sensitive information is housed properly secured?

The way an organization is structured can create vulnerabilities in multiple ways. Poor organizational structure may mean that no one individual can be responsible or accountable in the case of a realized threat because of communication problems. The reporting authority for cybersecurity executives can also be a vulnerability, if the top information security staff do not have access to the decision-makers, such as the CEO and Board of Directors. You can tell a lot from the reporting structure in an organization. Does the Chief Information Security Officer have access to the CEO and board? Does the Chief Information Security Officer have the appropriate background and experience for the job position? Are roles assigned so that responsibility for a cybersecurity function vests in one individual? The Equifax case highlights the importance of a cybersecurity-minded culture.

CASE IN POINT

There were many points of failure in the 2017 Equifax data breach. At least some of them can be traced to the organizational structure and culture. The US House of Representatives Committee on Oversight and Government Reform investigated and found that as Equifax moved aggressively to grow and enhance their bottom line, they failed to take security measures to protect the increase in data that accompanied the expansion. The report noted one cause of the breach was that "a lack of accountability and no clear lines of authority in Equifax's IT management structure existed, leading to an execution gap between IT policy development and operation" (Committee on House Oversight and Government Reform, 2018).

Poor Cyber Hygiene

We all know we should wash our hands frequently and brush our teeth at bedtime. Just as these personal hygiene measures provide us health protection, so do **cyber hygiene** measures protect against poor outcomes. Hygiene in any realm takes time and can slow us down, but we just need one bad experience, whether it be tooth decay or wiring money to a hacker, to demonstrate the importance of cautionary measures.

Cyber Hygiene The practices that users of computers and other devices follow, and the steps they take, to maintain system health and improve online security. These practices are often part of a routine to ensure the safety of identity and other details that could be stolen or corrupted.

While some users may commit a cybersecurity breach with malicious intent, it is more likely that the human vulnerability is due to error. This fallibility stems from poor cyber hygiene that includes a lack of security awareness. A phishing email need only find one person, who, either due to a lack of understanding or perhaps having just a bad moment, clicks on the malicious link, as happened in the following case.

CASE IN POINT

In 2016, a hacker group called Fancy Bear used spear phishing to obtain and exfiltrate emails from John Podesta's personal Gmail account. At the time, Mr. Podesta was chair of Hillary Clinton's 2016 US presidential campaign. Wikileaks obtained and published copies of the emails. These emails exposed unflattering details of the Clinton campaign. The phishing email purportedly said that Mr. Podesta's email had been hacked and that he needed to click on the included link to change his password.

Credential Stuffing A cyberattack where attackers use stolen username/password combinations to gain unauthorized access to multiple accounts.

Users within an organization can introduce malware purposefully in many ways, but they can also unintentionally create the potential for cybercrimes against their own and other organizations by not being careful. An example of this is the vulnerability created by **credential stuffing**, illustrated in the Figure 4-4. Without a password management system, individuals tend to use the same password to access many different sites and accounts. This means that a hacker who obtains logins and passwords through a data breach of one company can likely use those same credentials at many other sites, perhaps introducing malware many times over.

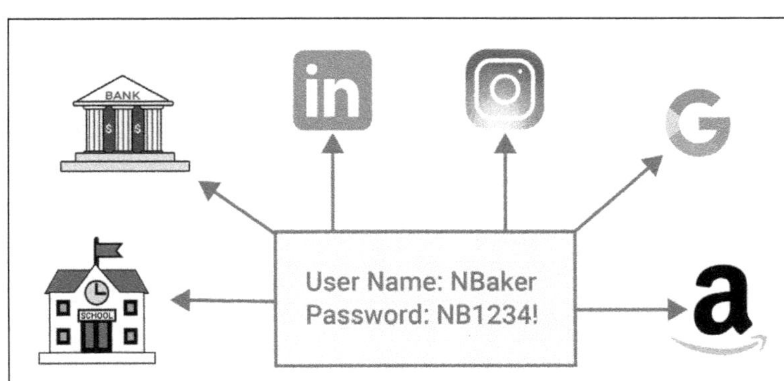

Figure 4-4. Credential Stuffing

Roles and Access

Login Name A unique identifier that a user enters to gain access to a system, network, or application.

Single Sign-On (SSO) An authentication process where the user authenticates once to gain access to all of the systems for which they are authorized based on their defined role.

Short Message Service (SMS) Attacks Same as "Smishing." A type of phishing attack carried out via SMS text messages, aiming to deceive users into providing sensitive information.

Each user should have an assigned unique **login name**. These names are frequently built around the user's actual name. Examples are jdoe@ or jane.doe@. One problem with this type of naming convention is that hackers can easily guess login names for any organization. Adding a middle initial (e.g., jane.c.doe@) can help. Each login name is associated with a user identification code that determines the level of access associated with that individual and their role. **Single sign-on (SSO)** allows users to log in just once to obtain access to a variety of related systems. This approach has benefits, such as reduced password fatigue, but it also can create a vulnerability because without enhanced security measures, the compromise of login credentials means more systems may be breached.

Password management and authentication techniques have necessarily become increasingly sophisticated over time. One author of this book recalls having a five-letter password for an Amazon account way back when. With brute force and dictionary threats, password requirements have evolved to longer strings of a mix of alphabetic and numeric characters. Login names, without appropriate password management or multifactor authentication (MFA), discussed in Chapter 6, create a frequently exploited vulnerability. However, as the following case demonstrates, when used in MFA, text messages themselves have their own vulnerability in **short message service (SMS) attacks**. In MFA, a frequent default option is for the application to text temporary codes to the user's phone. SMS security usually is not strong, which means an attacker need only access the message to read it and retrieve the temporary passcode.

CASE IN POINT

In 2021, thousands of customers of the cryptocurrency platform Coinbase had their funds stolen, partially because of a vulnerability in the company's SMS Account Recovery Process. As a result of the breach, Coinbase updated its Account Recovery Protocols to protect against the security flaw. It also recommended that customers use a robust password, such as a time-based, one-time password, to protect their individual accounts. Because text messages can be compromised, the use of SMS two-factor authentication has become less popular.

Least Privilege An information security concept in which a user is given the minimum level of access—or permissions—needed to perform their job functions.

Every information system user is another point on the attack surface. Each individual needs to be assigned an organizational role and the appropriate information system access that accompanies it. **Least privilege** is the technique of ensuring that assigned roles can *only* access and manipulate the data needed to perform their function. Without controls, user account access privileges present a vulnerability, as some users in a network are inappropriately given elevated access.

As an example, Table 4-1 illustrates the different data access needs of university students and faculty. Within each of these applications, roles would be further defined. For example, a student has access only to their own academic record, while a faculty member may access academic records of all students in their courses and all their advisees.

Table 4-1. Student and Faculty Data

Student Data and Applications	Faculty Data and Applications
Academic Records	Academic Records (Students and Advisees)
Campus Events/Student Organizations	Administrative Applications/Records
Career Services	Communications Systems
Communications Systems	Faculty Development/Training
Financial Information	Governance
Learning Management Systems	Human Resources
Library	Learning Management Systems
Student Support Services	Library

Some roles, such as system administrators, have superuser privileges. A system administrator assigns access rights to others and can access and modify almost all data within the network. Hackers who access an information system will seek to escalate privileges and assume a superuser role that allows them to have extended access and the ability to make changes to programs.

Third-Party Arrangements

Chapter 3 discussed threats from third-party relationships, including with customers, vendors, consultants, contractors, accountants and financial advisers, board members, and any other entities across the supply chain. The more relationships an organization has with external parties, the bigger the attack surface. Third parties represent both a threat and a vulnerability because of this increased attack surface, given that their security practices may not be as robust as those of the contracting entity. Organizations trying to mitigate risk from third-party vulnerabilities do so through access controls, assessment of the third party's security protocols and continuous monitoring, and a zero-trust framework that minimizes vulnerabilities. Requiring third parties to provide System and Organization Controls (SOC 2) reports, discussed in Chapter 8, is another way to protect against third-party vulnerabilities.

Third parties providing information systems resources, including software and cloud storage, present additional vulnerabilities. In-house software code can be protected through the security of your personnel and testing, but third-party software and software updates can introduce malware into your system, as exemplified by the SolarWinds hack, in which the Orion software update contained the bad code.

As more data storage migrates from in-house servers to public and private cloud service providers, the attack surface of an entity expands to these third-party providers. Many organizations today have migrated their software applications and associated data to the cloud. Cloud service providers, because of all the data they handle, are attractive targets for threat actors. As a result, they know that their business model relies on protecting that data. They must ensure that they control for vulnerabilities, such as software misconfigurations, access control weaknesses, and unpatched software. Contracting with a cloud service provider requires due diligence. As noted, one form of this may be a requirement for a SOC 2 report, discussed in Chapter 8.

Most organizations use a third-party email service, such as Google, MS Outlook, AOL, or Yahoo!. The following case further demonstrates the vulnerabilities introduced when third-party services are used and the contract entity does not have effective cybersecurity protections in place.

> **CASE IN POINT**
>
> **In 2021, the University of Colorado, among others, was subject to a ransomware attack that exposed mostly student data, about 300,000 records, but also some medical and financial information. The student data included transcripts, grades, ID numbers, donor information, and various status designations, such as visa status. The attack was on a vulnerability in third-party software from a firewall vendor. An investigation into the breach discovered multiple vulnerabilities in the file sharing application (Newman, 2021).**

Operational Vulnerabilities

Operational vulnerabilities are weaknesses in policies, procedures, and processes. They occur when operations are conducted suboptimally, creating security flaws and opportunities for threats to become realized. The following are examples of this type of vulnerability.

Operational Vulnerabilities Weaknesses in an organization's processes or systems that can be exploited to cause operational failures or security breaches.

Data Storage and Handling

Threat actors are after data, and poor data storage and handling practices create a vulnerability that lets them capture data at rest or in motion. Data should be classified and assigned storage protocols consistent with the degree of confidentiality and privacy required. The more sensitive the data, the more security measures needed. At a minimum, any data that an entity cannot share publicly should be encrypted, and encryption practices should be sophisticated and state of the art.

Failure to back up data does not by itself create a vulnerability, but it does impact the ability to recover from a data breach or ransomware attack. When threat actors make data inaccessible, having appropriate backup can mitigate the damage.

Encryption

Encryption is most often thought of as a way to protect data at rest or in motion. However, weak or poor encryption and faulty encryption practices create a vulnerability. Using out-of-date or legacy cryptographic algorithms is an example of weak encryption. Outdated encryption versions allow brute force attacks, where an attacker can generate enough guesses to ultimately crack the code. The following example demonstrates how encryption is evolving, creating vulnerabilities with failure to change.

> **EXAMPLE**
>
> **Rivest-Shamir-Adleman (RSA) has been a popular encryption algorithm for securing email messages, files, and digital signatures. However, as cryptography advances and quantum computing power become available, 1024-bit RSA keys have become vulnerable to cyber-attack.**

Poor encryption practice occurs when the encryption key is improperly stored or handled. Storing the encryption key in the database, in the file system, or in an application configuration file are all poor choices. Key management best practices dictate encrypting the keys and storing them in an alternate location. Other key management practices that address poor encryption vulnerabilities are changing keys periodically and using different keys for segmented data. Chapter 6 discusses encryption techniques in detail.

Security Certificates
Digital certificates used to verify the identity of websites or systems and to encrypt communications.

Expired Security Certificates

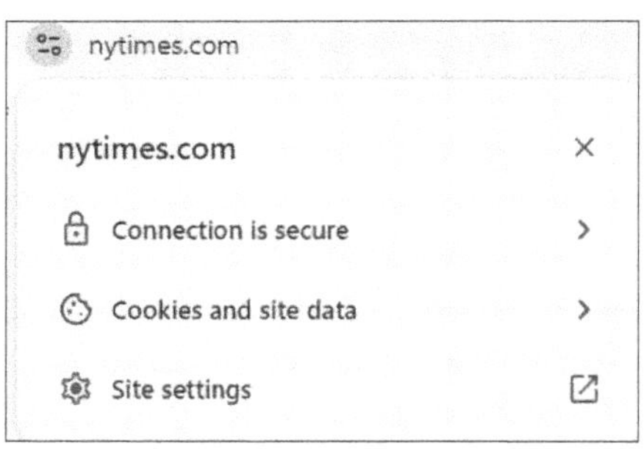

Figure 4-5. *New York Times* Website Security Certificate Information

Maintaining **security certificates** is another process in software management. They are data files issued by a trusted Certificate Authority to provide data about data. These data files verify identity with a unique signature. When browser software attempts to access a website, that site presents its certificate to your browser, which then checks it against the security certificate. If the certificate has expired or there is a discrepancy with the address or signature, you receive an error message, or the access to that site is blocked. You can see the security certificate of a secure (https://) website by clicking on an icon next to the URL, as Figure 4-5 illustrates. This action reveals information about the site and tells you who issued the security certificate and whether the connection is secure. When a security certificate has expired, the connection is not secure and is vulnerable to many web-based threats.

Monitoring and Reporting

Intrusion Detection and Prevention Software/System (IDPS)
Hardware and/or software tools that detect and respond to suspicious activities on a network or system, combining both detection and prevention functionalities.

Failure to continuously monitor communications and data in motion creates a vulnerability. Most networks are protected with antivirus or **intrusion detection and prevention software (IDPS)**. These applications track incoming communications and analyze them for anomalies that indicate a threat, such as phishing emails or spam communications containing malware.

The preferences on these programs may be set either strictly, so that some legitimate communications fail to get through security, or more liberally, so that some communications containing malware might get through. Security software needs to be kept up to date so that it can protect against new vulnerabilities and threats. Increasingly, these programs include behavioral analytics and other artificial intelligence tools to detect and reject communications that contain zero-day threats. Figure 4-6 shows important features in IDPS tools. The absence of any of these characteristics creates a vulnerability.

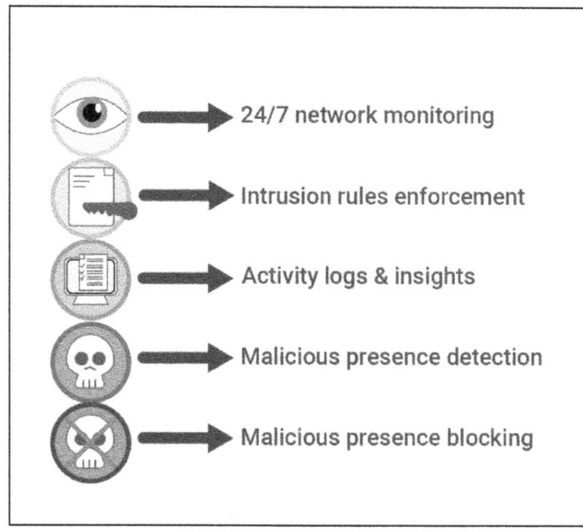

Figure 4-6. IDPS Tool Features (BasuMallick, 2022)

Public Charging and Public Wi-Fi

Another vulnerability is introduced when remote workers and travelers use public charging stations. Running out of juice? Plugging into a charging station at the airport might mean downloading malware. The hard line between work and home vanished for many following the pandemic, making it critical for devices to have access to safe charging.

Employees working outside the office may also be tempted to connect to public Wi-Fi. Coffee shops and airports are popular sites to log into a network from your laptop or mobile device and get some work done, but public Wi-Fi is highly vulnerable, as hackers can easily access it and steal user credentials. Accessing financial or confidential information over public Wi-Fi should be avoided at all costs.

Security Logs

IS staff should maintain **security logs** and follow up on them as needed. These logs might indicate patterns to intrusions, and they can provide insights and intelligence, including into user behaviors, anomalies, violations of policies, and system downtime. Logs that the system maintains but that security personnel fail to analyze present a security flaw. Analyses are looking for suspicious behaviors or patterns that are out of the ordinary, such as a privileged user logging into the system at an unusual time of day.

Security Logs Records of events and activities related to the security of a system, used for monitoring, analysis, and auditing.

CASE IN POINT

One way hackers might exfiltrate data or send commands to a remote access Trojan (RAT) is to use DNS (domain name server) tunneling. In a DNS query, Internet domain names are located and converted to IP addresses. A firewall might not block the query. In a DNS tunneling attack, the threat actor hides commands and data in the DNS protocol when communicating with the target computer. Information security personnel can detect a DNS tunneling threat by analyzing payloads for unusual data being sent and received or by analyzing traffic for unusual numbers of DNS domain requests. Failure to conduct these analyses would result in a successful attack.

Remote Access Trojan (RAT) A Trojan malware type that allows the attacker remote access to and control of the infected system.

Network security logs are an important resource in cybersecurity, and threat actors know it, so they may tamper with the log to prevent the recording of intrusion attempts. Insecure or unprotected logs are yet another vulnerability.

VPNs

Endpoint security and use of **virtual private networks (VPNs)** can mitigate vulnerabilities to some extent, but even those measures have weaknesses. A VPN strengthens security by protecting communications through encryption and providing a level of data privacy, but it does not protect against Internet malware infections. In addition, a VPN provides an entry into an organization's network, so any unpatched vulnerabilities in the VPN software can become potential entry points for attackers. VPNs that rely on outdated encryption algorithms or weak user authentication, which are also vulnerabilities, introduce security risks. You should ensure that you acquire your VPN from a legitimate provider, as a threat actor could potentially pose as a VPN vendor.

Virtual Private Network (VPN) A technology that creates a secure, encrypted connection over a less-secure network, such as the Internet.

Technical Vulnerabilities

Technical vulnerabilities relate to hardware and software. They frequently occur because of complexity and interactions between applications. This section describes several examples of each type.

Hardware Vulnerabilities

Computer hardware is all of the physical parts of a computer system. It includes external components, such as a screen or keyboard, and internal components, such as the motherboard and hard drive. Three types of hardware vulnerabilities are design flaws, outdated legacy hardware, and physical access, each of which we describe here.

Secure Boot A security feature that ensures a computer only boots using software that is trusted by the manufacturer or owner.

Legacy Hardware Older, possibly superseded hardware still being used by an organization.

- Design flaws. Several different kinds of design flaws can be present in computer hardware. One example is failure to design security features into the device, such as **secure boot**. When a user starts up a computing device, the firmware (bridge between the hardware and the operating system) checks to make sure that the operating system software running has a digital signature from a trusted source. This prevents unauthorized software from launching during the startup phase. Failure to update firmware regularly through patching creates another vulnerability.
- **Legacy hardware**. Hardware that is outdated is called legacy hardware, the term emphasizing that it is left over and has been superseded by newer equipment. This type of hardware is vulnerable because the manufacturer may no longer support it, and patching updates for vulnerabilities may not be available. In addition, the hardware may not have the security features included in more modern devices. Threats are evolving, and computer equipment protection needs to change with them.
- Physical access. The attack vector for physical access is easy to visualize. A threat actor can physically confiscate your phone or computer. By doing so, they may steal data, install malware, or simply reappropriate the device for their own use. The vulnerability is a lack of physical protection, such as CCTV cameras or security cables. Data centers housing multiple computer devices require all kinds of protection, including secrecy about their location. (Do you know where the data center on your campus is located?) Figure 4-7 describes some potential features of data center security.

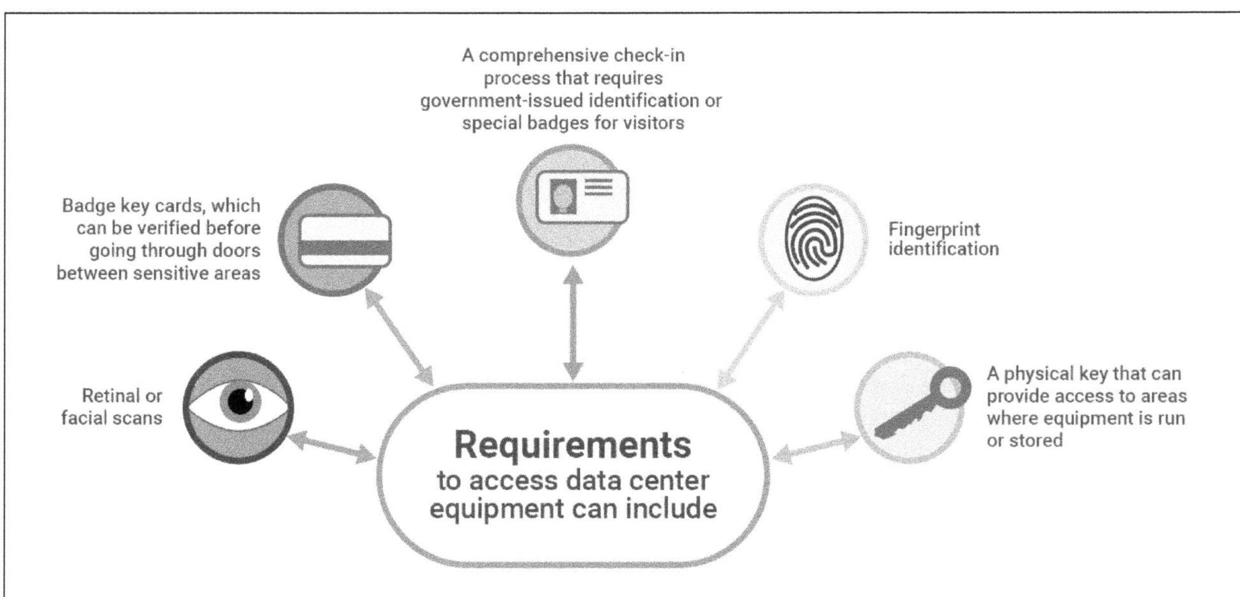

Figure 4-7. Data Center Access Controls (Fortinet, 2026)

Software Vulnerabilities

We are all familiar with software vulnerabilities, as they are often the reason behind those pesky software updates that take time and sometimes upend our settings, requiring reconfiguration or a change in how we navigate a device. Software vendors are almost always interested in bringing their products to market as quickly as possible, so they may not have discovered or protected against every vulnerability or bug prior to release. As new weaknesses are discovered, these companies offer **patches** to correct programming errors and other flaws.

Software Patch A piece of code released by developers to update or fix issues in a software program, such as bugs, security vulnerabilities, or performance problems.

Cybersecurity and Infrastructure Security Agency (CISA) A US federal agency responsible for protecting the nation's critical infrastructure from physical and cyber threats.

The **Cybersecurity Infrastructure and Security Agency (CISA)** developed a program based on the philosophy that software vendors should ensure that their software releases are not only functional but also Secure by Design. At the Black Hat USA meeting in 2024, Jen Easterly, then Director of CISA, explained that just as we expect airplanes to incorporate safety in their design prior to releasing new models, so should we expect software vendors to do so. Software developers who embrace the Secure by Design philosophy sign a voluntary pledge, promising to make good faith efforts to ensure that their software is "safe to fly" upon release. They can then display a logo (see Figure 4-8) to signify their commitment. Many of the world's largest software manufacturers, including Amazon Web Services, Google, and Microsoft, have signed the pledge and are included on CISA's website as signatories.

Figure 4-8. The Secure by Design Pledge Logo

In some cases, misconfigurations create software vulnerabilities, but there are other options for bad actors to attack through software. **Software vulnerabilities** occur from flawed or weakly written code, which might be in legacy software or the result of lazy coding habits. Computer software is complex, and as software programs interface or communicate with one another, this complexity increases exponentially. Computer programming bugs and unanticipated interactions between different software code are common vulnerabilities. The following are examples of software vulnerabilities.

Software Vulnerabilities Weaknesses or flaws in software that can be exploited by attackers to gain unauthorized access or cause damage.

- Backdoors. Programmers sometimes purposefully create vulnerable code. These **hidden backdoor program vulnerabilities** are pieces of code that let a programmer remotely access the application for technical reasons, such as to diagnose problems or provide support. Unfortunately, when users are unaware of this code, they create an opportunity for someone who does know about it to access the system and network.
- Operating system bugs. Operating system software can also be a source of software vulnerabilities. Given the complexity and functionality of this software, it is challenging for programmers to develop it without bugs. Many of these bugs are zero day, as they have not been previously discovered. Vendors release patches as updates as soon as they discover these vulnerabilities.

Hidden Backdoor Program Vulnerabilities (Backdoor) Security weaknesses in software that allow unauthorized access to a system, often intentionally left by developers for maintenance purposes.

> **EXAMPLE**
> **Apple's 2021 iOS (its mobile operating system) contained a vulnerability that could allow a hacker to compromise a user's iPhone.**

- **Parsing errors** and **stack overflow**. Parsing errors and stack overflow errors are two examples of coding flaws that may stem from legacy programming code. A parsing error is a mistake in code that keeps your program from converting code so that a program can run. Stack overflow errors happen when the program uses up the call

Parsing Errors Errors that occur when data is improperly interpreted or processed by a program or system.

Stack Overflow A type of software bug that occurs when a program tries to use more memory than is allocated, which can lead to crashes or vulnerabilities.

stack or buffer handling application requests. This type of error usually occurs because a program function is calling itself over and over again, in an infinite loop.

Remote Code Execution (RCE) A cyberattack where an attacker can execute code on a remote system.

- **Remote code execution (RCE).** Remote code execution is another type of software vulnerability. This weakness allows hackers to remotely exfiltrate data or execute malicious code. Attackers can exploit a vulnerability of this nature to gain control over compromised networks.

> **CASE IN POINT**
>
> **Apache Log4j is a popular logging program used in millions of Java-based applications. A zero-day RCE vulnerability was found in the code in November 2021 and publicly disclosed the next month. This vulnerability allowed attackers to exfiltrate data and run an attacker's code and commands. MITRE assigned the vulnerability, named Log4Shell, the designation CVE-2021-44228. Because the logging software is so widely used, this was a severe vulnerability that led to a significant number of cyberattacks in 2021.**

SQL Injection A type of attack where malicious SQL code is inserted into a database query, allowing the attacker to gain unauthorized access to data.

- **SQL injection errors**. As described in the previous chapter, SQL injections are a common cybersecurity threat that results from web application errors. SQL injection is the most exploited web application security risk because it can be relatively easy to execute, and the payoff—access to data—is significant. In SQL injection, the adversary enters a string of malicious code into the SQL program. Many SQL attacks occur when users are required to complete a form on the web. The hacker might populate that form with random characters that the web application does not validate or sanitize and instead passes them on to SQL queries. Improper validation of user input is a vulnerability that allows an adversary to launch a successful attack.

> **CASE IN POINT**
>
> **In 2019, an SQL injection vulnerability was found in Fortnite. Fortnite, owned by parent company Epic Games, is an online game played by hundreds of millions of users. The flaw in the log-in system allowed for an SQL injection attack. The hackers used stolen debit and credit cards to make in-game purchases until the vulnerability was discovered and patched.**

Syntax or Logic Error Mistakes in a program's code, in either its structure (syntax) or logic, that prevent it from functioning correctly.

- **Syntax or logic errors**. Many programming errors can be categorized as either syntax or logic errors. Syntax is the set of rules that the computer uses to read the code. Syntax errors are spelling mistakes that can stop an application from running. Errors in logic occur when the code is programmed properly for execution, but the mistake causes the application to return incorrect results.
- System misconfigurations. Information system hardware and software have a variety of settings and are organized in different ways. Those factors determine their system configurations. For example, Microsoft Windows has a tool called msconfig.exe. A user can launch that tool to change many of the settings within the software, including those that control how Windows starts and whether you want to load all or just some drivers and services. This software, like all software and hardware, is sold to users with default settings. The default may vary in the passcode length and allow the user to change it.

Computer networks include software, hardware, processes, and security controls. Network configuration or setup involves organizing these parts and their settings to accomplish communication and security goals.

Transmission Control Protocol (TCP) A communication protocol that establishes a reliable connection between two systems and ensures the correct delivery of data packets.

Computer networks also include communication protocols, which are the set of rules governing how communications occur. The **transmission control protocol (TCP)**, discussed in Chapter 2, is a critical part of a computer network, as it ensures that data are transmitted without any omissions or errors and in the appropriate order. The three-way TCP handshake is the process that establishes communication between a client and a server. The three steps in the handshake are SYN, for synchronization of the communication; SYN-ACK, where the server receives and acknowledges the communication; and ACK, where the acknowledgment is returned to the client. Figure 4-9 illustrates the three-way TCP handshake.

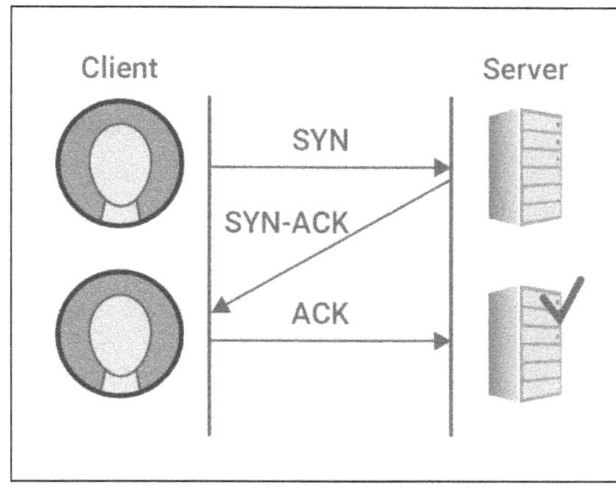

Figure 4-9. TCP Handshake

A **SYN flood** or half-open attack exploits a vulnerability in this handshake and overwhelms the server with messages to which it cannot respond. The SYN-ACK has nowhere to go because the IP address it tries to respond to is fake. These attacks can be prevented by establishing controls such as those that filter or limit requests.

System misconfiguration occurs when security and privacy control settings have not been set optimally to protect data and other resources. These misconfigurations create an exploitable vulnerability. An example of a misconfiguration is the setup of access rights. As mentioned earlier, one internal control in information systems security involves assigning each user the least amount of access needed for them to perform their function. Configuring what a particular user ID may see when they log into a system is an important task for a database administrator. Many breaches are the result of misconfiguration of hardware or software, including information security software such as firewalls. **Website errors** can result from either misconfigured server hardware or software, and they can create havoc for those trying to access a site or capture browsing data. Two attack types that may be enabled by a misconfigured system vulnerability are code injection and brute force password guessing.

SYN Flood A type of denial-of-service (DoS) attack that targets the TCP three-way handshake used to open connections. The attacker overwhelms a server with so many half-open TCP connection requests (SYN packets) that the server runs out of resources for legitimate clients—causing slowdowns or service outages.

System Misconfiguration Incorrect or suboptimal settings in a system that can lead to vulnerabilities or performance issues.

Website Errors Technical issues or malfunctions that prevent a website from functioning as intended, often caused by coding errors, server problems, or connectivity issues.

Network Segmentation The practice of dividing a computer network into smaller, isolated sections (segments) to improve security, performance, and manageability.

> **CASE IN POINT**
>
> **In 2017, a cloud storage misconfiguration led to the leakage of the personal information of about 400 million Time Warner Cable customers. Customers using the MyTWC app could have their data viewed by external sources because of a misconfiguration of the cloud storage by a third-party service provider. Researchers investigating an unrelated data breach discovered the unprotected data on an Amazon server. These cloud server misconfiguration user errors are not uncommon and have led to other breaches, including breaches at Verizon and Dow Jones.**

Another system misconfiguration vulnerability is lack of **network segmentation**. If a network is divided into smaller subnetworks, a threat actor who obtains access to one part of the network is not able to move laterally throughout the system.

- Web browser vulnerabilities. Web browsers are another type of software with frequently exploited vulnerabilities. Browsers are often hacking targets because so much sensitive data is communicated through them. A web browser has many subsystems in which a small coding error can lead to an opportunity to insert malicious code. Weak authentication code or flaws in application login security within a web browser allow cybercriminals and other adversaries unauthorized access to systems and databases. Some of these flaws are administrative, such as password protocols, but others may be traced to programming. As an example, a browser application that returns an error message that tells the user exactly what they did wrong provides hints to make it easier for the attacker to correct their approach.

Managing Vulnerabilities

We began this chapter with a discussion about how to identify vulnerabilities. Limitless resources are not generally available to close each and every vulnerability, so it is important to be able to rank vulnerabilities as part of a risk assessment. (Risk assessment and risk management are the subjects of the next two chapters in this book.) The National Cybersecurity Federally Funded Research and Development Center, operated by MITRE Corporation, manages and maintains the Common Vulnerabilities and Exposures (CVE) database (www.cve.org). The CVE is sponsored by both the US Department of Homeland Security and the Cybersecurity and Infrastructure Security Agency (CISA) that operates under it. This database is a publicly available record of known vulnerabilities, and there is no charge to use it. Businesses use the CVE as part of their vulnerability management program. Figure 4-10 shows a CVE discovered in 2024. *CNA* stands for the CVE Numbering Authority.

```
CVE-2024-38202

CNA: Microsoft Corporation
Summary Microsoft was notified that an elevation of
privilege vulnerability exists in Windows Update,
potentially enabling an attacker with basic user privileges
to reintroduce previously mitigated vulnerabilities or
circumvent some features of Virtualization Based Security
(VBS). However, an attacker attempting to exploit this
vulnerability requires additional interaction by a
privileged user to be successful. Microsoft has developed a
security update to mitigate this threat which was made
available October 08, 2024 and is provided in the Security
Updates table of this CVE for customers to download.
```

Figure 4-10. CVE Example

Common Vulnerability Scoring System (CVSS) A free and open industry standard for assessing the severity of computer system security vulnerabilities.

Once vulnerabilities are identified, they can be analyzed via the **Common Vulnerability Scoring System (CVSS)**. This industry standard allows an enterprise to assess the severity of security vulnerabilities by assigning numerical scores based on severity and prioritizing them by the threat they pose. Enterprise managers can then consider the vulnerability in light of their company's risk appetite and available resources. A score of 0–10 is calculated as a function of several metrics, including data CIA, authentication, exploitability, and target distribution or proportion of vulnerable systems in the environment.

> **CASE IN POINT**
>
> **CVE-2021-34423 was issued in late 2021. This CVE details a vulnerability found in Zoom's video conferencing software. The company quickly shipped patches for two security defects that could allow malicious hacker exploits. A researcher for Google Project Zero discovered and reported on the vulnerabilities. Zoom noted that the bug was a buffer overflow vulnerability with a CVSS score of 7.3.**

Threat-Risk Pairing The process of linking identified threats to specific risks to assess their potential impact.

As noted, the relationship between threats and vulnerabilities is an important one. It takes a vulnerability for a threat to become realized, as bad actors exploit that weakness. Threats without associated vulnerabilities do not carry risk. Many organizations perform a **threat-risk pairing** as an exercise to decide how to allocate resources in addressing vulnerabilities. For example, a vulnerability in authentication can activate a malware threat.

Summary

This chapter began with a discussion about the relationships among vulnerabilities, threats, and attacks and how to identify vulnerabilities. The chapter next described an expanding attack surface and how this creates a commensurate growth in cybersecurity vulnerabilities. Most of the chapter described three kinds of vulnerabilities: human, operational, and technical. Without a doubt, the human element is the greatest security weakness in protecting the confidentiality, integrity, and availability of data. Lastly, the chapter described briefly how vulnerabilities are managed. Since vulnerability management is also part of a risk management strategy, the chapters on risk assessment and risk management delve into this process in greater depth.

Chapter Review

Key Terms

Attack Surface
Attack Vectors
Automated Vulnerability Scanning
Common Vulnerabilities and Exposures (CVE)
Common Vulnerability Scoring System (CVSS)
Credential Stuffing
Cyber Hygiene
Cybersecurity and Infrastructure Security Agency (CISA)
Default Settings
DNS (Domain Name Server) Tunneling
Hidden Backdoor Program Vulnerabilities (Backdoor)
Internet of Things (IoT)
Intrusion Detection and Prevention Software/System (IDPS)
Least Privilege
Legacy Hardware
Login Name
Network Segmentation
Operational Vulnerabilities
Parsing Errors
Penetration (PEN) Testing
Remote Access Trojan (RAT)
Remote Code Execution (RCE)
Secure Boot
Secure by Design
Security Certificates
Security Logs
Short Message Service (SMS) Attacks.
Single Sign-On (SSO)
Software Patch
Software Vulnerabilities
SQL Injection
Stack Overflow
SYN Flood
Syntax or Logic Error
System Misconfiguration
Threat-Risk Pairing
Transmission Control Protocol (TCP)
Virtual Private Network (VPN)
Vulnerability
Website Errors
Zero-Day Vulnerability

Review Questions

1. What are some processes for identifying vulnerabilities?
2. Describe the phases involved in penetration testing.
3. What are attack vectors?
4. How can organizations reduce their attack surface?
5. Define *attack surface* and explain its significance in information security.
6. What is the principle behind "least privilege" in cybersecurity?
7. What is a common vulnerability associated with Internet of Things (IoT) devices?
8. What are some examples of software vulnerabilities?
9. How can weak encryption practices create vulnerabilities in data security?
10. What are some vulnerabilities remote work arrangements introduce, and how can organizations mitigate them?
11. Explain the significance of maintaining security certificates in software management.
12. What is the purpose of the Common Vulnerabilities and Exposures (CVE) database?
13. How does the Common Vulnerability Scoring System (CVSS) assist enterprises in assessing security vulnerabilities?

Multiple Choice

Identify the choice that best completes the statement or answers the question.

1. Which of the following accurately describes the relationship between vulnerabilities and exploits in cybersecurity?
   a. Vulnerabilities are synonymous with exploits.
   b. Exploits are a subset of vulnerabilities.
   c. Vulnerabilities create opportunities for exploits.
   d. Exploits are weaknesses susceptible to harm.
   e. Vulnerabilities are activated threats.
2. What is the purpose of penetration testing in vulnerability management?
   a. To find weaknesses and not attempt to gain unauthorized access
   b. To avoid communication protocols in the planning phase
   c. To conduct post-attack analysis before the attack itself
   d. To only identify vulnerabilities in purchased software
   e. To simulate unauthorized access attempts to assess system weaknesses
3. How can organizations mitigate the risk of vulnerabilities from third-party arrangements?
   a. Ignore third-party security protocols.
   b. Implement a zero-trust framework.
   c. Avoid access controls.
   d. Increase relationships with external parties.
   e. Decrease continuous monitoring.
4. What is a common reason for software vulnerabilities in operating system software?
   a. Legacy programming code
   b. Poor programming habits
   c. Built-in security protocols

   d. Lack of hardware complexity
   e. Syntax errors
5. What is a common reason behind information security breaches related to software management?
   a. Lack of clear software policies
   b. Inadequate change management
   c. Use of outdated hardware
   d. Poor data storage practices
   e. Failure to monitor communications
6. What is a primary goal of software management policies?
   a. Increasing network complexity
   b. Expediting software introduction
   c. Reducing software testing
   d. Minimizing software vulnerabilities
   e. Ignoring change management procedures
7. What tool do IT security personnel routinely use to identify network vulnerabilities?
   a. CVE database
   b. Cybersecurity and Infrastructure Security Agency (CISA)
   c. Bugcrowd
   d. Common Vulnerability Scoring System (CVSS)
   e. Vulnerability scanners

Case Studies

Yellow Subs

As discussed in a Chapter 3 case, Yellow Subs, a Herron client, has 400 sandwich shops in the United States. Some of the stores are company-owned, but the majority are owned by franchisees. The company has expanded rapidly during the past five years and has only recently become concerned about cybersecurity threats. Natalie has completed her assigned analysis of the threat landscape for Yellow Subs, and now they have contracted with Herron to do an assessment of their primary vulnerabilities.

Case Requirements

1. How would Natalie go about performing this type of work?
2. What data/information does she need to obtain from Yellow Subs to complete the assessment?
3. Many vulnerabilities are human. How can Yellow Subs best protect against these vulnerabilities?

Kim's Nail Salons

Kim Tran opened Kim's Nail Salons 10 years ago. The business began with one location, but now it has five stores within a 20-mile metro region. Each location has a local area network (LAN) that connects two clients, point-of-sale (POS) systems, printers, and other devices. Ethernet cable connects the devices in each LAN. The LANs are connected through a router into one wide area network (WAN). The business uses a packaged accounting software program that tracks sales and purchases. The software is cloud-based, and all data are backed up in the cloud.

Case Requirements

1. Identify one potential human, operational, and technical vulnerability that Kim needs to protect against to ensure confidentiality, integrity, and availability of client and employee data.
2. Kim has heard that another chain of nail salons was recently hit with a ransomware attack. She is concerned that this could happen to her business. What protective measures can she take to manage vulnerabilities to a ransomware attack?

Exercises

4-1. Visit the website for CVEs at www.cve.org. At the site, you can search CVE records to identify different types of common vulnerabilities and exposures. Required:
   1. Try typing "SQL Injection" into the search engine. How many CVEs were returned in the result? Next, try typing "SYN Flood." How many results did that return? What do the details of each CVE tell you about the vulnerability?
   2. Each CVE is assigned a CVSS score, ranging from 0–10. Can you identify a CVE with a score of 10?

4-2. Vulnerability scanners detect network and application weaknesses. Required:
   1. What are the different types of vulnerability scanners? Describe each type and what it does.
   2. How are vulnerability scanners different from antivirus software and firewalls?

4-3. Legacy hardware is outdated. Many companies have at least some legacy hardware because they cannot always afford to purchase the "latest" thing. Required:
   1. What are the security issues associated with legacy hardware?
   2. Research and identify a cyberattack that traces its cause to the presence of legacy hardware.

4-4. Programmers/coders sometimes create a "backdoor" in their application code so that they can more easily access the code for troubleshooting. Required:
   1. What are the advantages and disadvantages of these backdoors in coding?
   2. Hackers can also create backdoors. How do they do this, and how do they exploit them to launch a cyberattack?

References

BasuMallick, Chiradeep. (2022, March 4). Top 10 Intrusion Detection and Prevention System Software in 2022. Spiceworks. https://www.spiceworks.com/soft-tech/best-idps-software/

Committee on House Oversight and Government Reform. (2018, December). *The Equifax data breach.* https://oversight.house.gov/report/committee-releases-report-revealing-new-information-on-equifax-data-breach/

Fortinet, Inc. (2026). Data Center Security. https://www.fortinet.com/resources/cyberglossary/data-center-security

National Institute of Standards and Technology (NIST). (2025). Vulnerability. In *Glossary*. Retrieved September 26, 2025, from https://csrc.nist.gov/glossary/term/vulnerability

Newman, L. H. (2021, March 8). The Accellion breach keeps getting worse—and more expensive. *Wired*. https://www.wired.com/story/accellion-breach-victims-extortion/

Stahie, S. (2021, April 12). *Security researchers find critical zero-day exploit in Zoom Messenger*. Bitdefender. https://www.bitdefender.com/en-us/blog/hotforsecurity/security-researchers-find-critical-zero-day-exploit-in-zoom-messenger

Statista. Transforma Insights, & Exploding Topics. (May 19, 2025). Number of Internet of Things (IoT) connections worldwide from 2022 to 2023, with forecasts from 2024 to 2034 (in billions) [Graph]. Retrieved October 09, 2025, from https://www.statista.com/statistics/1183457/iot-connected-devices-worldwide/

Chapter 5

Information Security Risk and Risk Assessment

Chapter Outline

What Is Risk?
Types of Information Security Risk
What Do You Have That Someone Would Want?
How Much Risk Is Acceptable?
How to Assess Information Security Risk
Risk Assessment Models and Frameworks
Risk Assessment and Risk Management

Learning Objectives (LO)

After reading this chapter, you should be able to:

- LO1. Define risk and risk assessment.
- LO2. Understand operational, reputational, and compliance security risks.
- LO3. Know the difference between qualitative and quantitative risk assessments.
- LO4. Understand how models and frameworks can be of use in assessing cybersecurity risk.
- LO5. Be familiar with the FAIR Risk Model, ISO/IEC 27005, and NIST SP800-30 R1.
- LO6. Understand the link between risk assessment and risk management.

Introduction

You are much more likely to survive an automobile accident if you wear a seat belt. Just as you know that you can reduce your risk of injury by wearing one, enacting controls that prevent cyberattacks is a rational approach to managing cyber risk. Business leaders and accounting professionals are all familiar with the concept of **risk** as it impacts their organizations daily. A company's annual report includes a section on risk, requiring the organization to think about all the threats that might impact it, ranging from pandemics to natural environmental disasters. Since 2023, the 10-K report includes a separate section on cyber risk, recognizing the unique nature and scale of impact a cyberattack could have on the enterprise. The 2024 Allianz Risk Barometer found cyber risk to be business's number one risk concern (Allianz Commercial, 2024). Cyber risk affects every function in a business organization. When an attack occurs, the marketing function must handle the PR and communications, the accounting and finance teams need to deal with budgets and potential financial losses, IT must contain the incident and protect against future attacks, and the C-suite must manage the short- and long-term ramifications.

Risk The possibility that an event, action, or condition will occur that adversely affects objectives, assets, or outcomes.

When it comes to risk, there are two stages of actions: measurement and management. As the saying goes, "You have to measure it to manage it." This chapter concerns the measurement, or *assessment,* of risk. We begin by defining cyber risk and categorizing it in terms of operational, reputational, and compliance risks. We also discuss risk appetite and tolerance: How much risk is an organization willing to accept? What assets are at risk? The chapter considers both quantitative and qualitative cyber risk and offers some frameworks that are available as tools for risk assessment. The next chapter will be about how you control risks by managing them.

What Is Risk?

Risk is the chance that something bad will happen. The "bad thing" is the impact of an event, the "chance" is the likelihood of occurrence, and the risk is the result of multiplying the two.

Risk = Likelihood x Impact

> **EXAMPLE**
>
> **If a ransomware attack is 20% likely, and the impact of such an event is $5,000,000, then the risk exposure is $1,000,000.**

Impact may be mild or catastrophic. A mild information security risk could be temporary loss of noncritical data, such as product lists. Catastrophic impact might be permanent damage to business reputation or a service outage that lasts for some time. Infrastructure entities, including transportation, water supply, food processing and supply chains, and energy, are especially vulnerable to high-impact information security risks. In 2021, the Colonial Pipeline ransomware attack resulted in gasoline shortages along the East Coast of the United States. The impact was short-lived, so it fell short of being catastrophic, but it was significant. An attack on JBS USA Holdings, Inc., one of the world's largest meat processors, also highlighted the potential of catastrophic cybersecurity risks to affect anyone and everyone.

The specific threat, type of attack, motive, and exploited vulnerability all contribute to impact. Many impacts are financial, with either direct or indirect losses. A ransomware threat creates a direct financial loss, whereas an indirect financial loss could be a sales decline resulting from a customer data breach. Nonfinancial impacts may include data losses, service outages, business process interruption, regulatory or legal compliance failures, loss or erosion of intellectual property, brand and reputation damage, and even physical harm or death. Most of these impacts can be quantified but require some guesswork.

In considering risk management, and how much to expend to mitigate or change a risk, organizations need to consider the relationship between the threat (e.g., a ransomware attack) and the impact that such a threat would create. In our previous example, you would possibly be willing to spend as much as $1,000,000 to protect against the risk of a ransomware attack, through risk management.

Types of Information Security Risk

We can group information security risks into three categories: operational, reputational, and compliance. Each has its own distinct set of characteristics and challenges.

Operational Risks

Operational Information Security Risk The risk of loss resulting from inadequate or failed processes, systems, human errors, or external events relating to information security.

Operational risk is the impact times the likelihood of an adverse event affecting business operations. The core business functions may be impaired or disrupted so that the organization is not able to carry out its normal operating activities, such as serving customers or paying suppliers.

Figure 5-1 summarizes potential financial impacts from operational risks.

A prime example of an adverse event that presents operational risk is a ransomware attack.

Even when no ransom is paid, such attacks cause significant financial loss. According to Sophos, a cybersecurity product and services firm, the average loss from a ransomware attack was $2.73 million in 2024 (Sophos, 2024).

Nonfinancial impacts may be translated into financial effects over time, but generally an estimate of loss requires some speculation or judgment. Nonfinancial impacts from operational risks may range from customer dissatisfaction to productivity loss to physical harm or injury. In the latter case, consider disruption of our nation's electrical grid or a ransomware attack on a hospital as examples.

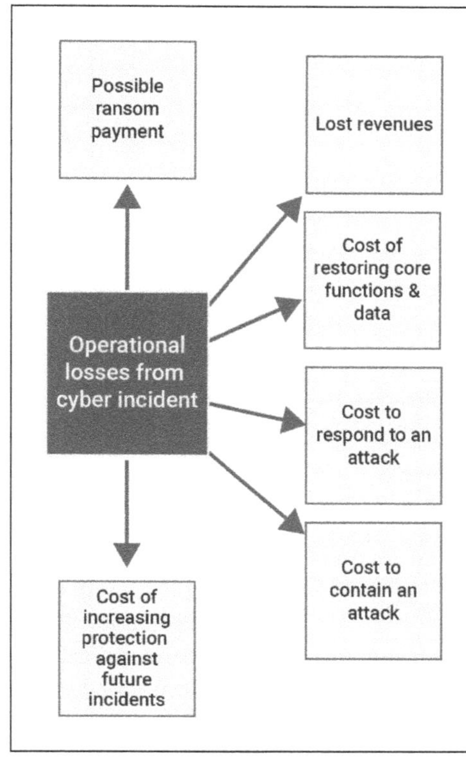

Figure 5-1. Operational Losses

CASE IN POINT

When Sony Pictures suffered a ransomware attack by the Guardians of Peace (GOP) in 2014, it literally sent the company back to the pre-computer era. The attackers stole data, made selected data public, and then destroyed the data. The publicly exposed data included private emails, scripts for forthcoming movies, and personally identifiable information (PII). Operations were disrupted, as employees had to communicate via messaging, could only accept cash for purchases in stores on the film lot, and issued paper checks to employees. A motive for the hack was thought to be the upcoming release of a movie, *The Interview*, that was critical of North Korea's leader. In addition to disruption of normal business processes, another operational impact was changes in film releases, including the release schedule for *The Interview*.

Reputational Risks

Warren Buffet has famously noted that "it takes 20 years to build a reputation and five minutes to ruin it" (Schwantes, 2020). **Reputational risk** as related to cybersecurity is the chance that a business reputation can be destroyed, perhaps in a nanosecond. Reputation loss can be spread among multiple stakeholders. We tend to think first of customers, but a decline in reputation can negatively affect relationships with stockholders, creditors, suppliers, current and potential employees, and regulatory agencies. Simply acknowledging that there has been a data breach sends a signal to stakeholders that a company is vulnerable to attack and perhaps is not well managed. The loss of consumer data can make customers and other stakeholders reluctant to continue to share data with the entity. Any breach or cyber incident has the potential to harm stakeholder relationships in the short run, and sometimes for much longer. However, it is important, and in many cases required, that companies be transparent about breaches that impact stakeholder data. The truth will out, as they say, and it is better for a business to proactively manage communications.

Reputational Risk
The potential for any event, controllable or otherwise, to damage your organization's reputation.

CASE IN POINT

Equifax is a credit monitoring company. Its business model is to collect financial data from banks, use proprietary analytical tools to analyze the data, and then sell the data back to financial institutions and consumers. It would seem, then, that it would have a higher level of security over its data and processes than just about any other business. However, in 2017, the company suffered a data breach that exposed PII for more than 147 million individuals. The financial ramifications for the company were significant, as it had to make available about $425 million to support individuals impacted by the breach. The company also suffered reputational damage, as trust was eroded by a breach that a congressional

investigation found had been preventable. Equifax's Buzz Score, an indicator of positive or negative consumer attitudes toward a business, fell 33 points (maximum is 100 points) in 10 days after the breach was announced. The company's stock price also decreased significantly in the aftermath of the breach, not recovering to pre-breach levels for more than a year and a half.

Reputation may be captured on the balance sheet as goodwill, but only if a company has been purchased. Otherwise, it is only reflected in, hopefully, superior revenues and profitability. However, reputation can be one of a business entity's most valuable assets, whether or not it is captured on a financial statement, so reputational risk can be a major threat.

Compliance Risks

Compliance Risk The potential for legal or regulatory sanctions, financial loss, or damage to reputation that an organization may suffer as a result of its failure to comply with laws, regulations, and standards.

General Data Protection Regulation (GDPR) A regulation in EU law governing data protection and privacy for all individuals within the European Union and the European Economic Area. It also addresses the transfer of personal data outside the EU and EEA.

Compliance risk is the chance of failing to meet legal and regulatory requirements. Every organization operates in a unique legal and regulatory environment that depends on its geographic location (country, state, locality), industry, and size, among other variables, such as complexity of operations and supply chain. This topic is discussed briefly here and in more depth in Chapter 9. In the United States, national laws exist regarding privacy and breach notification. Businesses operating in the United States may also be subject to international regulations, as well as those of their home country if they are not based in the United States. For example, although the **General Data Protection Regulation (GDPR)** is a European Union regulation, it applies to companies that operate in other countries and territories that serve customers in the EU. Huge penalties have been levied on companies for failure to comply with GDPR. These include a more than $250 million fine paid by Meta for a Facebook data breach and a fine of more than $800 million that Amazon paid to the Luxembourg Commission for Data Protection in 2021. In addition to country laws and regulations, many individual US states have governance over cybersecurity issues such as data breach notifications.

The consequences of failure to comply with laws and regulations go beyond paying fines. Stakeholders expect organizations to protect their data, and failure to comply with regulatory authorities, or even simple negligence, may result in lawsuits. For example, in 2022, the US telecommunications company T-Mobile settled a class action lawsuit, with $350 million awarded to claimants for a data breach that impacted millions of its subscribers.

Some industries are so critical that cybersecurity incidents involving them can affect broad populations. As a result, they are frequently governed by specialized regulations. These industries are typically associated with infrastructure, national security, transportation, and healthcare and include all the types of organizations depicted in Figure 5-2.

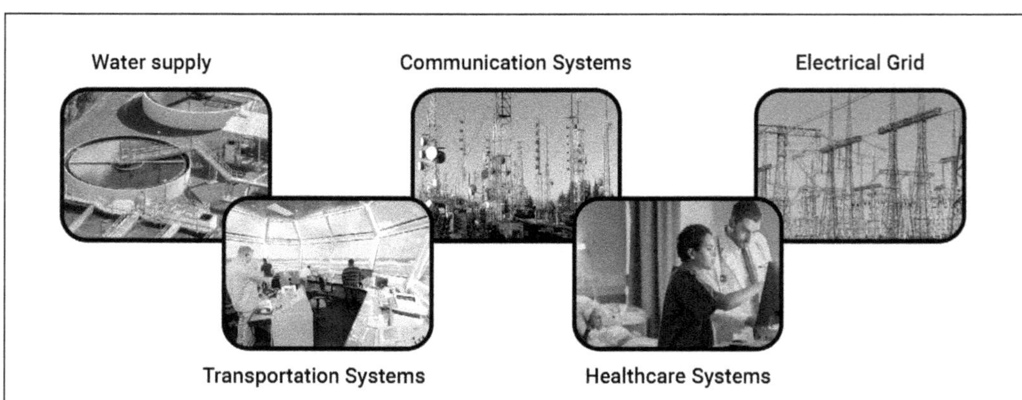

Figure 5-2. Critical Industries

CASE IN POINT

The Critical Infrastructure Protection Reliability Standards (CIP Reliability Standards) were created by the North American Electric Reliability Corporation (NERC) to protect North American power companies. These entities are required to comply with these standards, which are meant to protect the physical security and cybersecurity of those assets that are essential to the functioning of the nation's electrical infrastructure. Standards include a variety of security measures, beginning with the identification and categorization of cyber assets. They also include training, incident response, risk assessment, and system security.

What Do You Have That Someone Would Want?

To assess information security risk, you need to understand what it is you have that has value to someone. This entails **asset identification**, an essential step in determining **cyber value at risk (VaR)**. VaR is a metric that allows you to estimate financial losses by analyzing risk and risk management. Organizations have both monetary and nonmonetary assets. Financial resources are often a target, so companies tend to prioritize safeguarding monetary assets. However, business organizations have many other assets that may be just as or even more important. Some of these are **tangible assets**, while others are **intangible assets**. As the name implies, tangible assets are those you can touch. Figure 5-3 illustrates some tangible assets.

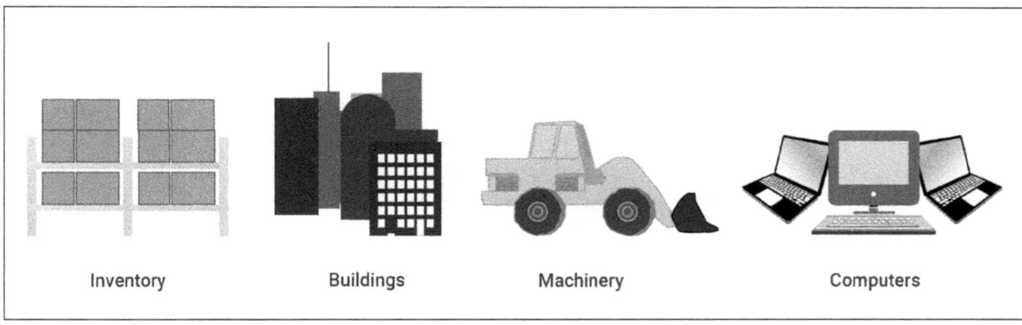

Figure 5-3. Examples of Tangible Assets

Often, safeguarding these assets includes physical security, but information security measures are important as well.

Intangible assets on company balance sheets have grown exponentially over the past few decades. In 1975, intangible assets made up just 17% of S&P company total assets. By 2020, that number had grown to 90% (Berman, 2021). Intangible assets include data, intellectual property, patents, copyrights, trademarks, and reputation.

Data are one of the most obvious assets requiring protection from threat actors. Criminals and other hackers may try to access data for many reasons. Adversaries may want the data because they can be converted into monetary assets. **Personally identifiable information (PII)** has value to criminals. Depending on what is taken, such as credit card numbers or Social Security numbers, databases of PII may be sold on the **dark web** to criminals, who will in turn use these identifiers to conduct transactions or steal identities. PII can also be used to embarrass companies; making the threat of publicly releasing an organization's PII can be used as leverage for ransom payments.

The following are data elements associated with PII:

- Name
- Address
- Phone number

Asset Identification The process of identifying and documenting all the valuable assets within an organization that need protection from risks.

Cyber Value at Risk (VaR) A quantitative risk management approach that applies the financial value at risk (VaR) concept to cybersecurity. It estimates the potential financial loss an organization could face from cyber incidents within a defined time horizon and at a specified confidence level.

Tangible Assets Physical items of value that an organization owns, such as buildings, equipment, and inventory.

Intangible Assets Non-physical assets such as intellectual property, goodwill, or brand value.

Personally Identifiable Information (PII) Information that, when used alone or with other relevant data, can identify an individual.

Dark Web A part of the Internet that is not indexed by traditional search engines and requires specific software, configurations, or authorization to access. It is often associated with illegal activities.

- Social Security number
- Credit card number
- Driver's license number
- Passport number
- Bank account number
- User ID (login)
- Password

> **EXAMPLE**
> **The COVID pandemic provides a good opportunity for us to think about risk appetite. You probably noticed that your friends and families had a variety of risk appetites when it came to their behavioral modifications in the pandemic. Some people would dine indoors and travel on airplanes. Others would eat only outside at restaurants, or avoid restaurants entirely. Some people stayed home and ventured out only when necessary. Businesses, too, had different risk appetites. At some companies, everyone still went to work on-site, and at others, employees mostly worked from home.**

Customer databases and transaction histories can also have value to threat actors. Because there are laws regarding protection of customer data, businesses may be extorted for ransom if hackers breach these data. Once the data are taken, they may be used for identity theft or in phishing attempts. Having a little knowledge about a target can take a criminal a long way toward obtaining unauthorized access to valuable resources.

Healthcare organizations have an extra burden of protection of data assets because, in addition to PII, they maintain databases of **protected health information (PHI)**. As you can imagine, these are data that require extra protection, if for no other reason than that there are legal and regulatory reasons to do so. Examples of these data include medical histories and prescriptions. Threat actors can use these data to obtain drugs or to file false medical claims with insurance companies. Because PHI is so important, it receives special protections from the Health Insurance Portability and Accountability Act (HIPAA), discussed in Chapter 9.

Protected Health Information (PHI) Any information about health status, provision of healthcare, or payment for healthcare that can be linked to an individual.

How Much Risk Is Acceptable?

Risk Appetite The amount of risk that an organization is willing to accept in pursuit of its objectives.

Risk Tolerance The acceptable level of variation in performance or outcomes that an organization (or individual) is willing to withstand in pursuit of its objectives.

People and businesses have **risk appetites**. This is the amount of risk that they are willing to take on. Closely related to risk appetite is the concept of **risk tolerance**, which refers to the variance from the risk appetite that the business can withstand.

In the COVID example, you might realize quickly that context matters. The behavioral modifications of individuals likely varied among different age groups, for people who had underlying health issues, and perhaps even with respect to political views. The same was true for companies. In some industries, such as healthcare, working from home was not an option. Businesses located in high-rises that needed social distancing in elevators couldn't have people all coming to work at once, so they had to set work-from-home policies accordingly. Situational analysis plays a role in setting the risk appetite for a business. This analysis includes an understanding of the technologies that the organization employs, the nature of the data held and processed, legal and regulatory requirements, entity size and scope, and the use of third parties.

Likelihood and impact of risk are both important. With COVID, the likelihood of hospitalization or death is much greater if you are 65 years or older. Dying is a pretty big impact, so you might expect that older people have done the calculation of impact times likelihood and determined that they should take precautions against the virus. Those precautions come with a cost, which must be weighed against the benefits of reduced risk.

Given the increasing ramifications of cybersecurity attacks, information system risk assessment should be part of an organization's strategic planning process. Decisions about expanding business to new geographic regions, consolidating or increasing branches, moving more business online, and storing data in the public cloud versus a private cloud or on-site are all examples of strategic decisions that will change cybersecurity risk.

How to Assess Information Security Risk

As noted at the beginning of the chapter, "You can't manage what you can't measure." When it comes to risk, it makes sense to recognize that some level of measurement or assessment can assist in managing or mitigating threats to the **CIA triad** of confidentiality, integrity, and availability, the elements of information security (Figure 5-4). An information security risk assessment exercise helps to identify sources of threats, potential threatening events, vulnerabilities, likelihood, and impacts. As noted earlier, risk is the product of the impact times the likelihood of that impact.

CIA Triad A model designed to guide policies for information security within an organization, consisting of three key principles: confidentiality, integrity, and availability.

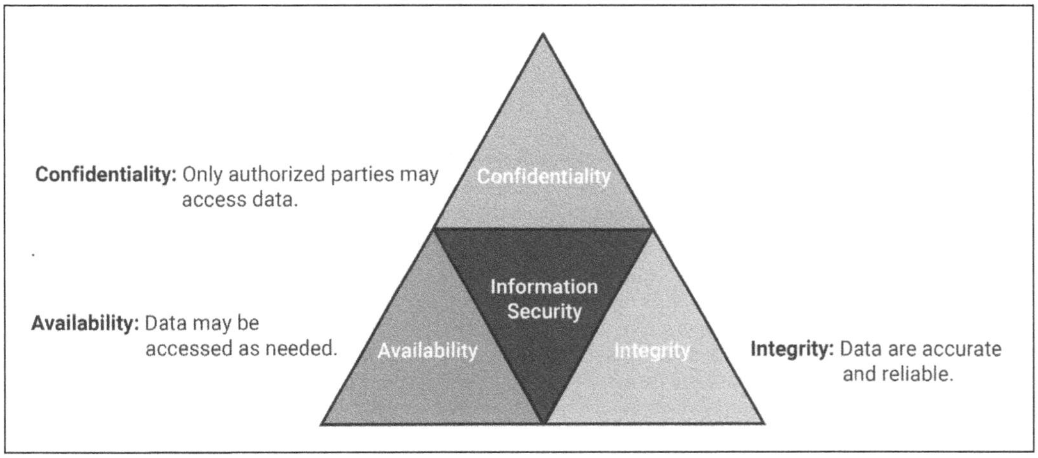

Figure 5-4. The CIA Triad

Risk assessment is invaluable in thinking about resource allocation. How much should you spend to mitigate a risk? You cannot really reduce a risk to zero, but you can change the likelihood and/or impact of a risk. For example, you might decide to move to a **multifactor authentication** model. How much will that cost? How does it change the likelihood and/or impact of the threat of unauthorized access and the risk of a data breach?

Multifactor Authentication (MFA) A security process that requires users to provide two or more authentication factors to verify their identity.

A business has many ways to assess its cyber risk. Considerations in selecting an approach include the complexity, cost, and time needed to perform the assessment. In-house information security personnel may conduct formal or informal assessments on a regularly scheduled or ad hoc basis. Alternatively, a firm might engage a consultant to do a cyber risk assessment. Consultants have a variety of tools, frameworks, and protocols available for this type of work, in addition to having broad, helpful experience working for multiple clients. In the best case, organizations will be proactive in assessing their cyber risk, doing so before they are affected by an adverse cyber event. However, in some cases, a breach or other cyber incident will occur and prompt a business to make this investment. Risk assessment of all kinds should be part of normal business operations and planning. Approaches to risk assessment can be both qualitative and quantitative; in many cases, an organization will use a combination of the two.

Qualitative Risk Assessment

Qualitative risk assessment can be as simple as asking the question of top leadership, "What keeps you up at night?" Qualitative risk assessment requires judgment and is somewhat subjective. This type of assessment should include discussions among a variety of stakeholders.

Qualitative Risk Assessment A method of evaluating risks based on subjective judgment and non-numeric criteria, such as severity or likelihood.

A structured approach to qualitative risk assessment involves the following steps (Table 5-1).

Table 5-1. Steps in Qualitative Risk Assessment

#	Step	Explanation
1	Identify potential threats	Sources of threats can be insiders or external stakeholders, including third parties and nation-states. Industry type, size of organization, location, operational complexity, and nature of data stored can all be factors in determining where a threat might originate. Another important factor in identifying threats is the value of assets to potential adversaries. For example, a company that has highly valuable intellectual property might be a target for data theft. A company with valuable PII might be the target of a data breach.
2	Identify vulnerabilities	Vulnerabilities can be in processes, software, hardware, or people. A vulnerability may be known, such as an unpatched software defect, or unknown. A **zero-day vulnerability** is one that has not yet been identified. It is so called because software developers have zero days to fix it. Unfortunately, it can take time to develop a fix, and during that interval, hackers may be exploiting it. Common vulnerabilities include lack of encryption, poor authentication processes, unpatched software systems, lack of or poorly configured firewalls, unsecured Wi-Fi, and human error. People are the biggest vulnerability, as human behavior is variable—more than 80% of cyberattacks occur because of human error or lack of security awareness.
3	Identify potential exploits	An **exploit** is a cyberattack event. It is the activation of a threat by using or *exploiting* a vulnerability. For example, a threat source might be a cybercriminal who wants to obtain a customer database. The database may be housed on a server that can be accessed through specific software. The criminal can launch a cyberattack through a phishing email that will exploit an unpatched software vulnerability.
4	Determine likelihood of exploits	In a qualitative risk assessment, the likelihood of occurrence of various exploits may be given a simple value such as high, medium, or low. This categorization still allows for security personnel to develop a plan to mitigate the most likely exploits and allocate resources toward the relevant vulnerabilities. For example, if the likelihood of a phishing attack that would result in a data breach is high, then resources might be directed toward a training and cyber awareness program for employees.
5	Determine the impact of a threat occurrence	As described earlier, the impact of a cyberattack can range from loss of data to business bankruptcy or loss of life. Many impacts might result from operational, reputational, and legal/regulatory risks. The risk assessment process will attempt to identify impacts that are most likely and those that can be mitigated in a cost-effective way. The problem with cybersecurity is that attackers need only one vulnerability to exploit, while information security that eliminates risk altogether would need to mitigate every vulnerability, including those that are unknown.

In a quantitative risk assessment, discussed next, these steps would be quantified with respect to likelihood and impact.

The following case shows how these steps played out in an actual cyberattack.

CASE IN POINT

All kinds of organizations may be victims of cyberattacks. In 2021, the Houston Rockets, a National Basketball Association team, suffered an attempted ransomware attack and data breach. The threat source was a hacker group called Babuk. The internal network was the target of the attack, but the organization had successfully protected its network to the extent that the attack had limited impact. The hacking group claimed to have taken contract information and financial data. In this example, the source of the threat was a cybercriminal group. The vulnerability was in the network system, and the exploit was a ransomware attack, where data and operations were threatened. The likelihood of such an exploit is the most difficult to estimate, but in an era when data breaches and ransomware attacks are rampant, it might universally be estimated as high. Fortunately, the Houston Rockets had risk management controls in place to minimize the impact, and there was no effect on operations.

Quantitative Risk Assessment

Quantifying risk can help to counter human bias. For example, we tend to have personal fears about certain threats, such as being attacked by a bear or being in a plane crash. According to the National Park Service, you are more likely to be killed by a bee than a bear. And we all know that your chance of dying in a car accident is much greater than death from a plane crash. In fact,

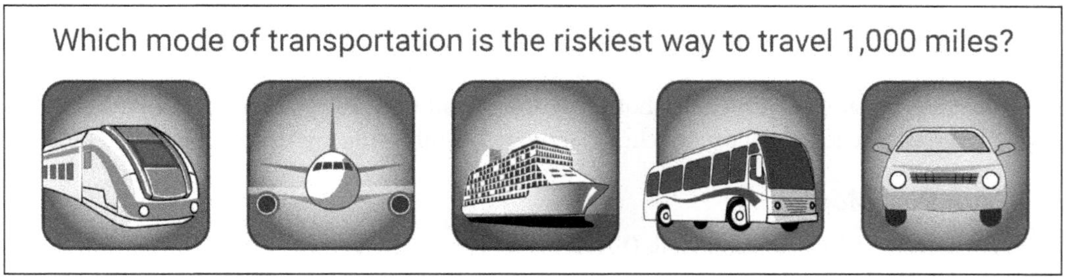

Figure 5-5. Risk and Transportation Modes

the odds of you dying from a plane crash are about the same as the risk of dying from a vending machine accident, and you'd be hard-pressed to find an individual with a phobia about that. Just knowing these statistics is useful, and when you have a limited budget to apply to risk management, quantified risk assessment can help you in allocating your resources effectively.

The answer to the question in Figure 5-5 is that it depends on a number of factors, including the weather, the experience of the operator, the location of the travel, the quality of the company, and more. All things considered, boats may be the least safe. Interestingly, individuals often fear these modes of travel over automotive travel, even though travel by cars is statistically the least safe.

Quantitative risk assessment requires assigning numerical values to impact and likelihood. One way to quantify impact is to identify company assets. Each asset has a value, and the impact could be the loss of that particular asset. As an example, a customer database can be assigned a monetary value. In other cases, there may be public data, perhaps industry-specific, regarding the dollar impact of various cyberattack scenarios.

Quantitative Risk Assessment A method of evaluating risks using numerical data, typically assigning values to probability and impact.

A quantitative approach to risk assessment has the advantage of allowing managers to prioritize risk factors. By ranking risk by potential loss, management can allocate resources toward information security more effectively. Both qualitative and quantitative risk assessment require judgment and estimates. With quantitative risk assessment, however, reducing risk to a number might provide a false sense of reliability. A quantitative versus qualitative approach also can require more effort, more expertise, and complex modeling.

> **EXAMPLE**
>
> **If the average cost to remediate a ransomware attack is estimated to be $2 million, and the likelihood of such an attack is thought to be about 40% for your company, then the risk value for that specific threat is $800,000. Knowing this value can help a company determine how much it should spend to defend against such an event.**

Addressing Risk

Treating risk is covered in the chapter on risk management, and it involves taking steps to avoid, control for, accept, or share the risk. Some risks are deemed acceptable. They are residual risks, those that are left over after controls and mitigation efforts are in place. Two ongoing processes are communication and monitoring. Organizations should be communicating continuously with stakeholders and consulting with technical experts throughout the risk processes. Risk assessment and management are not one-time exercises but need ongoing monitoring and review. Cyber threats are evolving and require continuous threat and vulnerability scanning. For example, the **Cybersecurity and Infrastructure Security Agency (CISA)**, under the US Department of Homeland Security, provides information alerts about vulnerabilities, exploits, and other security issues.

Cybersecurity and Infrastructure Security Agency (CISA) A US federal agency responsible for protecting the nation's critical infrastructure from physical and cyber threats.

Risk Assessment Models and Frameworks

Risk assessment models and frameworks can provide structure and guidance in estimating risk. In some cases, a framework includes both risk assessment and risk management. In other cases, frameworks include a general cybersecurity strategy. The following are three popular risk models.

The FAIR Risk Model

Factor Analysis of Information Risk (FAIR) A risk management framework used to understand, analyze, and quantify information risk in financial terms.

Monte Carlo Simulation A mathematical technique used to model the probability of different outcomes when there is uncertainty in the input variables. Instead of producing a single outcome, it runs a large number of trials (often thousands or millions), each time using randomly generated values within defined ranges, to estimate the distribution of possible results.

Factor Analysis of Information Risk (FAIR) is an international quantitative model that facilitates quantifying cyber risk. The model was adopted as a standard by the Open Group, an international consortium that offers standards and certification. FAIR includes a set of concepts that is composed of risks and risk relationships, methods to measure risk factors, a mathematical simulation of risk relationships, and a modeling construct. In addition to risk assessment, FAIR includes risk management, which is meant to "cost-effectively achieve and maintain an acceptable level of loss exposure" (RSI Security, 2020).

Implementing the FAIR framework as a quantitative model requires some sophisticated quantitative tools, such as **Monte Carlo simulation**. However, FAIR also includes a Basic Risk Analysis Guide. Figure 5-6 shows the basic steps that you should follow in completing a FAIR risk assessment.

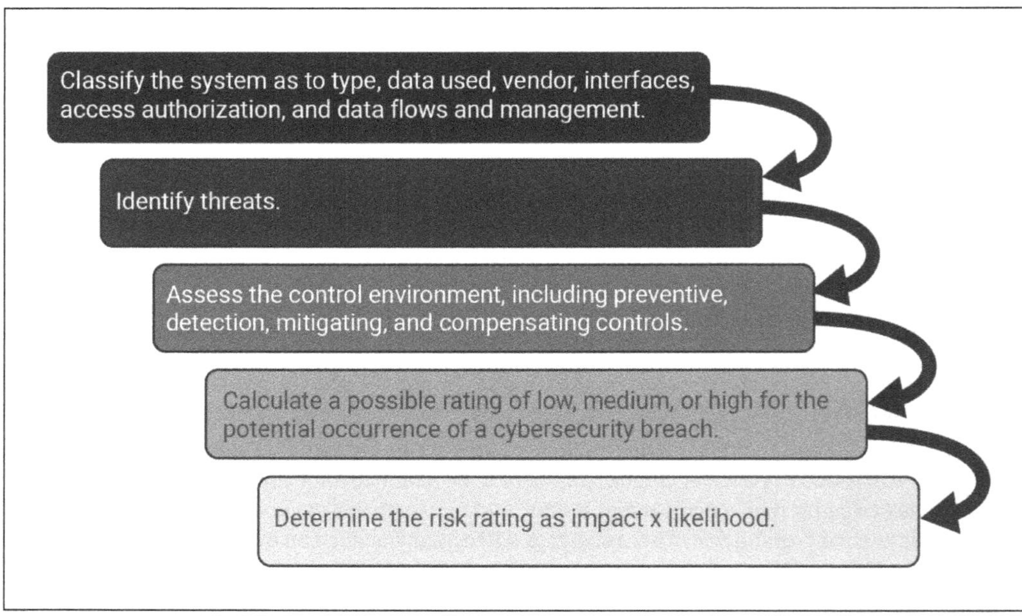

Figure 5-6. FAIR Risk Assessment Steps (NIST, 2012)

ISO/IEC 27005

ISO/IEC 27005 An international standard that provides guidelines for information security risk management.

Information Security Management System (ISMS) A framework of policies and procedures used to manage an organization's information security risks.

The International Organization for Standardization (ISO) and the International Electrotechnical Commission (IEC) published **ISO/IEC 27005**, the international standard that describes information security risk assessment processes that meet the requirements of ISO/IEC 27001, the international standard for an **information security management system (ISMS)** implementation. You can apply both standards to any type or size of organizational entity. ISO/IEC standards include technical, industrial, and commercial standards for enterprises globally.

Risk assessment and management are important components of an ISMS. ISO/IEC 27005 does not specify a methodology for risk management, but it does provide a framework for risk assessment and management. Note that risk assessment is a step in risk management, which we discuss in the next chapter.

The context or a situational analysis establishes the criteria for risk management. It includes determining the scope and organizational structure of risk management to ensure information security with respect to confidentiality, integrity, and availability. Risk criteria are dependent on the organization's risk appetite and context, such as industry and organizational objectives.

Risk assessment in ISO/IEC 27005 has three steps, as shown in Figure 5-7.

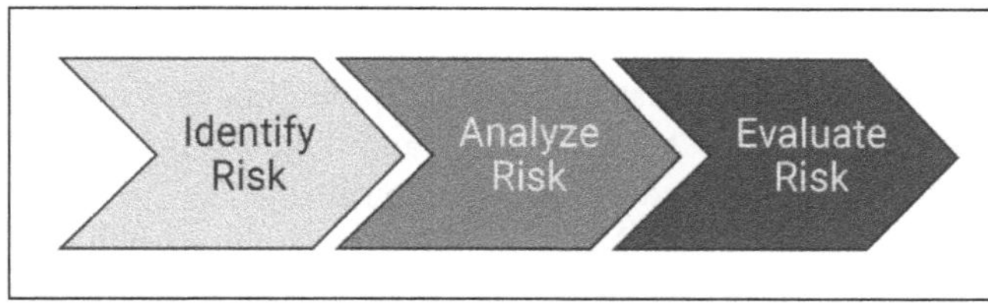

Figure 5-7. ISO/IEC 27005 Risk Assessment Steps

The first is to identify risks with respect to the sources and causes of risk, impact, and threats. To identify risks, you would look at historical data and industry trends and consult with experts. Analyzing risk is the step where you consider the likelihood or probability of threats occurring and their impact. Evaluation is where an organization determines if a risk is acceptable and what, if any, steps should be taken to mitigate it. This may include a cost/benefit analysis, whereby a weighted average of the risk (probability times impact) is compared with the cost to control the risk.

NIST SP 800-30 Revision 1

NIST SP800-30 is a guide for risk assessments published by the National Institute of Standards and Technology. While it is meant to be applied to federal government entities, it is also useful for other types of organizations, including businesses. The guide may be useful in estimating and prioritizing information risks on many levels, including to assets, individuals, operations, and national interests.

NIST SP800-30 A NIST special publication that provides guidelines for conducting risk assessments for information systems.

NIST SP800-30 states:

> The purpose of risk assessments is to inform decision makers and support risk responses by identifying: (i) relevant threats to organizations or threats directed through organizations against other organizations; (ii) vulnerabilities both internal and external to organizations; (iii) impact (i.e., harm) to organizations that may occur given the potential for threats exploiting vulnerabilities; and (iv) likelihood that harm will occur. (NIST, 2012)

There are several NIST publications on information security, including the NIST Cybersecurity Framework and NIST SP800-39 and NIST SP800-53, which provide guidance on risk management. SP800-30 provides readiness help for risk assessments, the conduct of risk assessments, communication of risk assessment results, and ongoing maintenance of risk assessments. NIST guidance on risk assessment describes models and approaches for assessing information security risk.

The process to conduct an information risk assessment within NIST SP800-30 involves the following steps:

- Identify threat sources and events.
- Identify vulnerabilities and predisposing conditions.
- Determine likelihood of occurrence.
- Determine magnitude of impact.
- Determine risk.

> **EXAMPLE**
>
> **If a ransomware attack is 30% possible next year, and the impact would likely be \$1,000,000, you would consider spending under \$300,000 to control for that risk.**

NIST SP800-30 points out that you can approach the analysis of risk in multiple ways. Although the same elements are involved in each case (threats, impact/assets, and vulnerabilities), the starting point is different. In a threat-based risk analysis orientation, you will first catalogue the sources and specific threats of concern. The impact of those threats and the vulnerabilities threat actors exploit to activate them follow. In an asset- or impact-oriented approach, the first step is to determine the assets more at risk and the consequences that could ensue from an information security exploit. Finally, a vulnerability-oriented approach begins with analysis of the entity's weaknesses and then continues to examine the threats and impacts that could ensue were an adversary to exploit them.

NIST SP800-30 contains several useful appendices, including a glossary of terms, a taxonomy of threat sources, examples of adversarial threat events, examples of adverse impacts, and a taxonomy of predisposing conditions.

Risk Assessment and Risk Management

The objective of cybersecurity is to manage digital risks to IT and data assets. The first step is for an organization to consider its risk appetite and risk tolerance. Identifying and assessing the level of risk for each threat and/or vulnerability is next. These risks may be documented in a risk register. The risk register is a structured approach to organizing each identified risk and noting its likelihood and impact to establish priorities for risk management. This documentation is useful in communicating risk strategies and for ongoing monitoring of risks.

In reality, calculating the likelihood and impact of cybersecurity risks is a complex process requiring expertise and a detailed understanding of both an organization's threats and its vulnerabilities. As described in this chapter, this calculation may be qualitative or quantitative.

Once risks are identified and assessed, they can be managed, accepted, mitigated, or transferred, as described in the next chapter. Note that risk assessment allows management to consider the costs and benefits associated with each identified risk and allocate resources appropriately.

Summary

Cybersecurity requires protection of the confidentiality, integrity, and availability of information assets. Assessing and managing risk are critical to accomplishing this goal. The chapter describes organizational, reputational, and compliance risks and how to assess them. This includes a description of information assets to be protected, risk appetite, and approaches to risk assessment. Many public and proprietary guides and frameworks are available to assist organizations in risk analysis and assessment. The chapter briefly describes the FAIR model, ISO/IEC 27005, and NIST SP800-30. Risk assessment is the first step toward information security risk management, which is the topic of the next chapter.

Chapter Review

Key Terms

Asset Identification
CIA Triad
Compliance Risk
Cyber Value at Risk (VaR)
Cybersecurity and Infrastructure Security Agency (CISA)
Dark Web
Exploit
Factor Analysis of Information Risk (FAIR)
General Data Protection Regulation (GDPR)
Information Security Management System (ISMS)
Intangible Assets
ISO/IEC 27005
Monte Carlo Simulation
Multifactor Authentication (MFA)
NIST SP800-30
Operational Information Security Risk
Personally Identifiable Information (PII)
Protected Health Information (PHI)
Qualitative Risk Assessment
Quantitative Risk Assessment
Reputational Risk
Risk
Risk Appetite
Risk Tolerance
Tangible Assets
Zero-Day Vulnerability

Review Questions

1. What is risk, and how does it relate to cybersecurity?
2. List the types of security risks and summarize each risk.
3. How would you respond to the statement "There's nothing on my computer that anyone would want to steal"?
4. Describe *risk appetite* and explain how it relates to cybersecurity.
5. What are examples of qualitative cybersecurity risks, and how do you assess them?
6. What are examples of quantitative cybersecurity risks, and how do you assess them?
7. List and describe the three main risk assessment models and frameworks.

Multiple Choice

Identify the choice that best completes the statement or answers the question.

1. What is *risk* defined as in the context of cybersecurity?
   a. The result of multiplying impact by likelihood
   b. The impact of a bad event
   c. The severity of a bad event
   d. The probability of a threat occurring
   e. The impact of a threat on an organization
2. What are *operational information security risks* defined as?
   a. The likelihood of an adverse event affecting business operations
   b. The financial loss incurred from a ransomware attack
   c. The impact of a threat on an organization's reputation
   d. The severity of a data breach
   e. The result of multiplying impact by likelihood
3. What is a potential financial impact of operational information security risks?
   a. Customer dissatisfaction
   b. Physical harm or injury
   c. Lost revenues
   d. Productivity loss
   e. Reduced information security costs
4. What is reputational risk in cybersecurity primarily concerned with?
   a. Financial losses incurred from data breaches
   b. The impact of a data breach on stakeholder relationships
   c. The severity of a cybersecurity incident
   d. The likelihood of a cyberattack occurring
   e. The operational disruptions caused by a data breach
5. What is compliance risk primarily concerned with in information security?
   a. Failing to meet legal and regulatory requirements
   b. Financial losses due to data breaches
   c. Operational disruptions caused by cyberattacks
   d. The reputational impact of a data breach
   e. The likelihood of a cyberattack occurring
6. What additional protection measures might businesses face if their customer databases are breached?
   a. Increased tax liabilities
   b. Enhanced cybersecurity insurance coverage
   c. Threats of legal action for data mismanagement
   d. Reduced operational costs
   e. Access to government subsidies for cybersecurity
7. What is the primary purpose of qualitative risk assessment?
   a. To provide a precise numerical value of risk likelihood and impact
   b. To identify potential sources of threats and vulnerabilities
   c. To completely eliminate all possible cyber threats
   d. To quantify the exact financial impact of a cyberattack
   e. To determine the exact timing of potential cyber exploits
8. What is one drawback of relying solely on quantitative risk assessment?
   a. It increases the likelihood of rare events occurring.
   b. It eliminates the need for judgment and estimates.
   c. It provides a false sense of reliability.
   d. It requires minimal effort and expertise.
   e. It simplifies the process of prioritizing risk factors.

Cases

Yellow Subs—Risk Assessment

Yellow Subs has been so pleased with the work that Herron has done to date that it has engaged the company to help in assessing its cybersecurity risk. Chris Mason is going to work with the company's management to determine its risk appetite and risk tolerance. Together, they will consider the work that Natalie has done in identifying threats and vulnerabilities to construct a qualitative risk assessment for the company.

Case Requirements

1. What factors will Chris consider in determining the risk appetite and risk tolerance for a retail food company operating in the United States? As a reminder, the company has 400 restaurants, most of which are owned by franchisees.
2. Taking into consideration the threats and vulnerabilities identified in the case in Chapters 3 and 4, how would you go about assessing a qualitative likelihood for each threat and vulnerability pair to determine the potential impact?

Hoyle Company—Cybersecurity Roles and Risks

The Hoyle Company is a small family business with about $10 million in annual revenue and 85 employees. The company sells garden decorations, which it imports from China and Vietnam. Hoyle is headquartered in Norfolk, Virginia, and has customers throughout the United States.

The company has three IT staff. One is the IT Manager, who sets the strategy, manages the budget, and troubleshoots. The Network Administrator manages the network and all the hardware and software. The third employee is an IT Technician, who provides support to employees throughout the company and manages user accounts and access. Responsibility for information security is spread across the IT staff. The IT Manager takes care of physical security for the data center. The Network Administrator manages the firewall, antivirus protection, network detection and response tools, and software patching. This role also maintains data backups and a disaster recovery plan. The IT Technician handles employee cybersecurity training, password security, and access control.

Jim Hoyle, the CEO, has been reading about several high-profile cybersecurity incidents. He is concerned that the company's IT staffing is not sufficient to provide protection against a ransomware attack.

Case Requirements

1. Using online research tools, including generative AI, explain why the three staff and their roles are adequate or inadequate to manage cybersecurity risk. Would you change any of the responsibilities of the three IT staff? If so, how?
2. Again, using online research tools, including generative AI, estimate the amount of risk that Hoyle has with respect to a ransomware attack.
3. In the event of a ransomware attack on Hoyle, describe specific operational risks, compliance risks (e.g., for this type of business), and reputational risk.

Greengate Furniture Company—Conducting a Risk Assessment

The Greengate Furniture Company is a midsize Midwestern US–based company that produces office furniture for businesses and homes. The company employs about 175 people and has annual revenues of $20 million. Greengate's president, Andy Johnson, is aware that one of the company's competitors has recently undergone a ransomware attack. As a result, he has engaged a regional accounting firm to conduct a risk assessment for Greengate.

Case Requirements

1. What do you believe are the most likely cyber risks facing Greengate?
2. What level of risk acceptance would you consider appropriate for this company?
3. Assume that you are the consultant assigned to conduct the risk assessment for Greengate. Identify the specific steps you would take to do so.

Exercises

5-1. Colleges and universities are frequent targets of threat actors. One of the reasons for this is that they have a variety of valuable assets. Required:
   1. Make a list of three tangible assets that could be taken from a university in a cyber incident.
   2. Intangible assets include data. What are five types of data your university maintains that could be of interest to adversaries?

5-2. Since 2023, the Securities and Exchange Commission has required companies to include a separate Cybersecurity Risk Disclosure in their annual 10-K filings. The following is an excerpt from Lululemon's 10-K filed on January 28, 2024:

   ITEM 1C. CYBERSECURITY
   Risk Management and Strategy
   Our business operations and relationships with customers and suppliers are heavily reliant on technology. We operate a cybersecurity program designed to assess our security risks and threats, to manage those risks and protect our technology systems and data, and to detect and respond to cybersecurity incidents.

   We manage strategic risks, including cybersecurity risk, through our Enterprise Risk Management program, which has direct involvement from the board of directors, the audit committee, and senior management. Through this process, we have identified cybersecurity as a risk management priority. Required:
   1. Compare Lululemon's Risk Management and Strategy statement in 10-K Item 1C with that of another public company retailer. How are the statements similar or different?
   2. Go to Lululemon's 10-K statement. Apart from the section on Risk Management and Strategy, what other information is included in Item 1C?
   3. Based on your reading of Lululemon's Cybersecurity Risk Disclosure, what would you estimate their risk appetite to be (high, moderate, or low risk appetite, or extremely risk averse)?
   4. Imagine that you are the CEO of Lululemon (cool job!). How would you go about determining your level of risk appetite?

5-3. In 2023, MGM Resorts suffered a cyberattack that affected its operations for more than one week. MGM Resorts is an international company that owns and operates casino resorts.

   Required:
   1. In the 2023 MGM cyberattack, there were several disruptions to operations. Describe three of these disruptions.

2. On September 11, 2023, MGM communicated that a cyber incident had occurred. What was the impact on the stock price the following day? One week later? One month later? One year later? Why do you believe the stock market reacted as it did throughout this period?

References

Allianz Commercial. (2024, January). *Allianz risk barometer: Identifying the major business risks for 2024*. https://www.allianz.com/en/economic_research/insights/publications/specials_fmo/2024_01_16-Allianz-Risk-Barometer.html

Berman, B. (2021, January 19). *Latest data show that intangible assets comprise 90% of the value of the S&P 500 companies*. IP CloseUp. https://ipcloseup.com/2021/01/19/latest-data-show-that-intangible-assets-comprise-90-of-the-value-of-the-sp-500-companies/

National Institute of Standards and Technology (NIST). (2012, September). *SP 800-30 Rev. 1, Guide for conducting risk assessments*. https://csrc.nist.gov/pubs/sp/800/30/r1/final

RSI Security. (2020, June 4). The basics to completing a FAIR assessment. *RSI Security Blog*. https://blog.rsisecurity.com/the-basics-to-completing-a-fair-assessment/

Schwantes, M. (2020, December 8). Warren Buffett explains how one mistake could ruin your reputation. *Inc.* https://www.inc.com/marcel-schwantes/warren-buffett-explains-how-1-mistake-could-ruin-your-reputation.html

Sophos. (2024). *The state of ransomware 2024*. https://assets.sophos.com/X24WTUEQ/at/9brgj5n44hqvgsp5f5bqcps/sophos-state-of-ransomware-2024-wp.pdf

Chapter 6

Information Security Risk Management

Chapter Outline

Managing Cybersecurity Risk
Classifying Cybersecurity Controls
Cybersecurity Controls by Type
Risk Management Models and Frameworks
Cybersecurity Insurance

Learning Objectives (LO)

After reading this chapter, you should be able to:

- LO1. Describe approaches to risk management strategy.
- LO2. Know the different types of cyber controls.
- LO3. Be able to describe examples of technical, administrative, and physical controls.
- LO4. Know the different cyber control functions.
- LO5. Be able to describe examples of preventive, detective, and corrective controls.
- LO6. Be familiar with the common risk management frameworks.
- LO7. Recognize the benefits of cybersecurity insurance.

Introduction

Cybersecurity professionals, accountants, managers in every functional area, and boards are among those tasked with managing information security risk. Once risk has been assessed and the parameters for risk appetite and risk tolerance are set, these stakeholders will tackle the job of instituting controls to manage risk to an acceptable level, based on the organization's risk management strategy. This strategy considers the costs associated with mitigating risk versus the benefits of doing so.

This chapter describes how to manage cyber risk and the related controls, frameworks, and planning efforts that can protect an organization from cyberattack. We describe cybersecurity controls along two dimensions: type and function. Types of controls are technical, administrative, and physical. The chapter provides examples of specific controls in each type. Control functions are preventive, detective, and corrective. Some controls are meant to prevent attacks or protect the information system, while others work to discover intrusions and correct for them. These functions are described within each type. The chapter also discusses two other components of risk management: control frameworks and cybersecurity insurance.

Managing Cybersecurity Risk

Residual Risk The remaining risk after all mitigation efforts have been implemented.

Acceptable Risk The level of risk that is deemed tolerable in an organization's operations, given the expected benefits and controls.

Chapter 5 defined risk and explained how to assess it. Once an organization has evaluated information security risk, leaders can consider how to manage it. Both senior management and the Board's attitude toward approaching risk will set the tone for risk management strategy. Even in an organization that prioritizes cybersecurity, not all risks will be controlled. There will always be some **residual risk**, where the cost of mitigation is simply too high to be worth the effort to eliminate or reduce that risk. There will also be **acceptable risk**, which falls within the organizational risk appetite. Table 6-1 describes approaches to cyber risk management.

Table 6-1. Cyber Risk Management Approaches

Risk Strategy	Action(s)
Accept	Allocate few resources to controlling for cyber risk.
Avoid	Take actions that avoid risk.
Mitigate	Allocate resources to control cyber risk.
Transfer	Purchase cyber insurance.

Technical Controls The measures that organizations put in place to safeguard their digital assets.

Administrative Controls The creation and enforcement of security policies, rules, procedures, and guidelines that are set by management.

Physical Controls The measures that organizations take to physically protect their assets.

Preventive Control A control put in place with the intention to avoid an incident or prevent it from occurring.

Detective Control A control put in place to detect and alert about an incident that has already occurred.

Corrective Control A control put in place to correct the damage done as a result of an incident that has already occurred.

Compensating Control Supplementary measures that are implemented to protect against identified risks or threats when primary controls are ineffective, infeasible, or inefficient.

To manage risk, an enterprise may choose to institute controls that reduce it to acceptable levels. Management might decide to avoid the risk altogether by changing processes so that it is no longer present—for instance, by altering vendor payment processes. If they cannot avoid the risk altogether, management may change either the likelihood of its occurrence or the consequences stemming from it, through changes in processes or by instituting controls. Another alternative is to transfer the risk, typically through purchase of cybersecurity insurance. As with risk assessment, a variety of frameworks, models, and guidelines are available to guide an organization in implementing risk management.

Organizations should have a policy and plan in place in case the risk management strategy fails and a cyber incident occurs. This should include a Cybersecurity Incident Response Plan. Even the best managed entities are likely to suffer an adverse cybersecurity event. Remember that protecting information resources requires plugging every vulnerability, including guarding against zero-day events, but it takes only one successful threat or one human error to compromise information security.

Classifying Cybersecurity Controls

Cybersecurity controls reduce or mitigate the risk to confidentiality, integrity, and accessibility of information assets and systems. As noted, one way to classify cyber controls is by type. Remember that we categorize them in this book according to three types: technical, administrative, and physical. **Technical controls** may also be defined as logical controls. These are hardware and software controls employed to protect information assets. **Administrative controls** include people and governance. These are the policies, procedures, processes, and practices that an organization uses to manage information security risk. **Physical controls** are tangible controls that prevent or detect unauthorized access, physical events, or actions that pose a threat to information system assets. Typically, physical controls will include access and monitoring controls.

As outlined earlier, a rubric for categorizing cybersecurity controls by function also has three categories: preventive, detective, and corrective. An organization may implement a set of **preventive controls** over information security to impede an attack from occurring. Examples of this type of control are multifactor authentication and encryption. **Detective controls** are intended to discover a threat and include intrusion detection systems. **Corrective controls** are those that occur following an adverse information security event and include communication strategies and incident response. In addition to these primary categories, **compensating controls** may be put in place when a primary control is not feasible. For example, in a small organization where a control objective is to protect servers, it may not be possible to physically prevent access. In that case, cameras or monitoring devices might be put in place as a compensating control. Table 6-2 shows how functions and types of controls can be classified in a matrix, providing examples of controls that fulfill each intersection of type and function.

Table 6-2. Control Matrix with Examples

Control Type	Control Function		
	Preventive	**Detective**	**Corrective**
Technical	Multifactor authentication	Intrusion detection systems	Software patches
Administrative	Cybersecurity awareness training	System change logs	Incident Response Plan
Physical	Security badge access	Closed circuit TV	Fire suppression

Cybersecurity Controls by Type

Controls over information security are constantly evolving. Earlier chapters discussed the relationship between threats and vulnerabilities. As threats evolve and escalate, the control landscape changes to counter these new threats. When new vulnerabilities appear, cyber controls are developed, often in the form of new software security patches. Understanding the different types of controls and whether they can prevent, detect, or correct cyber threats is key to developing a business's risk strategy. This section of the chapter lists the controls that an organization might use as part of its risk management strategy, broken out by type.

Technical Controls

Technical controls are automated software and hardware controls that protect data and information systems. This section details specific technical control categories and examples.

Access and Authentication

Logical access controls prevent unauthorized access to digital data and information systems. These controls are preventive because they deny access. Logical access controls manage who can get into the network, and they include both identifying users and verifying their identity. A user ID for logging into an information system is an example of a logical access control.

The verification process in digital access is called **authentication**. Earlier chapters have introduced the topic of **multifactor authentication (MFA)** to prevent threat actors from accessing digital resources. Authentication involves three factors: knowledge, possession, and inherence. Knowledge is something known to the user, possession is something you have, and inherence is who you are. Figure 6-1 shows examples of each of these factors. MFA requires a combination of two or more of them.

Logical Access Control Tools and/or protocols used to identify, authenticate, authorize, and hold accountable individuals trying to access a system.

Authentication The process of verifying the identity of a user or system.

Multifactor Authentication (MFA) A security process that requires users to provide two or more authentication factors to verify their identity.

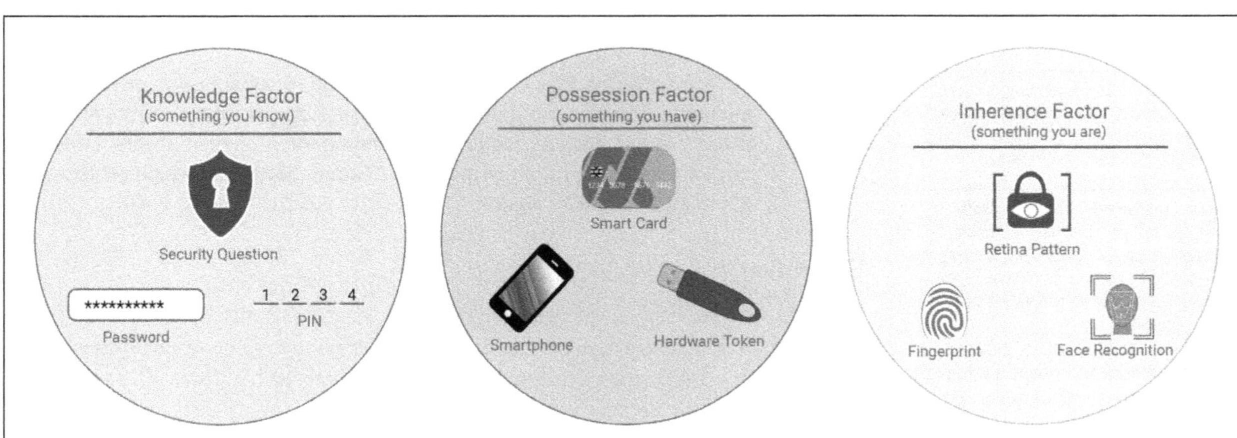

Figure 6-1. The Authentication Factors

Many access control systems rely on passwords for user authentication. Because they can be guessed or hacked through **dictionary** or **brute force attacks,** passwords have become increasingly complex over time. **Password management software** may be used to generate unique passwords and keep track of them. However, even highly complex passwords can be hacked with enough time and the right tools, making the case for MFA a compelling one.

In logical access control systems, the computer will check identification against an access control list of authorized users. These **access control lists** include levels of privilege that segment access. While authentication ensures that a person acquiring access is who they say they are, **authorization** allows the individual to perform certain functions. **Least privilege** is a control that ensures that each user will only be able to access data necessary to perform their job. As with access to different privacy levels of information (e.g., confidential versus highly confidential), authorization allows a user to perform different actions. For example, some users may be able to access data but not exfiltrate or edit them.

Authorization Models

An organization must select an appropriate authorization model to allow both access and authorization to perform functions within that segment of the network. The most common access control methods include **discretionary, mandatory, role-based, attribute-based, rule-based,** and **risk-adaptive** and answer the question "Now that I know who you are, what are you allowed to do"?

Antivirus Software

Virus protection applications scan incoming network traffic to detect and prevent malware, such as viruses, worms, Trojan Horses, and spyware, from accessing data and hardware. This control is both preventive and detective. The software scans incoming files or computer code against a database of known malicious code. **Antivirus software** can also provide real-time protection by running in the background and being activated whenever traffic is trying to enter the secured system. **Next-generation (next-gen) antivirus software** goes beyond earlier programs by detecting both known and unknown threats. Previous antivirus protection compared the incoming traffic against the signatures of known threats and prevented access when a match was found. Next-gen antivirus software uses machine learning and behavioral artificial intelligence to protect against zero-day malware threats.

> **EXAMPLE**
> **Norton AntiVirus Plus is available in a range of subscription prices, each including different protective features, such as parental controls and privacy monitoring. In addition to scanning for viruses, malware, and spyware, Norton's program includes AI scam detection, which protects users against online scams.**

Encryption

One of the most important controls for protecting data both at rest and in transit is **encryption.** Encryption entails encoding a message by using a **cipher** (algorithm) to scramble it into a different, generally unintelligible, form. The scrambled message can then be safely transmitted to a message recipient, as anyone who intercepts the message would not be able to interpret it without knowing the cipher.

Dictionary Attack A method of breaking a password or encryption by systematically entering every word in a predefined list of likely passwords.

Brute Force Attack Attempts to gain unauthorized access to a system by systematically trying all possible combinations of letters, numbers, and symbols for passwords or encryption keys.

Password Management Software Software that helps users generate, store, and manage complex passwords securely.

Access Control Lists Lists used to control who can access a network or system resource and what actions they can perform.

Authorization The process of determining which users or systems are allowed to access certain resources.

Least Privilege An information security concept in which a user is given the minimum level of access—or permissions—needed to perform their job functions.

Discretionary Access Control The owner of a resource decides who can access it and what they can do.

Mandatory Access Control Access is controlled by system-enforced rules based on classifications (e.g., Top Secret, Confidential). Users cannot change permissions.

Role-Based Access Control A type of control where permissions are assigned to roles (e.g., "Manager," "HR") and users inherit permissions through their role(s).

Attribute-Based Access Control Access decisions are based on attributes (user, resource, environment) and policies.

Rule-Based Access Control A type of control that uses predefined rules to grant/deny access, often based on conditions like time of day or IP address.

Risk-Adaptive Access Control A type of control that considers contextual risk and adjusts access dynamically.

Antivirus Software A computer program used to prevent, detect, and remove malware. Antivirus software was originally developed to detect and remove computer viruses, hence the name.

Next Generation (Next-Gen) Antivirus Software Antivirus software that uses a combination of artificial intelligence, behavioral detection, machine learning algorithms, and exploit mitigation, so known and unknown threats can be anticipated and immediately prevented.

Encryption The process of converting data into a coded format to prevent unauthorized access.

Cipher An algorithm for encrypting and decrypting information.

Symmetric encryption (Figure 6-2) relies on a shared key, meaning that both the message sender and the message recipient have the same key or algorithm. This secret key both scrambles the message and unscrambles it.

The Caesar cipher is one of the earliest uses of this method. Julius Caesar used a substitution method to scramble the messages sent to his generals in the field. If the messenger was intercepted, the enemy could not read the message. In a simple substitution cipher such as this, you simply substitute a letter of the alphabet for a different one, as in the following example.

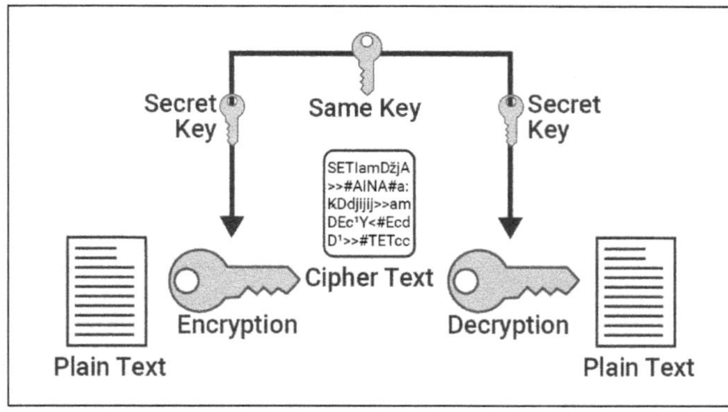

Figure 6-2. Symmetric Encryption

Symmetric Encryption An encryption method that uses a single shared key to encrypt and decrypt data.

Asymmetric Encryption An encryption method that uses a pair of keys (public and private) to encrypt and decrypt data.

EXAMPLE

In the following chart, you can see that the letter substitution has moved four spaces so that the letter "E" represents the letter "A," and so on. Try writing your first name using this cipher!

A	B	C	D	E	F	G	H	I	J	K	L	M	N	O	P	Q	R	S	T	U	V	W	X	Y	Z
E	F	G	H	I	J	K	L	M	N	O	P	Q	R	S	T	U	V	W	X	Y	Z	A	B	C	D

Asymmetric encryption makes use of two paired keys. One is public and shared, and the other key is private or secret. In asymmetric encryption, the individual sending a message obtains the public key of the intended recipient and uses it to encrypt the message. The message is unintelligible throughout transmission. Once the message is received, the individual uses the private or secret key from the key pair to decode and read the message. Encryption may be used to encode stored data and to protect data in transit. Figure 6-3 illustrates asymmetric encryption.

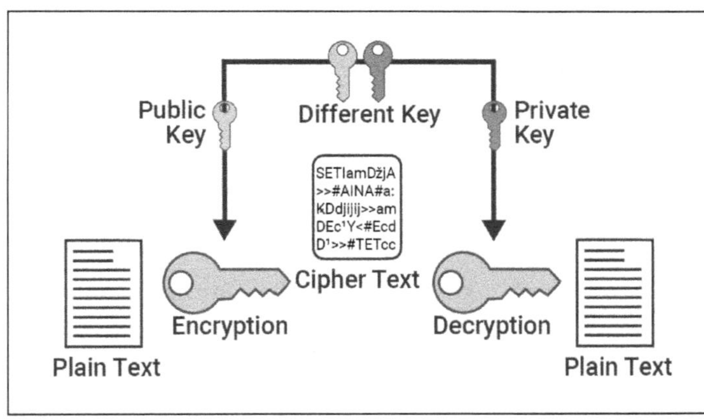

Figure 6-3. Asymmetric Encryption

Julius Caesar was able to use a simple substitution algorithm for encryption, but as computing power evolves, so, too, does the complexity of cipher algorithms. Many of these ciphers involve multiplying large prime numbers together, which is how to generate public keys. Quantum computing, discussed in Chapter 2, poses a threat to traditional encryption because the computing power makes it possible to crack the cipher. On the other hand, quantum computing may prove useful in creating ciphers that threat actors would be unable to break.

CASE IN POINT

Researchers at the National Institute of Standards and Technology (NIST) have been working on quantum-proof cryptography since early in this century, and in 2017 they initiated a public appeal for quantum-resistant public-key encryption algorithms that could be used for standardization. In 2024, the agency issued three Federal Information Processing Standards (FIPS) designed to be resistant to decoding by quantum computing. These standards described the mechanics for developing keys and digital signatures that can resist quantum computing attacks.

Digital Certificate An electronic document used to prove the ownership of a public key, issued by a trusted certificate authority.

Digital Signature A cryptographic technique used to verify the authenticity and integrity of digital messages or documents.

A **digital certificate**, issued by a Certificate Authority (an organization trusted to verify websites), provides verification that the encryption key pair used in asymmetric encryption belongs to the owner of the public key in the key pair. The certificate provides authentication of the party, creating a **digital signature** when encrypted using the private key. As we rely less and less on paper documents, these signatures have become critical for proof of identity in business transactions.

> **CASE IN POINT**
>
> **DocuSign is a Certificate Authority that processes digital signatures. The service is useful when individuals cannot sign a document in person. DocuSign uses a specific protocol to generate public and private key pairs. When an individual signs a document electronically, the signature is encrypted with their private key, creating the digital signature. The recipient of the signed document will also get a copy of the signer's public key, which may be used to verify the signature.**

Firewall A security system that monitors and controls incoming and outgoing network traffic based on predetermined security rules.

Firewalls

Firewalls may be included as part of an antivirus software application, or they may be installed independently. They act as information system barriers that monitor and control both incoming and outgoing network traffic based on a set of security rules. These rules are predetermined and are set to grant or deny access. Three basic types of firewalls are described in Table 6-3.

Table 6-3. Firewall Types

Firewall Type	Description
Packet filter	Small sets of data are sent in packets. The firewall checks the packets of data trying to enter the network against a set of identified threats. The packet is rejected if a match is found.
Proxy server	A proxy server prevents direct connections between incoming packets and the network. The proxy firewall is an intermediary sitting between an internal and external network. The proxy firewall verifies the authenticity of the incoming packet and forwards it, but it does not expose the internal network.
Stateful inspection	This type of firewall examines multiple elements in incoming data packets and compares them to a trusted database. Rather than denying access based on security rules, these firewalls grant access when a match is made against an authorized database of IP addresses or when the packets meet other characteristics indicating a safe communication.

Intrusion Detection and Prevention and Event Management Systems

Intrusion Detection and Prevention Software (IDPS) Hardware and/or software tools that detect and respond to suspicious activities on a network or system, combining both detection and prevention functionalities.

Attack Signatures Patterns or sequences of data that are indicative of a specific type of attack or malicious activity.

Chapter 4 described **intrusion detection and prevention software (IDPS)** that protects networks from threat actors by detecting unauthorized incoming network traffic and preventing it from coming into the network. IDPS is bundled software that includes both the detection and prevention functions.

The detection portion of these applications examines incoming network message packets to detect threats. These threats are identified by comparison with signatures of known attacks. **Attack signatures** are unique patterns of information associated with known vulnerabilities. For example, a suspicious IP address might be detected in the same way that your mobile phone warns you of an incoming call as a Spam Risk. A shortcoming of signature-based IDPS is that it is only able to detect threats that exploit *known* vulnerabilities. Anomaly-based IDPS can detect unknown threats by examining user behavior and monitoring for suspicious activity. Examples of anomalous behavior that IDPS may spot include a user logging into the system at an unusual time or from an unknown device. Once a threat is identified, the IDPS logs it and issues a security alert.

While IDPS warns of a potential adversarial exploit, the prevention portion stops the threat from reaching the network. Again, the process may be signature- or anomaly-based. Unfortunately, a by-product is that the process can create false positives, which is legitimate traffic prevented from entering the network and reaching the intended recipient. Most users have the experience of finding a message that they should have received in their email Junk or Spam folder, which happens when IDPS creates a false positive.

Another technical tool for detecting and preventing threat actor intrusions is **system information and event management (SIEM)** technology. SIEM applications collect and analyze log and event data about security alerts across network applications and devices. They then look for patterns and anomalies in real time and provide alerts and event reports. These systems are capable of advanced threat analysis and include incident response functions. SIEM systems might be best suited for larger businesses that have complex cybersecurity needs and regulatory requirements, such as GDPR or HIPAA.

Network Security

Networks, especially those connecting to the Internet, need to implement several technical cybersecurity controls for both preventing and detecting threats, colloquially known as **hardening** the system or network. Connecting to the Internet creates vulnerability exposures that might be mitigated by air gapping parts of the network. An **air gap** is a computer or a portion of an organization's network that stores data, may have private applications, and is not connected to the Internet. In ransomware attacks, having backup data on air-gapped machines isolated from the network is critical for recovery. Figure 6-4 illustrates an air-gapped computer network.

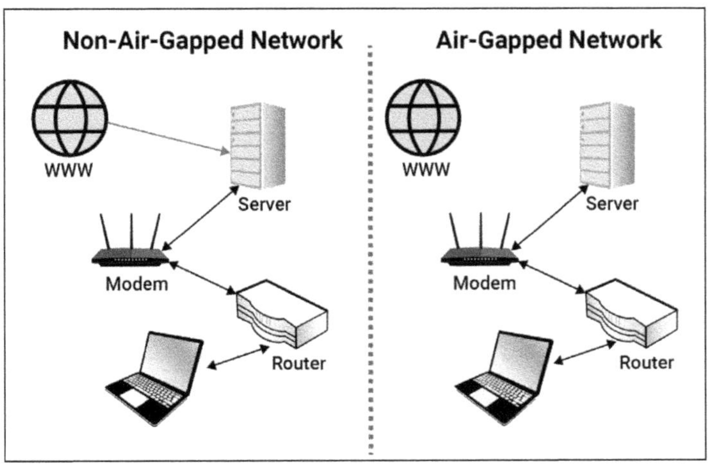

Figure 6-4. Air Gapping

Another network control is **port scanning**. Computer ports are the physical slots or virtual connection points on a device that allow it to connect and communicate with other devices. A USB port is an example of a physical port, and a web server port is a virtual one that allows connection with a web browser. Port scans check to see if ports are open. Network security staff run port-scanning software to proactively identify and secure open ports. Threat actors can use port scans to attack a system by finding an unsecured or vulnerable open (not in use) port that can send and receive data.

Yet another important technical tool in protecting networks is **secure configuration** of network devices, including routers and switches. For example, routers configured with Wired Equivalent Privacy (WEP), an outdated Wi-Fi security protocol, are more vulnerable than the more secure encryption offered by **Wireless Protected Access (WPA3)**. The main purpose of WPA is to prevent unauthorized network access, which it does through encryption and authentication processes.

Network engineers can also configure the network to be less vulnerable to external (and internal) attacks. **Isolation** (similar to air gapping) of portions of the network and logical **segmentation** of the network into subnetworks limits communication within a network, making it harder for attackers to move throughout the network once they breach it. To further secure a network, each of the segments or subnetworks should have an additional layer of security to provide depth to the security infrastructure. This will help to prevent a hacker from accessing the entire network once they breach the outer defenses.

System Information and Event Management (SIEM) A solution that provides real-time analysis of security alerts generated by network hardware and applications.

Hardened (System, Network, etc.) A system that has been configured and secured to minimize vulnerabilities and reduce the risk of unauthorized access or compromise.

Air Gap A security measure in which a computer or network is physically isolated from other networks, especially the Internet.

Port Scanning The act of systematically scanning a computer's ports to find vulnerabilities.

Secure Configuration The practice of configuring systems and software securely to reduce vulnerabilities.

Wireless Protected Access (WPA, WPA2, and WPA3) Security protocols used to secure wireless networks, with WPA2 offering stronger encryption than WPA and WPA3 stronger than WPA2.

Isolated Network A full network or segment of a network that has no connection with other sections of an organization's network.

Network Segmentation The practice of dividing a computer network into smaller, isolated sections (segments) to improve security, performance, and manageability.

Single Sign-On (SSO) An authentication process where the user authenticates once to gain access to all of the systems for which they are authorized based on their defined role.

Virtual Private Network (VPN) A technology that creates a secure, encrypted connection over a less-secure network, such as the Internet.

While these added defenses make it difficult for intruders to access the system, they also put up restrictions for individuals within the organization. To minimize the inconvenience to trusted individuals, systems security usually utilizes access control lists (ACLs) and **single sign-on** routines to allow access but still keep control over the system. Each of these controls works within the authorization models described earlier.

As with all controls, any security architecture always has weaknesses that work for the intruder as well as the employee. For example, if a hacker is able to authenticate as an existing employee and the organization uses single sign-on, the intruder now has access to all of the systems based on the role of the person hacked. Other controls that can be implemented to detect intrusion include review of logs and/or real-time monitoring to review access that is outside of normal behavior (on weekends, late at night, etc.).

Organizations use **virtual private networks (VPNs)** to allow remote users to connect to their company's network files as if their remote devices had a direct connection. VPNs make use of encryption to protect data communications in transit and prevent data leakage. A VPN also affords a user privacy and anonymity online by hiding the Internet Protocol (IP) address during online activity.

Patch Management

Patch An update to software.

Patch Management The process of managing updates to software, especially security software, to protect systems from vulnerabilities.

We have mentioned patching several times in this book, often in cases where security personnel failed to patch vulnerabilities. When software vendors discover a vulnerability in their software, they issue a **patch** (code change) to correct it. They may let users know about the patch through direct communication or through public release of the patch. Once an organization becomes aware that the software they are using has a vulnerability that might be exploited, they should apply the patch provided by the vendor as soon as possible. However, a patch might not be applied immediately, for several reasons. For example, security personnel may want to test it first in a controlled environment to make sure that it does not impact other hardware or software. The **patch management** process is critical in protecting information security. Unfortunately, many cybercriminals have been able to successfully exploit vulnerabilities due to failures in patch management. Automated tools can monitor for vulnerabilities and apply patches as appropriate. These software systems scan applications for missing patches and automatically download them from application vendors.

CASE IN POINT

While pushing patches to customers is critical as part of a protective mind-set, the testing of these patches needs to be thorough. In July 2024, CrowdStrike (a cybersecurity vendor) pushed out a patch to users of its Falcon platform. The Falcon platform is designed to stop cyberattacks using cloud technology. The software itself is located on the individual user machines, in addition to servers maintaining the network traffic. While CrowdStrike usually tests new patches extensively, a glitch in the company's validation check software allowed an untested patch to be pushed to all CrowdStrike Falcon users. This patch caused an "out-of-bounds" memory read, which made Windows machines crash and give users a **Blue Screen of Death (BSOD)** (Figure 6-5). To fix the problem, each user needed to boot their computer in safe mode, delete the bad patch file and the entry in the boot-up script, and restart the computer. The overall cost from the downtime and mitigation efforts has been estimated at more than $1 billion.

Blue Screen of Death (BSOD) An error screen displayed on Windows computers following a fatal system error.

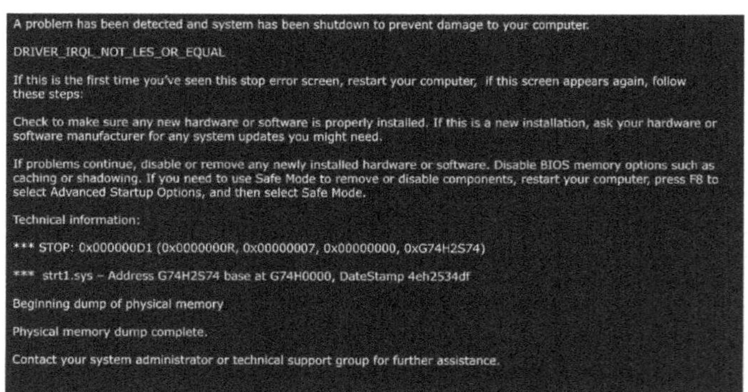

Figure 6-5. Blue Screen of Death (BSOD)

Administrative Controls

Administrative controls are the policies, processes, and governance in place to protect the CIA of information assets. The most important administrative control is the "tone at the top," or the attitude of top executives about security. We discussed this topic in Chapter 4. This attitude creates either a culture that highly values security and is willing to allocate resources and set policies appropriately to ensure the safety of data assets, or one that is lax and therefore highly vulnerable to exploits. A control environment that manages information security risk has many policies and procedures, which we will discuss.

Policies

A business organization's cybersecurity policies are an integral part of its risk management program. The range of policies is extensive, and it can be difficult to know where to start in crafting ones that best communicate the purpose and requirements of each individual control. Fortunately, many tools are available to help, such as the one described in the following case.

> **CASE IN POINT**
>
> **The SANS Institute (https://www.sans.org/information-security-policy/) offers templates for many information security policies. The templates provide language for an overview, the policy purpose and scope, and details for the policy itself. An example is the template for an Internet Usage Policy. This policy covers all internal and third-party users and states that Internet usage should be for business purposes only. Employees must acknowledge they have read and will comply with the policy, which also includes information about prohibited uses, such as gambling or posting libelous comments.**

Listed below are some of the more important cybersecurity policies a business should document and communicate. We will highlight and discuss in some depth a few that are the most critical: acceptable use, access control, data security, incident response, and remote access policies.

- **Acceptable Use Policy.** This policy provides guidance for how users are expected to use an organization's computer resources. The acceptable use policy may note the ownership of the organization's digital resources, and it protects against abuse of computer hardware, software, the Internet, and communication systems. This includes browsing and download activities.

  The acceptable use policy will include definitions of unacceptable use. Examples of these definitions include access to illegal or inappropriate content, abusive or discriminatory communications, hacking activities, and use of business resources for commercial purposes.

  Other components of an acceptable use policy are requirements for information security and data privacy and confidentiality. The acceptable use policy may inform users about how their online activities are monitored. The information security portion of the policy will note the importance of authentication protocols, a requirement to report security violations, and the need to verify external communications. The policy may remind users of the need to protect sensitive data and intellectual property.

Acceptable Use Policy
A set of rules that outline how a service or technology can be used.

- **Access Control Policy.** The purpose of an access control policy is to govern access to an entity's digital resources. The policy will define who it covers, which is likely to be employees and third-party partners. As described earlier, an important feature of access is least privilege, and the policy will reiterate that users should only be able to access those portions of the information system necessary for them to perform their functional activities.

  Another part of this policy is identifying and authenticating users. It is here that an organization describes its requirements for user identification and passwords,

Access Control Policy
A set of rules that determines who is allowed to access specific data, applications, or resources and under what conditions.

including their characteristics and management, and MFA requirements. The policy describes how user roles are defined and permissions are granted and rescinded. This includes who approves and manages access requests and how access is granted to business partners.

The Access Control Policy will describe how the system detects and logs unauthorized access attempts, as well as who reviews them. Clearly defined protocols should be established for logging and reporting these activities, including assigned responsibility for reviewing them.

Data Security Policy A formal document that outlines an organization's guidelines, rules, and standards for managing and protecting sensitive data assets.

Data Classification The process of categorizing data based on its sensitivity and the level of security controls needed.

- **Data Security Policy.** Data may be an organization's most valued asset. Protecting the CIA of data is what cybersecurity is all about. However, not all data are equal when it comes to requiring protection. The data security policy considers the levels of protection needed for different types of data. **Data classification** is a process organizations use to categorize data according to their value and privacy requirements. This classification determines associated risk and dictates the measures and costs to protect each data element. There are many ways to classify data according to sensitivity, such as secret and top secret, which is used extensively by the military. Table 6-4 illustrates a potential classification system for business data.

Table 6-4. Data Classification

Data Type	Description
Public	Available to anyone, including the public at large.
Internal only	Available to anyone within the organization. May include third-party relationships.
Confidential	Protected data, requires authorized access. May be protected by privacy legislation.
Restricted	Protected data, highly sensitive. May be protected by privacy legislation. Compromise of these data could cause severe harm to the organization.

As with other information security policies, the data security policy will describe its purpose and scope and provide specific guidelines. The policy should distinguish between security for data at rest and data in motion. Guidelines for data at rest will include encryption protocols, where data are stored, how long data are kept, and guidelines for data disposal. The Access Control Policy provides protocols for who may access data at rest. A data custodian should be assigned to each data element, and that individual bears responsibility for ensuring the data are appropriately protected, accurate, and available as needed by authorized stakeholders. Data in motion require protection during transmission. This most likely involves encryption and secure means of transmission, such as use of a VPN.

Incident Response Policy A formal document that outlines the procedures and responsibilities an organization follows to detect, respond to, and manage cybersecurity incidents.

Incident Response Plan (IRP) A set of procedures and guidelines that an organization follows when responding to a cybersecurity incident.

Computer Security Incident Response Team (CSIRT) A group within an organization that responds to security incidents and breaches.

- **Incident Response Policy and Plan.** The Incident Response Policy documents the strategy for detecting, responding to, and recovering from a cybersecurity incident. The policy may be created by senior level IS personnel with input from top executives. It guides the creation of the Incident Response Plan (IRP), which provides specific steps for responding to a cybersecurity event.

The first step in incident response is to create a cross-functional **computer security incident response team (CSIRT).** The CSIRT develops the IRP and coordinates people, processes, and resources that are involved if an attack occurs, with a goal to minimize the event's impact and maintain operations. The team analyzes incidents in the aftermath and provides guidance for avoiding future exploits. A senior IS executive, often the Chief Information Officer (CIO) or Chief Information Security Officer (CISO), should lead the CSIRT. Figure 6-6 describes best practices for creating a CSIRT.

The heart of the IRP is its actionable steps, which begins with preparation, as shown in Figure 6-7.

"Fail to plan and plan to fail" is a common phrase that is appropriate to mention here. The IRP does not begin with a discovered cyber breach but rather with planning and preparation. This includes assigning responsibilities to each member of the CSIRT.

Preparation and planning precede actual incidents. The next step is to detect and identify a system intrusion. Security breaches often go undetected when they first occur. In 2023, it took organizations an average of 204 days to identify a data breach (IBM, 2023). Once an incident is detected, responsible parties are assigned to analyze it. How did it happen? What systems are impacted?

1. Define CSIRT objectives
2. Select team members
3. Assign roles and responsibilities
4. Provide training
5. Establish tools and logistics
6. Conduct drills and reviews

Figure 6-6. Team Formation Steps

Figure 6-7. Steps in Incident Response

Next, the IRP outlines specific steps to contain, eradicate, and recover from the event. Containment is one of the most important steps to define. It requires isolating the affected system, perhaps by creating a **sandbox**, which is an isolated safe space for studying any potentially harmful code. The primary goal of containment is to prevent the attack from spreading, while a secondary goal is to preserve forensic evidence for later analysis. Once contained, the IRP describes how to rid the system of the intrusion (e.g., malware) and restore operations.

Sandbox An isolated environment used for testing and analyzing software without affecting the rest of the system.

The IRP process does not end with recovery. It is essential to conduct post-mortem activities to thoroughly understand the incident and how to prevent similar attacks in the future.

Finally, the IRP needs to be updated with the lessons learned, additional information, and new policies and procedures.

The IRP should specify a **communications protocol**. The communications protocol details who to tell, what to tell, and how and when to tell it. Who to tell includes both internal and external parties. Senior leadership, the board, employees, third-party partners, the press, law enforcement, and the legal and regulatory community are all potential recipients of security breach notices. Some parties need to be told immediately, particularly those who might be impacted by the breach itself. Others may be informed later, perhaps even after containment and restoration of operations. Increasingly, companies are required by law to inform customers and others about data breaches. The legal representative on the CSIRT will coordinate with the

Communications Protocol A formal description of the formats and rules for digital messages.

communications representative to ensure that legal and regulatory communication requirements are met.

It is important to strike a balance in communications between providing necessary information and avoiding overcommunicating. The best public statements are those that say briefly what happened and what is being done about it. Promises can create legal liability or expectations that may not be met.

Tabletop Exercise An informal, discussion-based session in which a team discusses their roles and responses during an emergency, walking through one or more example scenarios.

Incident Simulation A strategic exercise that replicates potential cyberattacks or security incidents to assess and improve an organization's ability to detect, respond to, and mitigate threats in a controlled environment.

An IRP that sits on the shelf is almost worse than useless. An IRP is a living, breathing document that organizations should update regularly. Years ago, a company suffered a worm attack and realized that the contact information for those who needed to be communicated with resided on the infected computers. This is just one example of a failure to test response and recovery plans. Two approaches to testing the IRP are **tabletop exercises** and **incident simulation.** A tabletop exercise involves having stakeholders discuss steps to take, including communications, under a variety of mock attack scenarios, as in the following example. These exercises may be conducted several times annually.

> **EXAMPLE**
>
> **A tabletop exercise can simulate a ransomware attack. The CSIRT and other members of management, including legal counsel and communications staff, participate. First the scenario is introduced, then each participant describes their role and actions. The exercise is likely to reveal weaknesses in response that are better addressed in the exercise than in a live attack.**

A more elaborate and resource-intensive test is to simulate an attack. The tabletop exercise involves discussion, whereas an attack simulation mimics a real incident.

Remote Access Policy A written document containing the guidelines for connecting to an organization's network from outside the office.

- **Remote Access Policy.** Many office workers continue to work remotely following the COVID pandemic; other business employees may work remotely as their job demands it. For example, a consultant is frequently at client offices, and pharmaceutical salespeople visit doctors' offices. Sitting at a desktop computer terminal that is linked to an internal network provides cybersecurity differently than being at a remote location and dialing in to the network. A **Remote Access Policy** provides protocols and requirements for accessing an organization's computer network and data resources. The scope of the policy will describe the stakeholders it covers. They are likely to be employees and third parties needing access to select parts of the network.

  The Remote Access Policy will specify how end users should communicate, most likely through a VPN. The policy will also prescribe the type of protection that should be on the devices accessing the network. For example, the organization's Help Desk might configure all company laptops with antivirus software.

  While the Access Control Policy controls access, remote or internal, the Remote Access Policy might include a **zero-trust** security model. In a zero-trust network, all access, whether inside or outside the network, must be continuously authorized, authenticated, and validated. Typically, IT network security has relied on trust within the network. With zero trust, each access attempt requires re-authentication. No one is deemed trustworthy.

Zero Trust A security model that assumes no implicit trust and requires verification for every access request.

Procedures

Risk management policies are useful as guidelines. They are complemented by ongoing procedures that evaluate management strategies, prepare for handling cyber incidents, and contribute to incident prevention. This section discusses audit and testing procedures, IT asset identification and tracking, and personnel practices.

Cybersecurity Audit A formal process for reviewing and assessing an organization's security policies, procedures, and controls.

- **Audits and Testing.** Organizations conduct regular **cybersecurity audits** to evaluate their level of information security and to ensure compliance with regulations such as privacy laws. In large organizations, internal staff may perform these audits on a regularly scheduled basis and use automated software to conduct continuous auditing (e.g., checking for updates in patch management). However, even in large companies, having an external party conduct the audit or at least perform some testing from time to time protects against any insider bias. Many organizations will use a combination of their own IT staff and third-party organizations as part of their cybersecurity assurance strategy.

A cybersecurity audit may include **penetration testing** (also called **pen testing**). In penetration testing, an organization contracts with an ethical hacker to try to breach their system security. This process exposes vulnerabilities and weaknesses in security and security policy so that the organization can fix them and prevent adversarial attacks. Pen testers are experts with experience that allows them to find vulnerabilities where those inside the organization may not be able to spot them. Figure 6-8 describes the stages of a pen test.

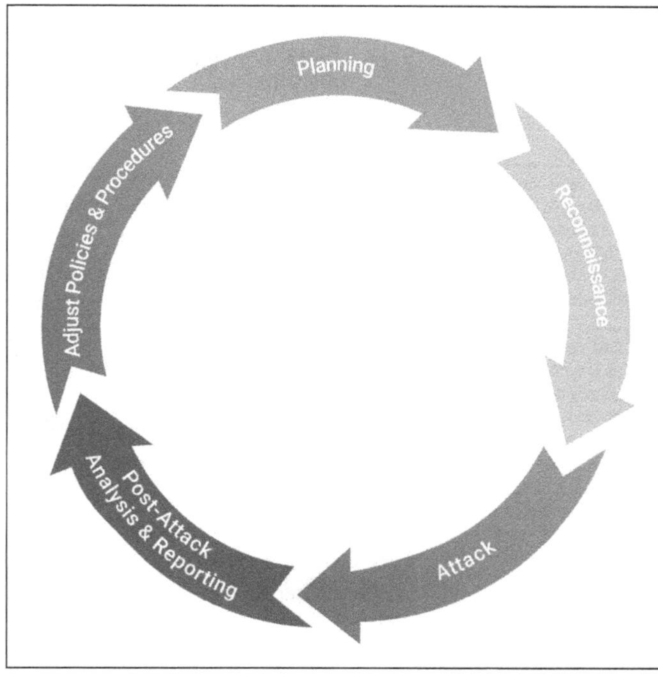

Figure 6-8. Phases in Penetration Testing

A **red team** engagement is like a pen test but conducts a more in-depth security assessment. Pen tests are usually limited engagements that test for known vulnerabilities and exploits. A red team assessment may follow pen testing, where most vulnerabilities have been discovered and patched. The red team will use every available technique to try to access a specified target. Red team attacks, like real cyberattacks, may take time to breach the system, so these engagements typically last longer than pen testing.

- **Backup and Recovery Procedures.** One of the most important processes in securing data is backing it up. An organization's backup strategy is the first line of defense against many types of cyberattacks, as illustrated in the following case.

CASE IN POINT

In 2024, Change Healthcare, a subsidiary of United Healthcare Group, suffered a severe ransomware attack that ultimately compromised PII for almost 200 million individuals. The company ended up paying a ransom of $22 million to be able to recover access to their information systems. The use of compromised credentials that lacked MFA were a direct cause, but a backup strategy failure contributed to the damage. The company lacked sequestered backups, which are those that cannot be altered, overridden, or deleted. The backup strategy also failed to include infrastructure gapping, which would isolate the backup from the primary data storage (Youngerwood, 2024).

There are different ways to back up data. A **full backup** is expensive and can be time-consuming, but it does include everything. An **incremental backup** copies the data since the last backup of any kind, and a **differential backup** copies all that has changed since the last full backup. The backup policy specifies the intervals at which each of these types of backup will occur.

Penetration (PEN) Testing A security exercise where a cybersecurity expert attempts to find and exploit vulnerabilities in a computer system. The purpose of this simulated attack is to identify any weak spots in a system's defenses.

Red Team A group that attempts a physical or digital intrusion against an organization.

Full Backup The most basic and comprehensive backup method in which a duplicate of all data is sent to another location for storage.

Incremental Backup A process that backs up all files that have changed since the last backup occurred.

Differential Backup A process that backs up only the files that have changed since the last full backup.

> **EXAMPLE**
> **A business might perform a full backup every 24 hours and an incremental back up every six hours. An alternative is to complete a full backup weekly, with incremental backups every six hours and a daily differential backup.**

Hot Site An alternate business location where the target environment is already up and running and can be immediately activated by a failover.

Warm Site An alternate business location where the target environment has some active equipment, requiring minimal setup before use.

Cold Site An alternate business location where the contracted facility has no active equipment, requiring extensive setup before use.

Failover The amount of time it takes to switch from production systems to backup systems.

Organizations may also establish off-premises sites for recovery, though much data today is backed up in the cloud. An organization's off-site backup center locations might be **hot sites** that are ready to go in a few hours. These locations are functional data centers that mirror the on-premise site, and they can be ready for use almost instantly. **Warm sites** that have hardware like the on-site data center may require some setup to become operational. **Cold sites** can be as simple as an empty building space, or they may have some basic hardware. There are trade-offs for each backup center option, with hot sites being best for quick recovery but also the most expensive to maintain. **Failover** is the backup process of changing to alternate redundant hardware or a network when primary components are compromised.

- **IT Asset Identification and Tracking.** Even small organizations are likely to have many IT assets to manage. Such assets include individual workstations, software, servers, mobile devices, and software licenses. Identifying and tracking these assets can be difficult because the hardware may be taken off-site or moved from office to office. Mobile phones, for example, can be almost impossible to trace and locate. Ensuring that the organization has processes and procedures in place for issuing hardware to new employees, maintaining hardware and software, and recovering these information assets when needed is necessary for risk management.

Fortunately, tools are available to help keep track of IT assets. Each asset should have a physical tag, such as a radio-frequency identification (RFID) tag that uses radio waves to allow tracking (Figure 6-9). Staff can scan the RFID tag and include the asset in an inventory database.

Staff should physically secure IT assets where possible and keep them in an appropriately controlled environment. This may entail temperature controls or raised flooring to protect assets from weather or natural overheating.

The asset database should include the asset description; its manufacturer, make, and model; the original cost; the designated custodian; and the location. Accountants are quite used to taking inventory of plant and equipment, including managing their depreciation schedules. IT asset identification and tracking may be a function within an organization's accounting or enterprise software.

Figure 6-9. Using RFID to Track Assets

- **Personnel Practices.** As we have noted several times, people are the weakest link in cybersecurity. Personnel policies and practices play an important part in ensuring the security of IT resources. Three personnel guidelines or procedures that are particularly relevant are hiring/firing policies, separation of duties, and security awareness training.

  Human resource functions within organizations have become increasingly complex and sophisticated over time. There are many reasons for this, such as laws ensuring equal opportunity in hiring and promotion and privacy and security concerns. Background checks and due diligence in hiring can prevent hiring errors. These actions can also protect against cyberattacks because insiders can be threat actors. It seems quite cold and heartless when we hear about security escorting a long-term employee out of a building when their job is terminated. However, these steps are a necessary part of risk management. Policies that employers conduct as due diligence when hiring and to ensure terminated employees immediately cede access to information resources and accounts are all part of a sound cybersecurity policy.

  - **Separation of duties** is a long-standing preventive internal control integral to controls in accounting. In accounting, the term means that no one person handles an economic transaction from beginning to end. We separate the recording, authorization, and custody of the asset or information. In information security, separation of duties has a similar meaning, in that functions and privileges should be divided among people so that no one individual has enough information or access to commit fraud or a cyberattack without collusion with others. In small organizations, separation of duties can be difficult to achieve and may require compensating controls such as monitoring or third-party audits. Separating duties requires identifying all IT roles and allocating each IT function into a distinct job so no one can impair the confidentiality, integrity, or accessibility of information purposefully or otherwise.
  - **Cybersecurity awareness training** should be part of every organization's information security strategy. Here is is a list of topics that employers may include in cybersecurity awareness training:
    - Antivirus software
    - Common scams
    - Device security
    - Incident reporting
    - Internet usage
    - Malware awareness
    - Password security
    - Phishing
    - Privacy regulations and policies
    - Remote working
    - Social media

    Security training can be offered to all employees, virtually or on-site, on a prescribed periodic basis, such as annually, and then can be accompanied with ad hoc or scheduled communications from IT. For example, if a known phishing email is circulating, the IT department may send out an alert email that warns employees about it.

Separation of Duties An organizational principle that divides tasks and responsibilities among multiple individuals to reduce the risk of fraud or error.

Cybersecurity Awareness Training Education programs designed to teach employees and users how to protect themselves and the organization from cyber threats.

> **CASE IN POINT**
>
> **KnowBe4 is a cybersecurity vendor offering security awareness training and what they refer to as a "human firewall." In addition to offering training, the company also provides clients with simulated phishing attacks and security culture and knowledge assessments. Kevin Mitnick, one of the world's most famous hackers, was KnowBe4's Chief Hacker. His**

security training focused on social engineering attacks (for which he was famous) and simulated phishing/spear-phishing attacks.

Physical Controls

Physical controls are infrastructure, hardware, and software put in place to control access to physical assets, control the environment where the physical assets are, and monitor those assets.

Environmental Controls

Equipment in data centers requires environmental protection against extreme temperatures, fire, flood, and natural disaster. Raised floors in data centers allow for cable placement, can help in controlling temperatures, and can let techs see if water leaks are occurring before they affect the hardware. Uninterruptible power supply (UPS) systems protect against outages and regulate power going to computer hardware. Maintaining remote off-site data centers outside a geographical region is another environmental control that mitigates risk associated with hardware, software, and data loss. For example, when a data center is located in a geographic area prone to hurricanes, the remote backup or recovery site would be in a region immune to that natural disaster.

Fire alarm systems should alert IT staff that a fire is present. Fire suppression systems put out the fire. Where there is computer hardware, liquid sprinkler systems can do more damage than good, so data centers have long been equipped with gaseous fire suppression systems. These systems may put out fires by interfering with chemical reactions or removing oxygen from the space. As with any automated system, hackers can stage attacks through fire alarm and suppression systems if they are not adequately secured both logically and physically.

EXAMPLE

Many data centers used to use Halon gas for fire suppression. Halon systems worked by removing oxygen and thus putting out the fire. However, Halon turned out to be bad for the ozone layer, and it has been replaced with Clean Agent Fire Suppression Systems. These systems work by reducing oxygen and heat levels and are environmentally friendly.

Monitoring Controls

Supervisory Control and Data Acquisition (SCADA) A system used for monitoring and controlling industrial processes and infrastructure.

Data centers and business offices are not always open for business. CCTV and other monitoring devices may keep watch over valuable IT assets when unattended. **Supervisory control and data acquisition (SCADA)** systems use sensors to get data about the operation of hardware devices, such as HVAC and access systems. SCADA systems act as supervisors, monitoring performance and the environment. They may be used in data centers and in industrial organizations.

Physical Access Controls

Physical access controls may be as simple as having locked offices that have computer hardware inside them. Data centers can and should be protected with a variety of access controls.

Smart Card Any ID card embedded with a microchip that allows secure processing and control over physical access, cryptographic keys, or digital signatures.

Decisions about how much to invest in physical protection require thinking about competing values of protection versus ease of access. For example, a campus building with a computer lab might warrant locking the lab, and perhaps the building that houses it, during off-hours. On the other hand, you might want students to be able to use the lab at any time. Use of ID cards or **smart cards** might allow those authorized, in this case students, to access the lab while keeping unauthorized visitors out. The level of sensitivity of data stored on the hardware, and the dollar value of the hardware itself, may require investments in more sophisticated identification and authentication, such as biometric readers. Another necessary investment may be for CCTV and security guards patrolling these areas randomly. Physical access protection by itself may not prevent a cyberattack, but it does send a signal about how security is valued.

> **CASE IN POINT**
>
> **In 2014, a team from a cybersecurity vendor visited the Sony Pictures Entertainment studio lot to meet with the company's cybersecurity management. They went through a security check at the lot gate, then went to the building housing the information security department. There was a sign identifying the department as such. They were able to enter the offices without an identification check. The offices had cubicles with computer hardware that could access the international data network. They found the office of the Senior VP for Information Security and waited by themselves to meet with him for about 15 minutes. Several weeks later, the company suffered a major cyberattack, and though unrelated to this access, the physical security controls did convey a message about the security culture (Elkind, 2015).**

As more of the workforce works remotely, ensuring the physical security of laptops and other mobile equipment gains importance. Much of this security is technical, like facial ID or fingerprint recognition for mobile phone access, but employees might also be advised to make sure they keep their computer equipment in safe places—for example, locked in a car trunk rather than visible to someone passing by the car window.

Risk Management Models and Frameworks

CIS Critical Security Controls

The Center for Internet Security (CIS) is a nonprofit organization dedicated to enhancing security over the Internet. Among their many products and services are the **CIS Critical Security Controls**. This set of controls provides specific, prioritized, actionable defense against cyberattacks. Many businesses, government agencies, and nonprofit organizations use CIS Critical Security Controls for managing cybersecurity risk and meeting various regulatory requirements. CIS controls are easy to understand and apply, so companies in the early stages of cybersecurity maturity may adopt them as a first step, before adopting a general cybersecurity framework. Version 8.1 of the controls map to the National Institute of Standards and Technology (NIST) Cybersecurity Framework 2.0 is discussed in Chapter 7.

CIS Critical Security Controls A risk management model made up of a prescriptive, prioritized, and simplified set of best practices that can be used to strengthen a cybersecurity posture.

> **CASE IN POINT**
>
> **The website for CIS Security Controls has testimonials from adopters and fans** (https://www.cisecurity.org/testimonials). **An example is this quote from Jim Long of the Long Law Firm, PLLC: "We use the CIS Controls to help our clients achieve compliance with state and federal cybersecurity regulations. The CIS 18 are prioritized, easy to understand, and extremely cost-effective for small to mid-size organizations looking to prove they are secure enough to do business in today's marketplace. I highly recommend starting with CIS in building your cybersecurity program."**

CIS controls represent best practices in cyber risk management. There are three implementation groups, described in Figure 6-10.

Each of the IGs corresponds to a recommended group of controls. For example, IG1 is the set of controls for essential cyber hygiene that all organizations should include to protect against common cyberattacks.

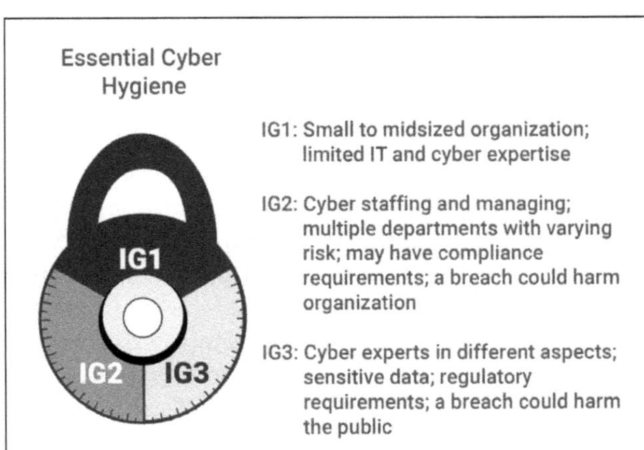

Figure 6-10. CIS Implementation Groups

CIS controls identify six asset classes: devices, software, data, users, network, and documentation. These asset classes help users identify their assets that are at the highest risk in a cyber incident. CIS controls include 18 controls and multiple safeguards (specific control activities) within each one. Here is an example:

> **EXAMPLE**
>
> **Control 3: Data Protection. Develop processes and technical controls to identify, classify, securely handle, retain, and dispose of data. The are 14 safeguards within this control. Two examples are Safeguard 3.6 Encrypt Data on End-User Devices and Safeguard 3.7 Establish and Maintain a Data Classification Scheme (Center for Internet Security, 2025).**

NIST Risk Management Special Publications

The National Institute of Standards and Technology (NIST) has several sets of standards and guidelines publicly available for information security, risk assessment, and risk management. They are harmonized where possible to reduce the administrative burden on entities that need to comply with both NIST and ISO/IEC standards. Chapter 5 described NIST Special Publication (SP) 800-30, which addresses risk assessment. Chapter 7 will dig deep on the NIST Cybersecurity Framework 2.0. NIST SP800-37, NIST SP800-39, NIST SP800-53, and NIST SP800-171 all concern risk management. Table 6-5 lists the titles, intended audience, and purpose of each SP.

Table 6-5. NIST Special Publication

NIST Special Publication	Title	Audience and Purpose
NIST SP 800-37	Risk Management Framework for Information Systems and Organizations: A System Life Cycle Approach for Security and Privacy	Risk management primarily for federal entities
NIST SP 800-39	Managing Information Security Risk	Risk management for federal and nonfederal entities
NIST SP 800-53	Security and Privacy Controls for Information Systems and Organizations	Security and control requirements primarily for federal entities
NIST SP 800-171	Protecting Controlled Unclassified Information in Nonfederal Systems and Organizations	Specific security controls and requirements for nonfederal environments

NIST SP800-37 provides a control framework for risk management with a series of steps. The framework is meant to be flexible so that steps and tasks following the Prepare step may be implemented in nonsequential order. Users can select controls and tailor them to meet specific organizational needs. In general, the framework in NIST SP800-37 provides a methodology to implement risk management at the information system level.

The purpose of NIST SP800-39 is to "provide guidance for an integrated, organization-wide program for managing information security risk to organizational operations (i.e., mission, functions, image, and reputation), organizational assets, individuals, other organizations, and the Nation resulting from the operation and use of federal information systems" (Joint Task Force Transformation Initiative, 2011). The guide is flexible and broad in scope. It defines the components of risk management as framing or context, risk assessment, risk response, and monitoring. These components form a risk management process intended to be integrated at the organization level, the mission/business process level, and the information system level.

NIST SP800-53, *Security and Privacy Controls for Information Systems and Organizations*, describes more than 1,000 security and privacy controls that are mandatory for federal information systems and agencies. Nonfederal entities that process, store, or transmit information can voluntarily adopt the framework. The controls described in NIST SP800-53 provide detailed guidance for organizations to protect systems and manage risk. The standard describes specific controls within each of 20 control groups. The guideline describes controls within each grouping in detail.

> **EXAMPLE**
>
> **AC-3 is Access Enforcement, which falls under the Access Control family. The specific control is: "Enforce approved authorizations for logical access to information and system resources in accordance with applicable access control policies." The control itself has many parts, including dual authorization and discretionary access controls.**

In addition to the controls themselves, NIST SP800-53 includes Discussion, Related Controls, and Control Enhancements.

NIST SP800-171 guides users in protecting **controlled unclassified information (CUI).** CUI is a category of government information that while not classified is still sensitive and requires safeguards. Examples include PII and proprietary business information. The guidance is used by third parties that conduct business with federal agencies and maintain and process CUI on their behalf. The standard is organized into control families that focus on a particular control area, such as access and contingency planning.

Controlled Unclassified Information Information that requires safeguarding or dissemination controls but is unclassified.

Cybersecurity Insurance

Depending on their risk appetite and risk assessment, organizations may choose to avoid or mitigate risk. Another option is for them to transfer that risk by purchasing **cybersecurity insurance.** Insurance can provide financial protection and other support in the event of a cybersecurity exposure. The insurance carrier and policy coverage that an organization chooses are important factors in its overall risk strategy. Making these decisions requires evaluating risk exposure against the cost of insurance premiums and comparing the terms of cyber insurance policies (coverage). Then to choose a specific carrier, the organization should consider reviews and consult with peers and experts as needed. The cybersecurity insurance industry is booming, and there are hundreds of carriers offering policies with a variety of coverage. They include well-known insurance carriers such as AIG, Chubb, and Liberty Mutual, and lesser known and more-specialized carriers.

Cybersecurity Insurance A type of insurance designed to protect businesses from Internet-based risks and risks related to information technology infrastructure and activities.

There are two basic types of cybersecurity coverage. First-party cybersecurity insurance covers the losses of the insured party itself. Third-party cybersecurity insurance covers the losses borne by third parties because of a cyberattack on the insured party. Table 6-6 summarizes the items that each type might cover.

Table 6-6. Insurance Coverages

First-Party Coverage	Third-Party Coverage
Communications costs (e.g., breach notifications)	Legal fees
Credit monitoring (for those affected)	Legal settlements with those harmed
Data restoration	Reputation restoration
Fines and penalties	
Incident response	
IT forensics	
Lost business income	
Network repair	
Public relations	

Policies typically do not include any losses associated with compromised intellectual property and potential future loss of revenues. They also likely do not include the costs an organization will incur in improving its cybersecurity defenses against future attacks.

CASE IN POINT

American International Group, Inc. (AIG), is a multinational corporation offering financial services and insurance. Insurance products include life, health, travel, and accident protection. In 1999, AIG added cybersecurity insurance to its product line. In 2014, AIG became the first carrier to offer services that would help prevent breach losses. In addition to insuring organizations, AIG also offers personal insurance coverage against online threats, such as extortion. AIG is currently one of the largest providers of cybersecurity insurance, and the corporation employs cybersecurity experts, artificial intelligence, and complex modeling to protect clients and provide financial relief and services.

Summary

Chapter 5 described risk and the process of risk assessment. This chapter explained how organizations manage risk, which might be avoided, mitigated, or transferred. Risk is managed through use of controls. These controls are technical, administrative, or physical, and managers and systems professionals can use controls to prevent, detect, or correct cybersecurity exploits. The chapter described many of the controls of each type. Incident response policies and plans are critical if a breach or exploit occurs. We discussed them as a type of administrative control. Many different risk management frameworks are available, and the chapter discussed a few of them. We ended the chapter with an explanation of cybersecurity insurance coverage.

Chapter Review

Key Terms

Acceptable Risk
Acceptable Use Policy
Access Control Lists
Access Control Policy
Administrative Controls
Air Gap
Antivirus Software
Asymmetric Encryption
Attack Signatures
Attribute-Based Access Control
Authentication
Authorization
Blue Screen of Death (BSOD)
Brute Force Attack.
Cipher
CIS Critical Security Controls
Cold Site
Communications Protocol
Compensating Control
Computer Security Incident Response Team (CSIRT)
Controlled Unclassified Information
Corrective Control
Cybersecurity Audit
Cybersecurity Awareness Training
Cybersecurity Insurance
Data Classification
Data Security Policy
Detective Control
Dictionary Attack
Differential Backup
Digital Certificate
Digital Signature
Discretionary Access Control
Encryption
Failover
Firewall
Full Backup
Hardened (System, Network, etc.)
Hot Site
Incident Response Plan (IRP)
Incident Response Policy
Incident Simulation
Incremental Backup
Intrusion Detection and Prevention Software (IDPS)
Isolated Network
Least Privilege
Logical Access Control
Mandatory Access Control
Multifactor Authentication (MFA)
Network Segmentation
Next Generation (Next-Gen) Antivirus Software
Password Management Software
Patch
Patch Management
Penetration (PEN) Testing
Physical Controls
Port Scanning
Preventive Control
Red Team
Remote Access Policy
Residual Risk
Risk-Adaptive Access Control
Role-Based Access Control
Rule-Based Access Control
Sandbox
Secure Configuration
Separation of Duties
Single Sign-On (SSO)
Smart Card
Supervisory Control and Data Acquisition (SCADA)
Symmetric Encryption
System Information and Event Management (SIEM)
Tabletop Exercise
Technical Controls
Virtual Private Network (VPN)
Warm Site
Wireless Protected Access (WPA, WPA2, and WPA3)
Zero Trust

Review Questions

1. Why do we need to manage cybersecurity risk?
2. Can all risk be eliminated? Why or why not?
3. What is the purpose of a cybersecurity risk strategy?
4. Summarize and explain the types of controls that can be used for cybersecurity.
5. Describe the different types of technical controls and give examples.
6. Describe the different types of administrative controls and give examples.
7. Describe the different types of physical controls and give examples.
8. How do risk models and risk frameworks work to improve cybersecurity in an organization?
9. What is the purpose of incident response planning?
10. What are the key parts of a cybersecurity Incident Response Plan?
11. What is the purpose of cybersecurity insurance?

Multiple Choice

Identify the choice that best completes the statement or answers the question.

1. What sets the tone for an organization's risk management strategy and cyber hygiene?
   a. The size of the organization
   b. The industry in which the organization operates
   c. Senior management and board attitude toward risk
   d. The technology the organization uses
   e. The number of IT staff in the organization
2. What do access control lists in technical access control systems define?
   a. The passwords used by all users
   b. The number of attempts a user can make to log in
   c. The hardware requirements for user access
   d. The software needed to access certain data
   e. Levels of privilege that segment access
3. What is the purpose of data classification within an organization?
   a. To rank employees by their job titles
   b. To categorize data according to their value and privacy requirements for protection
   c. To organize the company's book collection
   d. To classify the types of coffee available in the pantry
   e. To determine employee seating arrangements
4. Why is having an Incident Response Plan (IRP) crucial for organizations?
   a. It eliminates residual risk.
   b. It prepares the organization to effectively respond to and manage adverse cybersecurity events.
   c. It guarantees that cybersecurity events will not occur.
   d. It is required for compliance with all cybersecurity insurance policies.
   e. It ensures that all risks are transferred through insurance.
5. How does a comprehensive cybersecurity policy benefit an organization?
   a. It sets the agenda for annual general meetings.
   b. It lists the preferred vendors for office supplies.
   c. It consolidates all security programs and backup procedures to protect information assets.
   d. It outlines the procedures for office decoration.
   e. It determines the company's holiday schedule.
6. What is the primary goal of information security risk management?
   a. To transfer all risks to third parties
   b. To mitigate risks to an acceptable level
   c. To increase the organization's risk appetite
   d. To ensure that all risks are eliminated
   e. To ignore residual risks
7. What is the main function of intrusion detection and prevention software (IDPS)?
   a. To protect networks by detecting unauthorized incoming network traffic and preventing it from entering
   b. To enhance the speed of network traffic
   c. To generate new security rules automatically without human intervention
   d. To monitor the efficiency of the IT department
   e. To serve as the primary storage for network data

Case Studies

Herron Consulting—Risk Management

Natalie Magee and Chris Mason have completed a risk assessment for Yellow Subs. Chris would like to expand Herron's consulting practice to helping clients like Yellow Subs develop risk management programs. He has directed Natalie to do some research and select a framework in which they could develop specific expertise. They could then use that framework to help clients in managing their cybersecurity risk.

Case Requirements

1. Two sets of guidance for risk management are the CIS Critical Security Controls and NIST SP800-53. Both are useful in identifying current controls and gaps in them versus targeted risk management. Compare and contrast these two sets of guidance.
2. Assume that Natalie and Chris decide to use CIS Critical Security Controls to help Yellow Subs develop its risk management program. CIS Critical Security Controls has

multiple implementation groups. Which group best suits Yellow Subs?

a. Once you determine the appropriate group (i.e., IG 1, 2, or 3), choose five safeguards that will be most critical for the company to employ to protect its data from unauthorized access. Provide a rationale for each selected safeguard.

Libby Office Supplies, Inc.

Libby Office Supplies is a publicly traded US national retailer that sells all kinds of office supplies, from paper and pencils to digital devices. The company has stores in 30 US states, primarily in the western and middle parts of the country. The company's data center is in San Diego, California, where the company is headquartered. Drew Scott, the company's president, recognizes that this location faces unique threats from natural disasters, including fires and earthquakes.

Case Requirements

1. Describe three environmental controls that Libby should employ to protect the data center. Do any of these address the unique situational threats related to geography?
2. If you were to devise a backup location for Libby, where would you physically locate the center? Would you recommend a hot, warm, or cold site for backup? Provide a rationale for your answer.
3. Libby has an Incident Response Policy and recently developed an Incident Response Plan. Given the company's size, industry, and geographic reach, are there any special features you believe it should incorporate in the plan? What are they?

Exercises

6-1. Many cybersecurity professionals believe passwords are passé. That said, they are still the most used method of authentication. Required:

1. Consulting AI or other online resources, what is the best practice for passwords in terms of length, characters, and how often they should be changed?
2. Check five of your personal accounts, including at least one financial account, and note the passwords you are using to access them. Do they conform to best practices? (Skip this part if you use a password manager, and good for you!) How many of these accounts require MFA?

6-2. Table 6-2 in this chapter provided an example of each control type and function. Required:

1. Create a table like Table 6-2 and complete each cell by providing a different example from the one given in the chapter.

6-3. Administrative controls include both policies and procedures. Managing cybersecurity risk requires that an organization have multiple policies in place to protect data CIA. Required:

1. Visit your institution's website and find the link to Information Services. What policies are listed there? (If your school does not have a listing, check for another school that does.)
2. Find your institution's incident response policy. (If your organization does not have one, find a sample online.) What job titles are included on the CSIRT or have a defined role in the response within this policy? How does the policy define an incident?

6-4. The CSIRT includes individuals with a variety of technical expertise in detecting, handling, and analyzing cybersecurity incidents. The team is typically led by the individual responsible for cybersecurity (e.g., the CISO) and may include legal counsel, a public relations or communications director, a human resources officer, an incident manager, and a security analyst. Required:

1. Form small groups within your class. Each group should represent one of the roles on a CSIRT. Assume that a ransomware attack has occurred. Each group should briefly meet and then report on the actions they would take immediately following the attack.
2. Why is it important to have legal counsel, an HR officer, and a public relations representative on the CSIRT?

Reference

Center for Internet Security. (2025). *CIS Critical Security Controls v8.1.* https://learn.cisecurity.org/cis-controls-download

Elkind, P. (2015, June 25). Sony Pictures: Inside the hack of the century. *Fortune.* https://fortune.com/longform/sony-hack-part-1/

IBM. (2023, July). *Cost of a data breach report 2023.* https://d110erj175o600.cloudfront.net/wp-content/uploads/2023/07/25111651/Cost-of-a-Data-Breach-Report-2023.pdf

Joint Task Force Transformation Initiative. (2011, March). *NIST SP800-39, Managing information security risk: Organization, mission, and information system view.* National Institute of Standards and Technology. https://csrc.nist.gov/pubs/sp/800/39/final

Youngerwood, D. (2024, September 30). Five backup lessons learned from the UnitedHealth ransomware attack. Continuity Blog. https://www.continuitysoftware.com/blog/five-backup-lessons-learned-from-the-unitedhealth-ransomware-attack/

Chapter 7

Cybersecurity, Governance, and Risk Management Frameworks

Chapter Outline

Models, Frameworks, and Information Security
NIST Cybersecurity Framework 2.0
ISO/IEC 27001
Control Objectives for Information and Related Technology (COBIT)
MITRE ATT&CK
Proprietary Cybersecurity Frameworks

Learning Objectives (LO)

After reading this chapter, you should be able to:

- LO1. Understand the purpose of cybersecurity models and frameworks.
- LO2. Be knowledgeable about the NIST Cybersecurity Framework 2.0 and its core, profiles, and implementation tiers.
- LO3. Be knowledgeable about ISO/IEC 27001 and its components.
- LO4. Know how an organization might implement the NIST CSF or ISO/IEC 27001.
- LO5. Evaluate the key components of NIST CSF 2.0 and ISO/IEC 27001.
- LO6. Be familiar with COBIT and MITRE ATT&CK frameworks.
- LO7. Be familiar with industry-specific cybersecurity frameworks.

Introduction

Imagine you have just been hired as a Chief Information Security Officer (CISO) for a company. During the interview process, you likely will ask questions about its current cybersecurity profile. You will want to know if it has adopted a cybersecurity framework. If it has, then understanding that framework application can provide you with knowledge about what you would want to do to ensure a high level of information security. If the company has not adopted a framework, you might consider one so you can have a structured approach to managing cyber risk.

Cybersecurity frameworks are useful tools for assessing an information security posture and developing a plan to achieve an appropriate level of security that is commensurate with an organization's risk management needs and preferences. They are also helpful for communicating a cybersecurity strategy or maturity. Boards and executives in all functional areas need to know where an organization stands with respect to securing information assets and protecting them from a cyberattack. Corporate boards today are especially aware of the cyber threats that can harm their companies financially and operationally. Cybersecurity frameworks such as NIST CSF 2.0 and ISO/IEC 27001 can provide boards with structured approaches to managing cyber risk. Through dashboards and graphics related to the frameworks, information security leaders can more easily discuss the challenges they face, and the resources needed to confront them.

We have discussed other models for risk management in Chapter 6 and for risk assessment in Chapter 5. Chapter 7 provides detailed descriptions of the NIST Cybersecurity Framework 2.0 and the ISO/IEC 27001 framework. Both of these frameworks are widely used by governments and businesses as tools to build a comprehensive cybersecurity program within an organization. In addition, we describe COBIT, a framework for IT governance and management. We also discuss MITRE ATT&CK, a framework that focuses on threat actors and their attack methods. Many industries have created specialized information security frameworks to address their specific needs. We discuss examples at the end of this chapter, including HITRUST CSF, NERC CIP, and PCI DSS.

Models, Frameworks, and Information Security

Many different models and frameworks are available for structuring security for digital assets. They may be divided into three categories: risk assessment, risk management, and general cybersecurity. Table 7-1 shows the categories and specific frameworks within each one. Earlier chapters have described frameworks for risk assessment and risk management. Companies might adopt a risk management framework, such as CIS Critical Security Controls, discussed in Chapter 6, as a prelude to implementing a comprehensive risk strategy. Chapter 8 discusses the COSO frameworks, focusing on the COSO Internal Control–Integrated Framework.

Table 7-1. Frameworks for Control Types

Type	NIST	ISO/IEC	AICPA	Other
Risk Assessment	SP800-30	27005		FAIR (quant)
Risk Management	SP800-37 SP800-39 SP800-53 SP800-171	27002 27004	Risk Management Reporting & SOC Reports	• CIS Critical Security Controls • COSO
General Cybersecurity and IT Governance	Cybersecurity Framework 2.0	27001		• COBIT • MITRE ATT&CK • Proprietary Industry (e.g., HITRUST CSF, NERC CIP, PCI DSS)

NIST Cybersecurity Framework 2.0

In early 2013, President Barack Obama tasked the National Institute of Standards and Technology (NIST) to develop a cybersecurity framework that would help protect critical US infrastructure. NIST is an independent agency within the US Department of Commerce. The presidential executive order required the framework development to include support from both industry and academia. The collaborative nature of this development process resulted in a voluntary framework that can assist critical infrastructure entities in information security risk management.

The first version of the framework was launched in 2014. NIST has continued to collaborate with industry in refining and enhancing the NIST Cybersecurity Framework (CSF). Today, the NIST CSF, which includes standards, guidelines, and practices, is broadly used by business, government, and NGOs, particularly across the United States and Canada. The framework is flexible and allows adopters to customize their security profile, targets, and outcomes. It is a risk-based approach, and users may adopt the framework to build on or complement other NIST specialized guidelines for risk assessment and risk management or work with other general or proprietary cybersecurity standards. NIST launched Version 1. of its framework in 2018. In February 2024, NIST released Version 2.0 after collaborating with industry and government users of the framework to improve it and more accurately reflect the current challenges in the cybersecurity landscape.

Estimates of the framework's usage vary. In 2023, one survey claimed that 50% of companies used the NIST CSF (Spakowski, 2023). Many cybersecurity consulting firms, including the Big Four public accounting firms, assist organizations in adopting the framework. Using it is voluntary, and a business might opt to use just a part of it, such as one or more of the core functions.

> **CASE IN POINT**
>
> **The NIST website includes many resources for implementing the organization's framework, along with many special publications. It also features Success Stories, documenting how various organizations have implemented the framework. For example, the University of Chicago's Biological Sciences Division implemented the NIST CSF across its 23 departments. The framework allowed it to coordinate its risk management strategy across these different departments by adopting a common set of target outcomes (NIST, 2019a).**

Components of the NIST Cybersecurity Framework

The NIST CSF 2.0 has three components: Core, Implementation Tiers, and Profiles. The Core has six functions that detail specific outcomes. The Implementation Tiers help organizations determine their desired level of cybersecurity risk management, which is dependent on their appetite for risk, financial resources, industry, threat landscape, and mission. The Profiles assist an entity in aligning its risk management posture with the Core's desired outcomes. An entity can use the Profile to set outcome priorities.

The Framework Core

The NIST CSF's Core, shown in Figure 7-1, is appropriately named because it is the heart of the model.

Within the core functions are all the activities, potential outcomes, and references that are the basis of any information security system. They are described within the framework's six functions, 22 categories, and 106 subcategories. The categories define the cybersecurity objectives, and each category includes outcome statements or subcategories. These subcategories are like controls, but rather than thinking of "how," as you would do with a control, they focus on the "what" of the desired outcome.

The framework core also has cross-references for each subcategory. These informative references will cite how that particular outcome corresponds with an item in a different framework, such as COBIT or ISO/IEC 27001.

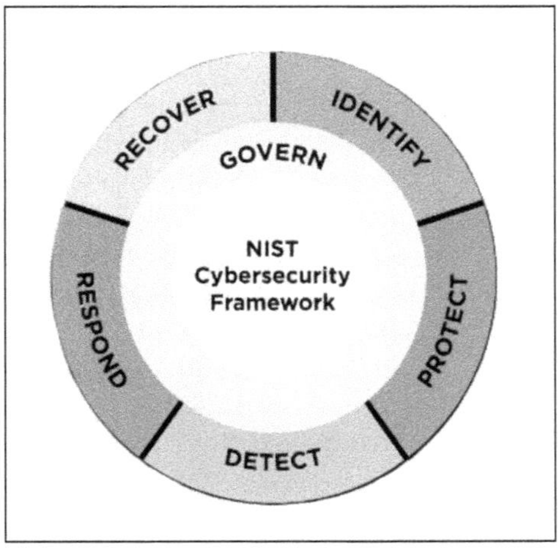

Figure 7-1. NIST CSF 2.0 Functions (NIST, 2024)

Here is a detailed explanation of the six functions:

- **Govern.** This is the first of the six functions within the core. This function, new in Version 2.0, is overarching, as shown in Figure 7.1. It has outcomes that guide organizations to prioritize and achieve the outcomes in the other five functions, given the company's mission and the expectations of the stakeholders. Governing is important and necessary to incorporate cybersecurity into the organization's broader risk management strategy.

  The Outcome Categories within the Govern function are Organizational Context; Risk Management Strategy; Roles, Responsibilities, and Authorities; Policy; Oversight; and Cybersecurity Supply Chain Risk Management. Each category has multiple subcategories associated with it. These subcategories are the desired outcomes.

> **EXAMPLE**
>
> **Within the Govern (GV) function is the category Risk Management Strategy (GV.RM). This category includes seven subcategories or outcomes. One of these is "GV.RM-02: Risk appetite and risk tolerance statements are established, communicated, and maintained" (NIST, 2024).**

- **Identify.** In this functional area, an organization names (identifies) its business context and resources. These resources include people, data, information systems, and assets. Understanding each resource for a particular entity is a first step in developing a risk management plan. For example, with respect to business context, the size of the organization and its industry sector can help in assessing risk by identifying potential threats. Large organizations in critical infrastructure industries are likely to have more risk than small retail operations.

  The Outcome Categories within the Identify function are Asset Management, Risk Assessment, and Improvement. Asset Management requires the identification of all assets, including data, hardware, and people. Risk Assessment requires that the organization understands its cybersecurity risk. The Improvement category recognizes the ongoing nature of cybersecurity and the need to identify security improvements. This category includes the need to maintain, communicate, and improve an Incident Response Plan (IRP).

> **EXAMPLE**
>
> **Within the Identify (ID) function is the category Asset Management (ID.AM). There are eight subcategories or outcomes within this category. One of these is "ID.AM-01: Inventories of hardware managed by the organization are maintained" (NIST, 2024).**

- **Protect.** This function is meant to ensure protection of organizational assets by preventing or limiting the impact of a potential cybersecurity attack or breach event. As noted in multiple places in this book, to prevent adverse cybersecurity events, an organization must plug every vulnerability, whereas a bad actor need exploit only one vulnerability to activate a threat. As a result, even if an organization adopts all available protection techniques and activities, an attack, such as a zero day, can still occur. However, by thinking in terms of desirable protection outcomes, organizations can reduce risk and mitigate adverse impacts.

  The five Outcome Categories within the Protect function are Identity Management and Access Control, Awareness and Training, Data Security, Platform Security, and Technology Infrastructure Resilience. The following case illustrates a category and subcategory within Protect.

> **EXAMPLE**
>
> **As an example, PR.AT is the Awareness and Training Category within the Protect (PR) function. This category is meant to ensure that all the entity's stakeholders are educated and trained as appropriate for their role. Subcategory PR.AT-01 addresses the knowledge that all personnel should have regarding cybersecurity. An organization's stakeholders should receive cybersecurity awareness and practices training to perform their general tasks while being mindful of information security.**

- **Detect.** The purpose of this function is to make sure that an attack or breach is identified on a timely basis. Ideally, that would be as soon as the incident occurs, but unfortunately negative cyber events are often discovered weeks or months after the bad actor has had the opportunity to broaden the intrusion.

The Detect function is a part of a defense-based approach to cybersecurity. Protect and Identify form a prevention-based strategy where organizations emphasize resource allocation toward blocking a negative cyber event. Prevention is seen as a proactive approach, while detection is a reaction. Most entities employ both prevention and detection approaches to information security. Increasingly, however, information security professionals have placed greater emphasis on detection because of the increasing frequency and sophistication of cyberattacks. This recognizes that it is not so much a matter of "if" an attack will occur but "when." Detection is important for identifying and minimizing damage from exploits. This awareness is highlighted in the following case.

CASE IN POINT

A variety of software tools are available for detecting cyber intrusions. They include endpoint detection and response (EDR) software. One such product is CrowdStrike Falcon, which incorporates artificial intelligence and machine learning to detect and block malicious software. A potential problem with any EDR software is that it may detect and block legitimate traffic. Such errors are called "false positives," which CrowdStrike Falcon advertises that it minimizes (CrowdStrike, 2025).

Endpoint Detection and Response (EDR) Technology that continuously monitors endpoints for evidence of threats and performs automatic actions to help mitigate them.

The Detect function within the NIST CSF is organized similarly to the other core functions.

EXAMPLE

There are two Outcome Categories within the Detect (DE) function: Continuous Monitoring, and Adverse Event Analysis. The purpose of the Adverse Event Analysis category, abbreviated DE.AE, is to ensure that unusual events are detected and the threat they pose is understood. Two of the five desired outcomes within DE.AE are the subcategories "DE.AE-04: The estimated impact and scope of adverse events are understood" and "DE.AE-08: Incidents are declared when adverse events meet the defined incident criteria" (NIST, 2024).

- **Respond.** Following detection of an activated threat, the cybersecurity risk management program includes appropriate response actions. They may range from simply rejecting an email containing malware to mitigating damage by activating response plans. It is critical that organizations have comprehensive response plans to deal with cyberattacks. Ransomware attacks occur frequently. Businesses need to know what to do when they arise. The NIST Respond function is intended to help organizations contain and minimize the impact of an adverse information security event.

EXAMPLE

The Respond (RS) function category RS.MI describes Incident Mitigation as "Activities are performed to prevent expansion of an event, mitigate its effects, and resolve the incident" (NIST, 2018). The three subcategories within Mitigation are containment, eradication, and "RS.MI-3: Newly identified vulnerabilities are mitigated or documented as accepted risks" (NIST, 2024).

- **Recover.** Of course, the point of having strategies to manage risk is to ensure that an organization can recover from a cyberattack as quickly as possible with minimal damage. A fast recovery is particularly important for critical infrastructure industries, or in the case of healthcare organizations. Physical harm and even death can result from attacks on hospitals when critical services cannot be delivered.

Two Outcome Categories within Recover are Incident Recovery Plan Execution and Incident Recovery Communication (RC.CO). Note that Communication was also a category within the Respond function, which stresses the importance of this activity. For both functions, that category encompasses internal and external communications or messaging to all stakeholders. Within Recover, however, it is also concerned with public relations messaging, as reputational damage is one of the primary risks of a cyberattack.

> **EXAMPLE**
>
> **RC.CO involves internal and external stakeholders being kept aware of the ongoing recovery efforts. The two subcategories or desired outcomes within Communications are "RC.CO-03:* Recovery activities and progress in restoring operational capabilities are communicated to designated internal and external stakeholders" and "RC.CO-04: Public updates on incident recovery are shared using approved methods and messaging" (NIST, 2024).**

Implementation Tiers

Organizations vary in the amount of risk they face from cybersecurity threats. For a small retailer with limited online sales, the risk may be easily managed. A water utility, on the other hand, requires a significant level of resources devoted to cybersecurity. The NIST Implementation Tiers allow an organization to think about where it is positioned on a spectrum of risk management, and where it might aspire to at a future date, perhaps as it grows and becomes more complex. The tier chosen considers industry context, risk appetite, available financial and staffing resources, mission and goals, and potential threats. The tiers are not intended to be a cybersecurity maturity measure but instead are used to indicate how an organization believes its cybersecurity policies and procedures support its risk management strategy.

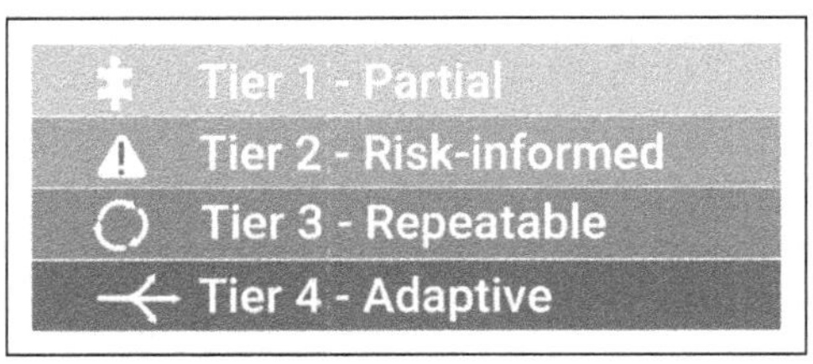

Figure 7-2. NIST Implementation Tiers (NIST, 2024)

Organizations may select from four tiers, as shown in Figure 7-2.

- **Partial.** The partial implementation tier is appropriate when there is little understanding of possible cybersecurity threats and a lack of a cybersecurity plan or coordinated effort around information security risk management and governance. The approach to threats is reactive. This may sound like a poor approach to cybersecurity, but for small businesses, it may well be a starting point.
- **Risk-Informed.** An organization that is in the risk-informed tier does not have an overall policy for cybersecurity but does have some risk management practices in place. The organization's management is aware of cybersecurity risk but has not established an entity-wide risk management strategy. As the name of this tier implies, cybersecurity governance and risk management are based on perceived risks. As an example, a business unit within an organization might view a specific cyber threat, such as unauthorized access to a sensitive database, as a risk requiring mitigation and might incorporate controls to manage it. There is no formal plan, but informal sharing of information security measures may occur.
- **Repeatable.** An organization adopting the repeatable implementation tier has formal risk management policies and an organization-wide cybersecurity risk management

* The subcategories within NIST CSF 2.0 do not always start numerically with 01. Subcategories that have been moved or incorporated into other categories with the update from Version 1.1 to Version 2.0 retain their place in the framework and indicate the topic to which they have been moved. For example, RC.CO-01 ("Public relations are managed") in the framework indicates that it has been withdrawn and incorporated into RC.CO-04 ("Public updates on incident recovery are shared using approved methods and messaging").

system in place. Management continuously monitors threats and communicates frequently about threats and vulnerabilities. On a broader basis, the entity at this tier is communicating with internal and external stakeholders about cyber risk management.

- **Adaptive.** The most aggressive cybersecurity program is associated with the adaptive Implementation tier. This level of security has all the characteristics of the repeatable tier but is also constantly evolving cybersecurity policies and practices to deal with an ever-changing threat landscape. An entity-wide cybersecurity risk governance and management strategy guides policies and procedures. Organizations adopting this tier are proactive and are ready to respond quickly to a negative cybersecurity event. Information security threats and risks are communicated broadly throughout the organization and to third-party partners.

Profiles

Based on the Implementation Tier, the Profiles component of the NIST CSF allows organizations to consider their current and desired information risk management state. The profiles include risk appetite, resources available and committed to security, objectives, and requirements. These focal points are aligned based on the desired outcomes from the Framework Core.

The current profile is the set of security controls (outcomes) aligned with a specific organization's needs. To develop a current profile, management identifies business objectives and then evaluates the information system security program against the framework outcomes to identify how many of the outcomes the organization is currently achieving.

> **EXAMPLE**
>
> **Within Govern and Risk Management Strategy is the outcome "GV.RM-02: Risk appetite and risk tolerance statements are established, communicated, and maintained" (NIST, 2024). If this is consistent with business objectives as a desired outcome, management considers to what extent the organization has accomplished the outcome.**

A target profile specifies the desired outcomes that an organization has selected and prioritized to achieve its cybersecurity risk management objectives. This target considers risk appetite and anticipated changes to the organization's cybersecurity posture, including new requirements, the adoption of new technologies, and emerging threat intelligence trends. Once you have established both the current and target profiles, security personnel can describe and analyze the gaps between the two. This process enables management to allocate resources to cybersecurity in an effective and rational manner. After completing this process, a plan can be developed to reduce cybersecurity risk, which includes a timeline, budget allocations, assigned responsibilities, and accountability.

The NIST Cybersecurity Center of Excellence is a space for government, industry representatives, and academics to collaborate in finding solutions to cybersecurity challenges. NIST CSF community profiles are an example of this collaboration. They are templates for specific industries. The framework also includes mapping to other cybersecurity models and frameworks, including CIS Controls and ISO/IEC 27001.

> **CASE IN POINT**
>
> **The NIST Cybersecurity Framework Version 2.0 Semiconductor Manufacturing Profile (NIST, 2025) is an example of a community profile that has been created to address the needs of a specific industry sector. The profile recognizes the complexity of semiconductor manufacturing and the specific cyber threats companies face in this process. The profile identifies the industry business objectives and applies them to create an industry profile with appropriate functional categories and outcomes within the NIST CSF 2.0.**

Implementing the NIST Cybersecurity Framework

A key feature of the NIST CSF is its flexibility. An organization may choose to implement the framework in its entirety; use parts of it in conjunction with other NIST guides (e.g., SP800-53), specialized industry frameworks, or general guidelines; or select one or more functions from the core, such as Protect or Detect. NIST resources include Quick Start Guides that make it easy for users to adapt components of the framework.

This chapter began with the scenario of a newly hired CISO. A helpful way for this individual to think about the job at hand is to consider the context of the organization's information security. If the NIST CSF is already in place, the CISO might consider if the organization is aligned with the appropriate implementation tier, assess the current profile and compare it with a target profile, and plan for implementing a risk management program that will result in desired outcomes. As an example, if the Implementation Tier should be Adaptive and the organization is still at Reactive or below, the CISO will need to arrange for the funds needed to achieve cybersecurity goals and map the timeline for implementing various controls to achieve the next tier. If the NIST CSF is partially adopted, the CISO can consider implementing additional functions. To adopt the NIST CSF from scratch, the implementation process will follow the steps in Figure 7-3.

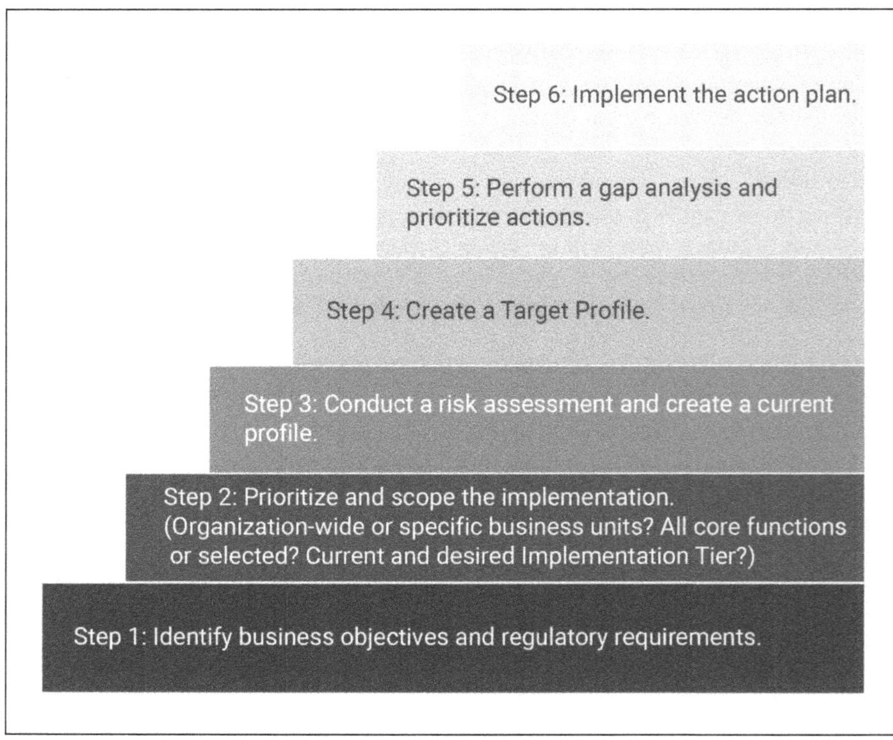

Figure 7-3. NIST Adoption Steps

As noted earlier, the NIST CSF has broad industry appeal, particulary within the United States and Canada. The NIST CSF 2.0 website (www.nist.gov/cyberframework) has resources available to assist users, including Quick Start Guides and templates. The site also provides examples of successful implementations. The following case presents a summary of one success story.

CASE IN POINT

Higher education is an industry that is particularly threatened by and vulnerable to cyberattacks. The University of Pittsburgh adopted the NIST CSF 1.1 in conjunction with NIST SP800-171. It receives government funding in the form of financial aid and research

> **grants, requiring it to meet a variety of regulatory requirements. Using the NIST CSF in conjunction with NIST SP800-171 provided a coordinated approach to information security and regulatory compliance. The adoption process included a risk assessment, mitigation plan, and periodic NIST SP800-171 assessments. This process resulted in compliance and a broad understanding of security risks across decentralized units (NIST, 2019b).**

ISO/IEC 27001

The International Organization for Standardization (ISO) partnered with the International Electrotechnical Commission (IEC) to develop ISO/IEC 27001, a standard for **information security management systems (ISMSs)** (ISO, 2022). The 27001 standard was introduced in 2005 and was updated in 2013 and again in 2022.

Information Security Management System (ISMS) A framework of policies and procedures used to manage an organization's information security risks.

The 27001 standard is part of the ISO information security suite, known as the ISO/IEC 27000 family, which has a dozen supporting standards. These standards are all meant to protect information security. ISO/IEC 27001 is the standard through which organizations may obtain certification for their ISMS. Chapter 5 discussed ISO/IEC 27005 as a guide for risk assessment.

ISO/IEC 27002, "Information Security, Cybersecurity and Privacy Protection," is another member of this family. This standard focuses on cybersecurity controls and procedures and may be used as a support for ISO/IEC 27001 certification. Organizations in any industry and of all sizes can use ISO/IEC 27002 as a guide to identify and implement cybersecurity controls. The standard organizes these controls into the categories of organizational, people, physical, and technological controls. ISO 27004 is another complement to the 27001 standard. This guidance measures the effectiveness of the ISMS.

ISO/IEC 27001 includes specific requirements for scope, references, organizational context, leadership, an information security policy, planning, support, operation, and performance evaluation and improvement. Annex A of the standard lists 93 information security controls derived from ISO/IEC 27002. Organizations seeking certification will indicate whether each control is in place, and if not, why.

Implementing ISO/IEC 27001

ISO provides a structured approach to organizing security controls to protect information. The standard requires management to develop a framework for ISMS, conduct risk assessments, and implement information security controls. Many consultants offer clients a checklist to help them prepare for ISO/IEC 27001 certification. Figure 7-4, created by UnderDefense Cybersecurity, is an example of one such checklist.

To obtain the ISO/IEC 27001 certification, an interested organization must complete a Statement of Applicability. This document describes the information security controls in Annex A chosen to mitigate information security risks. The organization provides the rationale for the controls chosen to be implemented. The auditors providing the certification will consider this document in evaluating whether those controls are appropriate for the entity to obtain certification.

There are two stages in the ISO/IEC 27001 certification process. In the first stage, an external auditor conducts a review of policies and processes to ensure that they meet the standard requirements and that an ISMS has been implemented. The external auditor in this instance is associated with an independent, accredited certification body and has completed the ISO/IEC 27001 Lead Auditor course or similar certification training. In the United States, a CPA might support the process, but the certification must come from an accredited certification body.

ISO 27001 Checklist

- ❑ Create a dedicated team
- ❑ Establish scope
- ❑ Conduct a risk assessment
- ❑ Organize and document Information Security Management System (ISMS)
- ❑ Implement ISMS policies and controls
- ❑ Conduct employee training
- ❑ Monitor and measure controls
- ❑ Undergo an internal audit
- ❑ Resolve issues
- ❑ Gain management approval of ISMS and ensure that it aligns with business goals
- ❑ Engage in continuous improvement

Figure 7-4. ISO 27001 Checklist

The first stage is essentially a documentation review. In the second stage of the ISO/IEC certification process, the auditor conducts an assessment on-site to check that the organization's ISMS complies with ISO/IEC 27001. This is an implementation and effectiveness assessment, which includes interviews and a review of business processes. Once certification is obtained, it is valid for three years, but the process requires annual audits for two years and a recertification audit in the third year.

An organization that obtains ISO/IEC 27001 can demonstrate to its partners that it has met a rigorous standard for information security. This certification is akin to an international quality seal that an independent expert has assessed either the entire organization's or a business unit's ISMS and found it to meet best practice.

CASE IN POINT

Enterbrain Software GmbH is a German fundraising software company. It manages PII and payment data for nonprofit organizations, and protecting that data is critical. Since 2019, the organization has had its information security management system certified with ISO 27001. The company has used the framework to ensure the CIA of data it holds and processes. A framework requirement for regular employee awareness and cybersecurity training helped the company to avert a cyber incident in 2020. Although the employee accidentally opened an infected email from a customer, they quickly reported it, and the emergency team was able to isolate the malware and prevent its spread (Saeckel, 2024).

Comparing NIST CSF 2.0 and ISO 27001

The NIST CSF and ISO/IEC 27001 are both popular general cybersecurity frameworks that have several commonalities and differences. They are both risk-based, with the goal of protecting information assets. But while NIST is free and open, the ISO 27001 standard includes an audit that results in certification. ISO charges a certification fee that varies with entity personnel size. As with NIST, ISO/IEC 27001 requires risk assessment and management, but it primarily emphasizes the core tenets of information security: confidentiality, integrity, and availability. The ISO standard has a more international focus and might be used by a governmental, private, or not-for-profit entity that is concerned with data protection. It is also possible for one or more business units within an entity to adopt 27001. Like the NIST framework, 27001 is voluntary, though suppliers or other stakeholders might require a trading partner to hold certification. Many of the controls within both standards overlap. NIST and 27001 may be adopted separately, or an organization may migrate from one to the other. The NIST framework is more flexible, and an organization might adopt it earlier in the entity's life cycle, migrating to ISO certification as it matures. Table 7-2 provides some comparisons of the two frameworks.

The NIST CSF and ISO/IEC 27001, though different, can map to each other. The controls within the standards can also map to other frameworks, including COBIT (discussed next), CIS

Table 7-2. Comparing NIST and ISO Frameworks

	NIST CSF 2.0	ISO/IEC 27001
Adopters	Any size organization and industry, global	Any size organization and industry, global
Certification	No; self-assessment	Yes; available from external accredited providers
Cost	Free	The ISO 27001 standard typically costs around $125, and the related implementation guidance standard ISO 27002 is around $225, bringing the total to approximately $350 for both documents (Karthik, 2025).
Purpose	Voluntary framework for managing cybersecurity risk	An international standard with specific requirements for an information security management system (ISMS)
Structure	Flexible, with core functions, categories, subcategories, Implementation Tiers, and Profiles	Structured, with clauses and an Annex that provides a list of controls

Controls (discussed in Chapter 6), SOC 2 (the subject of Chapter 8), and a few proprietary industry frameworks.

Control Objectives for Information and Related Technology (COBIT)

COBIT is an IT governance framework that includes principles of cybersecurity and can complement dedicated cybersecurity frameworks, such as the NIST CSF and ISO 27001. While not in itself a cybersecurity framework, COBIT's use in managing IT processes contains elements of risk management. The Information Systems Audit and Control Association (ISACA) first developed the Control Objectives for Information and Related Technology (COBIT) in 1996 as a set of control objectives for information systems auditors to use in managing information technology. The latest version, COBIT 2019, has a broader scope than the NIST and ISO cybersecurity frameworks in that it is a valuable resource for IT governance, management, and control. The guidelines within COBIT are applicable to all industries and types of organizations.

COBIT offers some advantages to adopters, including maximizing benefits from investment in IT resources. For IT managers and anyone concerned with risk management, COBIT provides guidance on structuring IT, identifying risks, ensuring appropriate internal controls, and providing risk management assurance. The model has multiple governance principles and key concepts. The governance and management objectives are sorted into five domains, which form the COBIT Core Model.

IT managers implement the COBIT model to create a governance system for IT that ensures not only security, but also best strategic direction and use of resources. Figure 7-5 illustrates governance within the COBIT framework.

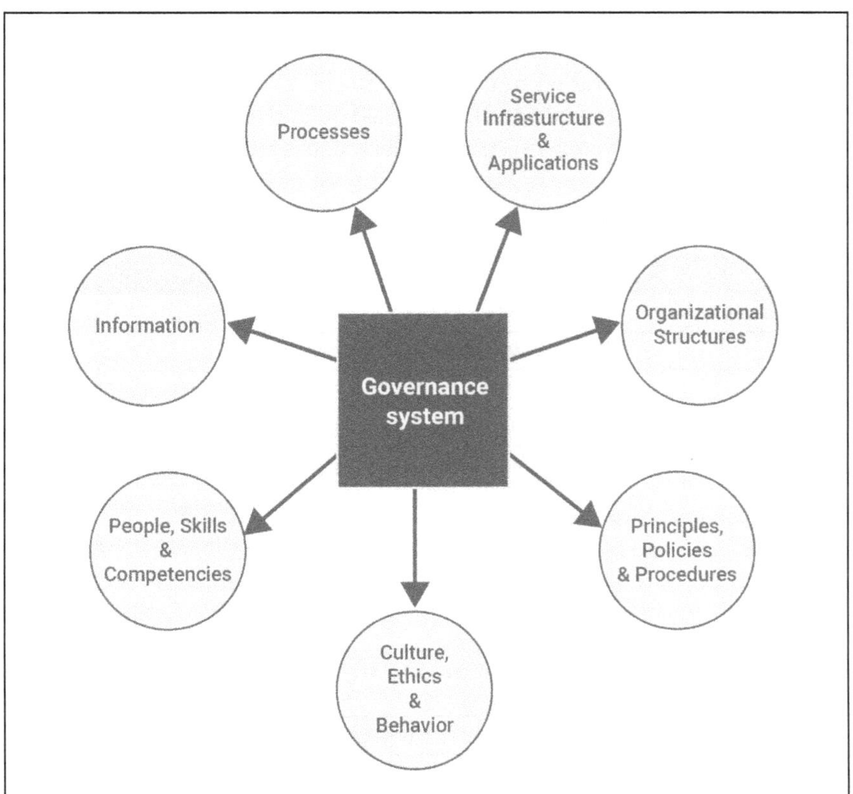

Figure 7-5. COBIT 2019 Governance System

ISACA provides an Implementation Guide for COBIT 2019 that recognizes that information and technology (I&T) are intertwined with business goals and activities. The model substitutes I&T for IT to emphasize that information and technology have impact throughout an enterprise, rather than being confined to an IT department. As noted, you can use COBIT 2019 in conjunction with other cybersecurity frameworks. For example, the NIST CSF subcategories provide a cross-reference to specific objectives within COBIT.

The ISACA organization provides case studies of successful COBIT adoptions as part of the resources it offers to users. The following case is an example of COBIT's use in IT governance and management.

CASE IN POINT

The National Bank of Angola used COBIT in 2019 to improve its governance and management of information and technology. The bank was guided by the Southern African Development Community guidelines, which were meant to coordinate and improve central bank governance and information and technology management throughout South Africa. Using COBIT, the bank was able to align its operations with external requirements, train employees, and document IT governance and management practices (Soares, 2023).

MITRE ATT&CK

The MITRE Corporation is a nonprofit organization managing federally funded research and development centers that support several US governmental agencies. One area of focus and support is cybersecurity. Many cybersecurity frameworks focus on cybersecurity threats. MITRE ATT&CK takes a different approach by being centered on the tactics and techniques of threat actors. The framework or guide is a knowledge base created from experts' observations. The MITRE ATT&CK Matrix for Enterprise is available at no charge (https://attack.mitre.org/).

The MITRE ATT&CK Enterprise matrix is organized around 14 tactics or potential goals of an adversary, such as defense evasion or lateral movement. Within each of these tactic columns are multiple techniques. One example of a tactic would be Initial Access, where the threat actor is establishing a foothold in the victim's system. The techniques associated with this tactic are listed in Table 7-3. These techniques are approaches an adversary might use to launch an attack upon initial access.

Table 7-3. MITRE ATT&CK Initial Access Tactics

Initial Access Tactic	Description
Content injection	Gain access and inject malicious content
Drive-by compromise	Gain access through a user visit to a compromised website
Exploit public-facing application	Exploit a weakness (e.g., a software bug) in an Internet host or system
External remote services	Leverage these services (e.g., a VPN) to access or persist within a network
Hardware additions	Introduce computer hardware or accessories (e.g., printer) to gain access
Phishing and variations	Phishing message and spear phishing, where social engineering is used to message victim
Replication through removable media	Potentially attack an air gapped system through removable media, such as a USB stick
Supply chain compromise and variations	Manipulate product or service delivery en route to final consumer
Trusted relationship	Compromise relationships with trusted third parties to gain access
Valid accounts	Obtain and compromise credentials of authorized users

Some of the techniques, such as phishing, may have more than one subtechnique, as discussed in the following example.

> **EXAMPLE**
> **One subtechnique for phishing is Spearphishing Attachment, where the threat actor attaches a file with malware to an email message. Another is Spearphishing Voice, where the threat actor uses a phone call to try to gain information or access. MITRE ATT&CK provides specific examples of the use of these techniques or subtechniques to launch an attack. The Spearphishing Attachment subtechnique was used in the 2015 Ukraine electric power attack by the adversary group Sandworm.**

MITRE provides several resources for users adopting the matrix. These include a blog post, use cases, and a road map (Figure 7-6).

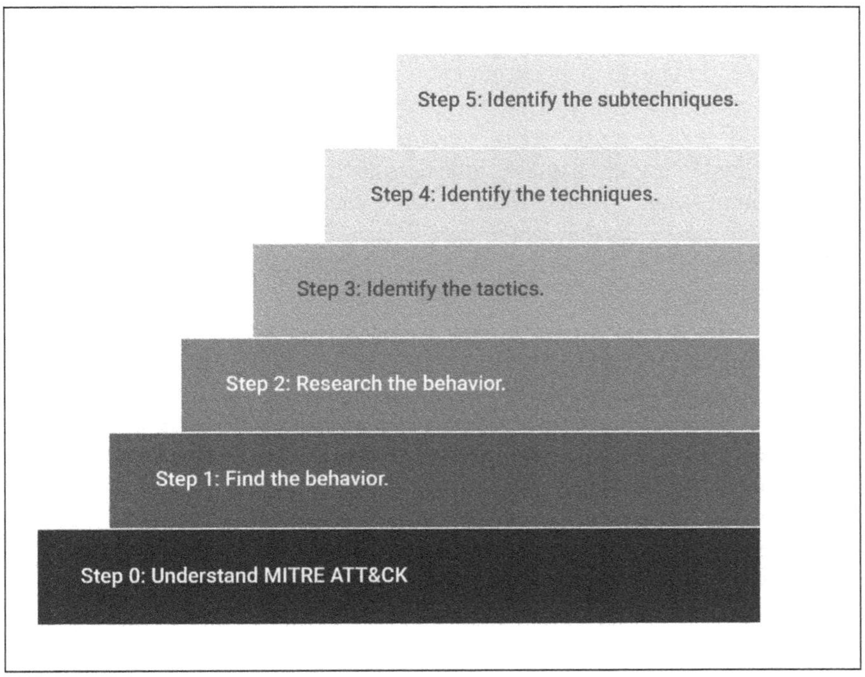

Figure 7-6. Steps in MITRE ATT&CK Road Map

Proprietary Cybersecurity Frameworks

There are several types of proprietary cybersecurity frameworks. Some are nation-specific standards or industry guidelines, while others may be individual company-developed frameworks or requirements. Some cybersecurity consulting firms have developed frameworks that they use for their clients, as illustrated in the following case.

> **CASE IN POINT**
> **The Big Four public accounting firms have been offering consulting services in information security for decades because of their expertise in assurance and risk management. Currently, each of these firms maintains a cybersecurity practice. Deloitte's practice, Cyber and Strategic Risk, resides within its Risk and Financial Advisory Practice. Deloitte has developed a proprietary Cyber Strategy Framework that incorporates risk management, IT governance, threats, capabilities, and specific business needs and objectives. An online**

> **platform and dashboards support the framework to help clients understand their cyber resilience (Deloitte, 2025).**

Another type of proprietary framework is industry-specific frameworks. Three examples are HITRUST CSF for the healthcare industry, NERC CIP for electric utilities, and PCI DSS, which is applicable to the credit card industry.

HITRUST CSF

The Health Information Trust Alliance (HITRUST) is a nonprofit organization founded to ensure information security in the healthcare industry. In 2007, it developed the HITRUST Common Security Framework (CSF). HITRUST has updated its CSF many times since then, releasing Version 11 in 2024. The framework can be adopted and lead to certification that can demonstrate compliance with regulatory requirements, such as the Health Insurance Portability and Accountability Act (HIPAA), discussed in Chapter 11. Although the framework was originally intended specifically for the healthcare industry, it has evolved such that other industries may also find it useful. As with many information security frameworks, HITRUST CSF maps to others, including the NIST CSF and multiple NIST special publications.

HITRUST CSF focuses on the security and privacy of protected health information (PHI). As we have described in earlier chapters, these data are particularly sensitive, as threat actors can use them to obtain prescription drugs and to commit identity theft. The CSF requires adopters to implement multiple safeguards that will protect PHI.

The HITRUST CSF has 14 control categories, 49 control objectives, and 156 control references (specific controls).

> **EXAMPLE**
>
> **Control Category 1 is Access Control. Access Control has seven Control Objectives. One of these is Control Objective 01.03 User Responsibilities. Within this Control Objective are three control specifications or references: Password Use, Unattended User Equipment, and Clear Desk and Clear Screen Policy.**

HITRUST certification requires a validated assessment by an external assessor. The multistage certification process can take several months and be somewhat costly, but the end result is a certification that demonstrates a high level of security for sensitive data assets.

NERC CIP

The Federal Energy Regulatory Commission (FERC) is an independent federal agency within the US Department of Energy that is charged with regulating energy services, including oil, natural gas, and electricity. The agency oversees the National Electrical Reliability Council (NERC), created in 1968 to ensure a reliable supply of electricity to consumers in North America. In 2003, the United States sustained a significant power blackout in the Northeast, caused by various errors and malfunctions. NERC was subsequently granted the authority to create regulations that would protect against future electrical power disruptions. In 2008, NERC issued the first set of FERC-approved mandatory regulations, called Critical Infrastructure Protection (CIP).

The NERC CIP standards have evolved over time with multiple versions. As cybersecurity concerns have increased, so, too, have the regulatory protections included in the CIP standards. For example, NERC now requires entities critical to its mission to report attacks and breach attempts. They also require background checks of employees.

NERC CIP v5 defines the standards required for entities that own or manage US or Canadian facilities that supply electrical power. Failure to comply with the standards can result in significant fines and penalties. Each standard includes a set of applicable systems, requirements, and measures, and, as with many proprietary industry standards, NERC CIP can map to the NIST CSF and ISO 27001.

> **EXAMPLE**
>
> **Part 1.1 of NERC CIP-004-7 applies to medium- and high-impact Bulk Electric System (BES) cyber systems. Security awareness training is required at least once each calendar quarter to reinforce cybersecurity practices for those with a particular kind of access to these cyber systems. For measures, the standard describes, as an example, documentation that the quarterly training has been provided.**

Chapter 12 of this book discusses the importance of protecting a nation's power grid. The following case is a chilling example of the importance of managing cybersecurity threats to critical infrastructure.

> **CASE IN POINT**
>
> **The 2015 Russian hack of the Ukraine power grid brought an increasing awareness of the damage that could result from such attacks. The Russian-Ukrainian war in 2022 heightened concerns about attacks on the US power grid. In spring 2022, there was some evidence that the power grid in the state of Texas was being scoured for weaknesses by Russian hackers. Attempted breaches of electrical utilities are always ongoing, but this time there was evidence of elevated efforts, warranting high alert (Ferman, 2022).**

PCI DSS

Specialized cybersecurity standards are created when an industry or business sector faces relatively more risk than others. Critical infrastructure, such as the power grid, is an obvious example. Another is the financial services industry. Threats increase when monetary assets are being stored or transferred. Businesses that handle credit or debit cardholder information need to comply with the Payment Card Industry Data Security Standard (PCI DSS). This standard applies to any organization, regardless of size, that accepts, stores, processes, or transmits data about cardholders.

The PCI DSS standard is not mandated by government, it is required by the major credit card brands that established the Payment Card Industry Security Standard Council. American Express, Discover, JCB International, MasterCard, and Visa founded the Council in 2006. They developed the standard as the industry saw a rise in the number of credit card data breaches, such as the Target breach in 2013 that exposed the data of more than 40 million customers.

Table 7-4. PCI DSS Compliance Levels

Merchant Level	Applicability
Level 1	More than 6 million card transactions annually
Level 2	1 million–6 million card transactions annually
Level 3	20,000–1 million e-commerce transactions annually
Level 4	Less than 20,000 e-commerce transactions, or up to 1 million transactions total

While all entities that process credit card transactions need to be PCI DSS compliant, the level of compliance varies with the volume of transactions, as shown in Table 7-4.

A Level 1 merchant must have an annual audit by a Qualified Security Assessor, while those merchants in the other levels may require an annual Self-Assessment Questionnaire and perhaps regular scans of their network by an approved party. While compliance requirements vary with levels, PCI DSS 4.0 has 12 general data security requirements that apply to all card merchants. These are the general categories for these requirements:

- Build and maintain a secure network and systems.
- Protect account data.
- Maintain a vulnerability management program.
- Implement strong access control measures.
- Regularly monitor and test networks.
- Maintain an Information Security Policy.

PCI DSS can map directly to the NIST CSF. Organizations may use ISO 27001 as support in their PCI DSS compliance, though the ISO standard is a comprehensive ISMS framework. Proprietary frameworks have objectives different from those of the more general cybersecurity frameworks, but in recognizing requirement redundancies, most standards provide opportunities to leverage others.

Summary

Developing policies and practices that ensure information security requires a structure. Organizations seeking to protect their data and information resources may make use of a general, proprietary, or specialized cybersecurity framework, or a combination thereof. The NIST Cybersecurity Framework and the ISO/IEC 27001 standards are the most widely adapted because they are both flexible and comprehensive. This chapter discusses these two standards in some detail. COBIT is a framework for governance of information and technology, and the chapter describes its benefits and structure. We also discuss MITRE ATT&CK, a framework that is built on adversarial tactics and techniques. Many proprietary frameworks are available for industries and other segments. The chapter describes three significant ones: HITRUST CSF, NERC CIP, and PCI DSS.

Chapter Review

Key Terms

Endpoint Detection and Response (EDR)

Information Security Management System (ISMS)

Review Questions

1. What is the purpose of having a cybersecurity framework?
2. What are the differences among the types of cybersecurity frameworks?
3. What is the NIST Cybersecurity Framework?
4. List and summarize the five NIST core functions.
5. List and summarize the NIST Implementation Tiers.
6. What is the purpose of the Profile in the NIST Cybersecurity Framework?
7. Explain the functions of the ISO and IEC 27001 frameworks.
8. How is an organization certified with ISO or IEC 27001?
9. What is MITRE ATT&CK? Describe the major areas of this framework.
10. Summarize the purpose of COBIT.
11. Explain the need for/function of proprietary cybersecurity frameworks.
12. Summarize each of the proprietary frameworks described in this chapter.

Multiple Choice

Identify the choice that best completes the statement or answers the question.

1. What does ISO/IEC 27001 certification primarily emphasize?
   a. Cybersecurity framework outcomes
   b. Physical security measures
   c. Confidentiality, integrity, and availability of information
   d. Technical cybersecurity controls
   e. Organizational behavior and culture

2. What was the purpose of President Barack Obama's Executive Order regarding the development of the NIST Cybersecurity Framework?
   a. To mandate the use of the NIST Cybersecurity Framework by all federal agencies
   b. To create a mandatory set of cybersecurity practices for private businesses
   c. To establish a voluntary framework to assist in information security risk management for critical infrastructure entities
   d. To launch a cybersecurity awareness campaign
   e. To replace all previous cybersecurity guidelines with the NIST Framework
4. Which of the following best describes the relationship between COBIT and other cybersecurity frameworks like the NIST Cybersecurity Framework?
   a. COBIT is incompatible with other cybersecurity frameworks.
   b. COBIT is a subset of the NIST Cybersecurity Framework.
   c. COBIT can be used in conjunction with other cybersecurity frameworks.
   d. COBIT replaces the need for any other cybersecurity framework.
   e. COBIT is solely focused on the financial aspects of IT governance, unlike other frameworks.
5. What are the three main components of the NIST Cybersecurity Framework?
   a. Core, Implementation Tiers, and Profiles
   b. Functions, Categories, and Subcategories
   c. Implementation Tiers, Profiles, and Subcategories
   d. Profiles, Functions, and Implementation Tiers
   e. Core, Functions, and Categories
6. What triggers the need for an organization to comply with the Payment Card Industry Data Security Standard (PCI DSS)?
   a. Operating within the financial services industry
   b. Experiencing a cybersecurity incident
   c. Having an online presence
   d. Accepting, storing, processing, or transmitting cardholder data
   e. Being part of the global supply chain
7. What is a key feature of the NIST Cybersecurity Framework that makes it appealing for various organizations?
   a. Its flexibility to be used in its entirety, in parts, or alongside other frameworks and guidelines
   b. Its exclusive focus on US federal government entities
   c. Its mandate for private sector businesses to adhere without exceptions
   d. Its requirement for all organizations to fully implement every aspect
   e. Its strict guidelines that leave no room for customization

Case Studies

Sammy Company—Adopting NIST CSF 2.0

Sammy Company is a midsize beverage manufacturing company that makes a variety of flavored seltzer water canned beverages. The company has grown over the past three years and now has annual revenues of more than $500 million and 700 employees. Cybersecurity has become an increasing concern for top management, and the company's CISO, Kristine Tsui, is considering adopting the NIST Cybersecurity Framework. She has contacted Herron Consulting to assist her with this project. You can access the NIST CSF 2.0 at https://nvlpubs.nist.gov/nistpubs/CSWP/NIST.CSWP.29.pdf.

Case Requirements

1. Discuss the NIST CSF 2.0 Implementation Tiers and provide guidance to Kristine on selecting the company's current tier. Given the level of cybersecurity threats to a midsize manufacturer, what level of Implementation Tier would you recommend as a target and why?
2. The NIST CSF is flexible, and a company may adopt any or all core functions: Govern, Identify, Protect, Detect, Respond, and Recover. Kristine would like to begin with the Detect function. Describe the importance of this function and provide examples of three outcomes associated with Detect.
3. Kristine has selected the Detect function as a starting point in adopting the NIST CSF. An argument might be made that it would be better to begin with the Prevent function. Make a case for each option.

Drew Corporation—Choosing a Cybersecurity Framework

Drew Corporation is a manufacturer and seller of athletic shoes and other athletic apparel. The company, founded in 1921, sells its products globally and has annual revenues of $5 billion. The company's CISO, Sam Sutton, reports to the CIO, Jenny Martin. The company takes cybersecurity very seriously and has allocated financial resources to manage cyber risk effectively to date. The company complies with regulatory standards, including PCI DSS, but it has not formally adopted a general cybersecurity framework. Jenny and Sam are concerned about escalating cyber risk and wonder if a more structured approach to risk management would make sense for the company.

Case Requirements

1. How would a general cybersecurity framework help Drew in managing its cyber risk?
2. How can Jenny and Sam use a cybersecurity framework in ensuring that appropriate resources are allocated to a constantly evolving cyber threat landscape?

3. Sam and Jenny have been discussing the benefits of various frameworks. They are not sure if it is better to adopt MITRE ATT&CK or go for ISO 27001 certification. They have also considered adopting the NIST CSF. Which would you recommend and why?

Exercises

7-1. The biggest change to the NIST CSF 2.0 from the earlier version was to add the Govern function. Required:
   1. Why do you think NIST added the Govern function?
   2. COBIT differs from other cybersecurity frameworks in that it emphasizes IT governance and management. What attributes of COBIT's focus on governance are different from the NIST framework that includes the governance function?

7-2. MITRE ATT&CK considers threat actors and their behaviors in creating a cybersecurity risk management program. Building a strategy for managing cyber risk could begin with an analysis of threats, assets at risk, vulnerabilities, or threat actors. Required:
   1. How does the focus on threat actors make the MITRE ATT&CK approach so different from a framework that considers threats and vulnerabilities rather than the bad guys?
   2. What are the advantages to an organization in choosing the MITRE ATT&CK approach versus adopting either the NIST CSF or ISO 27001?

7-3. This chapter described three proprietary sets of cybersecurity standards: HITRUST CSF, NERC CIP, and PCI DSS. Each serves a particular industry. Required:
   1. Research and find another industry that has a proprietary set of cybersecurity standards or guidelines. Describe the reason(s) why that industry needs specialized standards.
   2. For the standards or guidelines selected, explain how the model addresses cybersecurity risk in ways that are different than for other industries. (For example, the NERC CIP has special guidance about the physical security of electric facilities, and PCI DSS has compliance requirements that vary with the volume of credit card data handled or processed.)

References

CrowdStrike. (2025). *Secure the endpoint. Stop the breach.* Retrieved September 26, 2025, from https://www.crowdstrike.com/en-us/platform/endpoint-security/

Deloitte. (2025). *Cyber strategy framework.* Retrieved September 26, 2025, from https://www.deloitte.com/an/en/services/risk-advisory/services/cyber-strategy-framework.html

Ferman, M. (2022, March 31). Texas power grid, energy sectors facing elevated Russian cyber threats during war in Ukraine. *The Texas Tribune.* https://www.texastribune.org/2022/03/31/texas-energy-grid-russia-cyberattack-hackers/

International Organization for Standardization. (2022). *ISO/IEC 27001:2022.* https://www.iso.org/standard/27001

Karthik, S. (2025, January 1). *ISO 27001 certification cost breakdown: Plan your compliance budget.* Sprinto. https://sprinto.com/blog/iso-27001-certification-cost/

National Institute of Standards and Technology (NIST). (2018, May 21). *Cybersecurity framework: Respond.* https://www.nist.gov/cyberframework/respond

National Institute of Standards and Technology (NIST). (2019a, February 21). *Success story: University of Chicago Biological Sciences Division.* https://www.nist.gov/cyberframework/success-stories/university-chicago

National Institute of Standards and Technology (NIST). (2019b, February 21). *Success story: University of Pittsburgh.* https://www.nist.gov/cyberframework/success-stories/university-pittsburgh

National Institute of Standards and Technology (NIST). (2024, February 26). *The NIST cybersecurity framework (CSF) 2.0.* https://nvlpubs.nist.gov/nistpubs/CSWP/NIST.CSWP.29.pdf

National Institute of Standards and Technology (NIST). (2025, February). *Cybersecurity framework profile for semiconductor manufacturing.* Retrieved September 26, 2025, from https://www.nccoe.nist.gov/projects/cybersecurity-framework-profile-semiconductor-manufacturing

Saeckel, A. (2024, May 3). Lessons learned from ISO 27001: A case study of Enterbrain Software. *DQS in Dialog.* https://www.dqsglobal.com/en/explore/blog/lessons-learned-from-iso-27001-a-case-study-of-enterbrain-software

Soares, B. H. (2023, May 24). *Improving governance at a national bank with COBIT.* ISACA. https://www.isaca.org/resources/isaca-journal/issues/2023/volume-3/improving-governance-at-a-national-bank-with-cobit

Spakowski, K. (2023, September 6). NIST cybersecurity framework (CSF) controls fundamentals. *The AuditBoard Blog.* https://auditboard.com/blog/fundamentals-of-nist-cybersecurity-framework-controls

Chapter 8

System and Organization Controls

Chapter Outline

Attestation Standards
COSO
AICPA Trust Services Criteria
System and Organization Controls (SOC) Reports
SOC 2 Reports
SOC 2 Report Sections
SOC 2+ Engagements
Using SOC 2 Reports

Learning Objectives (LO)

After reading this chapter, you should be able to:

- LO1. Be familiar with the evolution of attestation services.
- LO2. Be knowledgeable about COSO frameworks and their components.
- LO3. Be knowledgeable about AICPA Trust Services.
- LO4. Be familiar with AICPA SOC 1, SOC 3, SOC for Cybersecurity, and SOC for Supply Chain reports.
- LO5. Be knowledgeable about SOC 2 reports from an auditor and user perspective.

Introduction

Certified Public Accountants (CPAs) have been providing assurance for **financial reporting** since the 19th century. In 1887, the American Institute of Certified Public Accountants (AICPA) formed as a professional membership association and began developing accounting standards and guidance for attestation services. The types of services that CPAs offer have varied, from public company audits to tax compliance and consulting to a broad array of advisory services, all of which have changed since the inception of the AICPA based on business circumstances, the legal and regulatory environment, and technology.

Financial Reporting A systematic process of recording and representing a company's financial data.

This chapter focuses on a specific type of attestation service, System and Organization Controls (SOC) reports (not to be confused with the other cybersecurity SOC—security operations center), and the AICPA Trust Services that underlie them as they relate to cybersecurity. CPAs currently provide the following SOC reports:

- SOC 1 Report
- SOC 2 Report
- SOC 2+ Report

- SOC 3 Report
- SOC for Cybersecurity
- SOC for Supply Chain

Internal Controls Processes used in an organization that ensure the integrity of the organization's records, systems, and processes as well as ensure regulatory compliance.

A SOC 1 report focuses on **internal controls** over financial reporting. SOC 2 reports provide assurance to user entities that an organization is protecting its data. SOC 2+ expands the scope of SOC 2 reports to incorporate other assessments. SOC 3 reports focus on cybersecurity but are a scaled-down general-use report versus SOC 2. The SOC for Cybersecurity is a comprehensive enterprise-wide cybersecurity risk management framework. SOC for Supply Chain is the newest entry to the suite of services, and these reports are meant to show that an organization's supply chain risk management program and controls are appropriate. SOC 2, 2+, 3, SOC for Cybersecurity, and SOC for Supply Chain all rely on the AICPA Trust Services as a foundation. They are detailed in this chapter.

Assurance The level of confidence an auditor provides that information they have examined is free from material misstatement and conforms to applicable standards.

Although CPAs provide **assurance** in SOC reports, many of an organization's functional areas are involved, and management of the organization issuing the report has the ultimate responsibility over its content. Users of the report must verify that it is accurate and suits their needs. This chapter concludes with the user perspective and how users should go about this verification.

Attestation Standards

Attestation A formal statement that confirms the accuracy or authenticity of a company's processes or reports, often in the context of compliance.

Attestation standards have evolved over the decades, moving from purely financial reporting on financial institutions to more currently encompassing cybersecurity, AI, cryptocurrency, and privacy. The evolution of the attestation standards can be seen in the timeline in Figure 8-1.

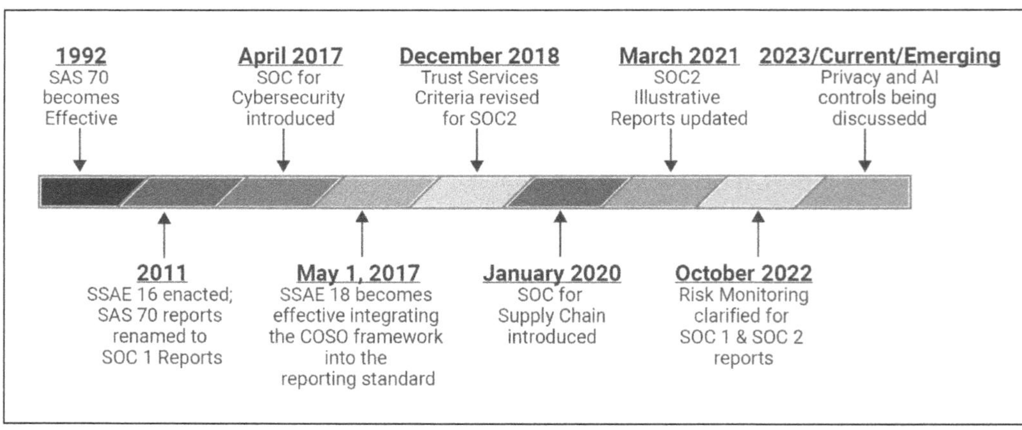

Figure 8-1. Attestation Reporting Standard Timeline

SAS 70

Statement on Auditing Standards (SAS) Official guidelines issued by the Auditing Standards Board (ASB) of the American Institute of Certified Public Accountants (AICPA). The SAS provides rules and procedures that auditors must follow when conducting audits of companies in the United States.

Software as a Service (SaaS) A cloud computing service model that delivers software applications over the Internet on a subscription basis.

In 1992, the AICPA published **Statement on Auditing Standards** (SAS) 70, a set of standards for auditors to use in reporting on service organization internal financial controls and processes. Service organizations are third-party entities that an organization might use in outsourcing some of its functions, such as payroll or billing, though they may simply be vendors or customers that share data. Service organizations can offer information systems functionality, such as cloud computing or **Software as a Service** (SaaS). The internal controls of these third parties become part of the internal control system of the contracting company because the service organization's failure to protect data can impact the entire supply chain.

SAS 70 reports adapted over time as both information technologies evolved and businesses began outsourcing more of their services that impacted financial reporting and information security.

SSAE 16

Statement of Standards for Attestation Engagements (SSAE) 16 was enacted in 2011 as a new service auditor reporting standard that somewhat matched the International Standard on Assurance Engagements 3402, issued by a board of the **International Federation of Accountants (IFA)**, and superseded SAS 70 reporting. Similar to SAS 70 engagements, the deliverable of this type of attestation was a report on the controls and processes of the assessed entity for financial reporting. Compliance with SSAE 16 required management to write an attestation about their control system, detailing the controls currently in place along with a description of the suitability of those controls. SSAE 16 provided for two different types of auditor reports. A Type 1 attestation verifies the following:

International Federation of Accountants (IFA) A global organization for the accountancy profession.

- Management-provided description of existing controls
- Management-provided description of the design of existing controls
- Management's attestation to the effectiveness of existing controls
- Auditor attestation about the existence of management controls and the effectiveness of those controls (auditors test the controls) at a specific point in time

A Type 2 SSAE 16 attestation provides all of the features of a Type 1 report but expands the testing of controls over a specified period of time (usually three to six months) to give the user of the report a sense of the operating effectiveness of the existing controls over time. The reports issued under SSAE 16 are called Service Organization Control (SOC) 1 reports.

SSAE 18

SSAE 18 was enacted in 2017 and supersedes SSAE 16. SSAE 18 expands beyond the narrow focus of financial reporting to include vendor management, risk assessment, and monitoring controls, as well as specifying additional reports (SOC 2 and SOC 3) that a CPA can issue. Around this time, the SOC for Cybersecurity was also introduced. The new standard also included a name change for SOC reports from Service Organization Control reports to System and Organization Controls (SOC) reports. Today, the standard covers a suite of six types of SOC reports, discussed later in this chapter. SSAE 18 provides requirements and guidance for examining controls at third parties that provide services, where the internal controls of the third party are relevant to the service organization. For example, a vendor that will maintain and process data for a customer may need to assure the customer that their data will be appropriately protected. SSAE also added the requirement for covered entities to perform annual risk assessments.

COSO

In 1985, five organizations joined together to create the Committee of Sponsoring Organizations (COSO) of the Treadway Commission. These founding organizations were the American Accounting Association (AAA), the American Institute of Certified Public Accountants (AICPA), Financial Executives International (FEI), the Institute of Internal Auditors (IIA), and the Institute of Management Accountants (IMA). The objective of these professional associations was to gain an understanding of the causal factors behind fraudulent financial reporting.

COSO

In 1992, COSO released its "Internal Control—Integrated Framework" (ISACA, 1992). This framework included five components of an internal control system, along with tools for evaluating the effectiveness of such systems. The framework was updated and rereleased in 2013, integrating technology and changes in business models and practices. The COSO cube, shown in Figure 8-2, provides a visual representation of the COSO framework.

The front side of the cube lists the components of the COSO internal control model. The control environment is the "tone at the top," including ethics and governance. Risk assessment is the process of identifying and analyzing risks. Control activities include internal control policies

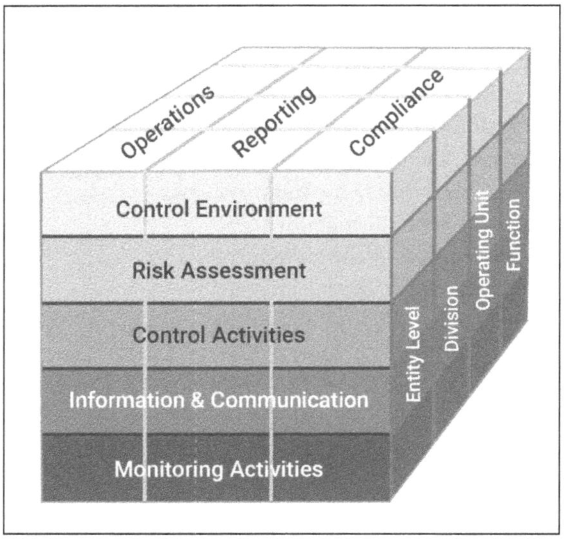

Figure 8-2. The COSO Cube

and procedures for managing risks. Information and communication are the information systems and information flows across them, including stakeholder communications. Monitoring activities are the evaluation of internal control system performance quality over time.

The top part of the COSO cube (columns) shows the three categories of internal control objectives. They are the effectiveness and efficiency of an entity's operations; ensuring that financial and nonfinancial reporting are reliable and accurate; and compliance with laws, regulations, and internal policies.

The cube's side illustrates the integration of the components and objectives across an organization, from the entity level to functions within operating units.

Within the five components of the COSO framework are a total of 17 principles, listed in Table 8-1. These principles guide managers and auditors in evaluating an organization's internal control system.

Table 8-1. COSO Components and Associated Principles

Component of Internal Control	COSO Principles
Control Environment	1. Demonstrates commitment to integrity and ethical values 2. Exercises oversight responsibility 3. Establishes structure, authority, and responsibility 4. Demonstrates commitment to competence 5. Enforces accountability
Risk Assessment	6. Specifies suitable objectives 7. Identifies and analyzes risk 8. Assesses risk of fraud 9. Identifies and analyzes significant change
Control Activities	10. Selects and develops control activities 11. Selects and develops general controls over technology 12. Deploys control activities through policies and procedures
Information and Communication	13. Uses relevant information 14. Communicates internally 15. Communicates externally
Monitoring Activities	16. Conducts ongoing and/or separate evaluations 17. Evaluates and communicates deficiencies

CASE IN POINT

Many public companies have implemented COSO's Internal Control—Integrated Framework. The Coca-Cola Company is an example of a company that references the framework in its annual report. The auditor opinion on Internal Control over Financial Reporting in Coca-Cola's 2023 10-K states, "We have audited The Coca-Cola Company and subsidiaries' internal control over financial reporting as of December 31, 2023, based on criteria established in Internal Control—Integrated Framework issued by the Committee of Sponsoring Organizations of the Treadway Commission (2013 framework) (the COSO criteria)" (Coca-Cola Company, 2024).

The COSO framework has been an effective tool for managers to use in developing and maintaining internal control systems to manage risk. The original intent of COSO—to understand the reasons behind financial fraud and fraudulent reporting, has led to the creation of a robust internal control framework used globally by both publicly traded companies and private firms alike. Public accounting firms use the framework in helping their clients assess their internal controls over business processes. This is often done to ensure compliance with the **Sarbanes-Oxley Act (SOX)** of 2002. Sections 302 and 404 of that legislation respectively call for the disclosure and assessment of internal controls.

In addition to its internal control framework, in 2004, COSO issued its **Enterprise Risk Management** (ERM) framework, which focuses on enterprise-wide risk management. The COSO ERM framework has five components: Governance and Culture; Strategy and Objective Setting; Performance; Review and Revision; and Information, Communication, and Reporting. COSO has developed guidance for organizations in several areas of ERM, including cloud computing, cyber risk, and artificial intelligence (see Table 8-2). Each of these guidance documents explains how to adapt the framework components to the particular risk or technology.

Sarbanes-Oxley Act (SOX) A US federal law that sets requirements for public company boards, management, and public accounting firms to protect investors from fraudulent financial reporting.

Enterprise Risk Management A structured, organization-wide approach to identifying, assessing, managing, and monitoring risks that may affect a company's ability to achieve its objectives.

Table 8-2. Selected COSO ERM Guidance

Publication	Purpose
Enterprise Risk Management for Cloud Computing	Enterprises can apply the ERM framework to establish governance over cloud computing. The guide shows how to apply the framework's components and principles to a cloud computing environment. Included is an appendix describing a roadmap to cloud computing.
Managing Cyber Risk in a Digital Age	Enterprises can use the ERM framework as a foundation to build a cyber security program by leveraging the components and principles in the framework to help the enterprise identify and manage cyber risks.
Realize the Full Potential of Artificial Intelligence	Enterprises can use the ERM framework to use and leverage artificial intelligence (AI) consistent with AI risks and their business strategy.

AICPA Trust Services Criteria

In 2017 the AICPA's Assurance Services Executive Committee (ASEC) explicitly integrated COSO's framework and components to create the AICPA Trust Services Criteria.* These criteria are the underpinning of all of the reports that SSAE 18 governs and are used in the assessment of controls for all reporting engagements. The criteria delineate the outcomes an organization's controls should meet to achieve organizational objectives. Both the COSO Framework and the Trust Services Criteria are used to evaluate internal control. The AICPA updated its Trust Services Criteria in 2022 with additional points of focus. Trust Services Criteria include COSO's five components, and they incorporate the principles within these categories (see Table 8-3) as **Control Criteria**. The Trust Services Criteria are meant to provide control over the **CIA triad** of Confidentiality, integrity (Processing Integrity), and Availability, which are called Trust Services Categories (Table 8-3). Two additional categories are Security and Privacy. Security is for the protection of information and systems against unauthorized access, unauthorized information disclosure, and any systems damage that might compromise the CIA. Privacy dictates that stakeholder information always be protected during its creation, use, processing, transmission or transfer, and storage. Confidentiality focuses on protecting any information an organization considers secret or restricted to a specific group. In sum, Privacy is about personal information, while Confidentiality refers to organizational information.

Control Criteria Standards or benchmarks used to evaluate the design and effectiveness of internal controls in an organization. They define what "good" controls should look like and provide a framework for assessing whether controls are achieving their intended purpose.

CIA Triad A model designed to guide policies for information security within an organization, consisting of three key principles: confidentiality, integrity, and availability.

AICPA & CIMA

* AICPA Trust Services Criteria are available with registration at www.aicpa-cima.com.

Table 8-3. Trust Services Criteria Categories	
Trust Services Categories	**Description**
Confidentiality	Protects confidential (secret or restricted) information
Processing Integrity	Ensures that processing is complete, accurate, timely, valid, and appropriate in meeting entity objectives
Availability	Ensures that information and systems are available and open to meet entity objectives
Security	Protects information and systems against unauthorized access, disclosure, and damage
Privacy	Protects personal information

As listed in Table 8-1, the five COSO components included 17 principles. The Trust Services Criteria incorporate them as well, naming them Common Criteria. In addition to these Common Criteria, the categories of Confidentiality, Processing Integrity, Availability, and Privacy also include supplemental criteria. They are included to augment "COSO Principle 12: *The entity deploys control activities through policies that establish what is expected and procedures that put policies into action*" (AICPA, 2022). Table 8-4 describes these supplemental criteria.

Table 8-4. Trust Services Supplemental Criteria	
Supplemental Criteria	**Description**
Logical and physical access controls	Controls that restrict logical and physical access and prevent unauthorized access
System operations	Management of system operations and detection and mitigation of deviations
Change management	Controls related to identifying needed changes, established process for changes, and preventing unauthorized changes
Risk mitigation	How the organization identifies, selects, and creates risk mitigation processes and procedures

Each Trust Services Criteria component includes common criteria, supplemental criteria (except for Security), and points of focus related to each criterion.

EXAMPLE

The Control Environment (a COSO component) includes five criteria, each corresponding to a COSO principle. CC1.1 (a control criteria) of the Trust Services Criteria correlates with COSO Principle 1: The entity demonstrates a commitment to integrity and ethical values. CC1.1 incorporates points of focus from the COSO framework, such as Sets the Tone at the Top. An additional point of focus for CC1.1 is Considers Contractors and Vendor Employees in Demonstrating Its Commitment (AICPA, 2022). In some cases, additional points of focus are related to a specific category.

EXAMPLE

CC1.5 is COSO Principle 5. Multiple points of focus are connected to the COSO framework, and one additional point of focus is added when using the Trust Services Criteria for the privacy category. This additional point of focus is: Takes Disciplinary Action (AICPA, 2022).

Figure 8-3 summarizes AICPA Trust Services categories and criteria.

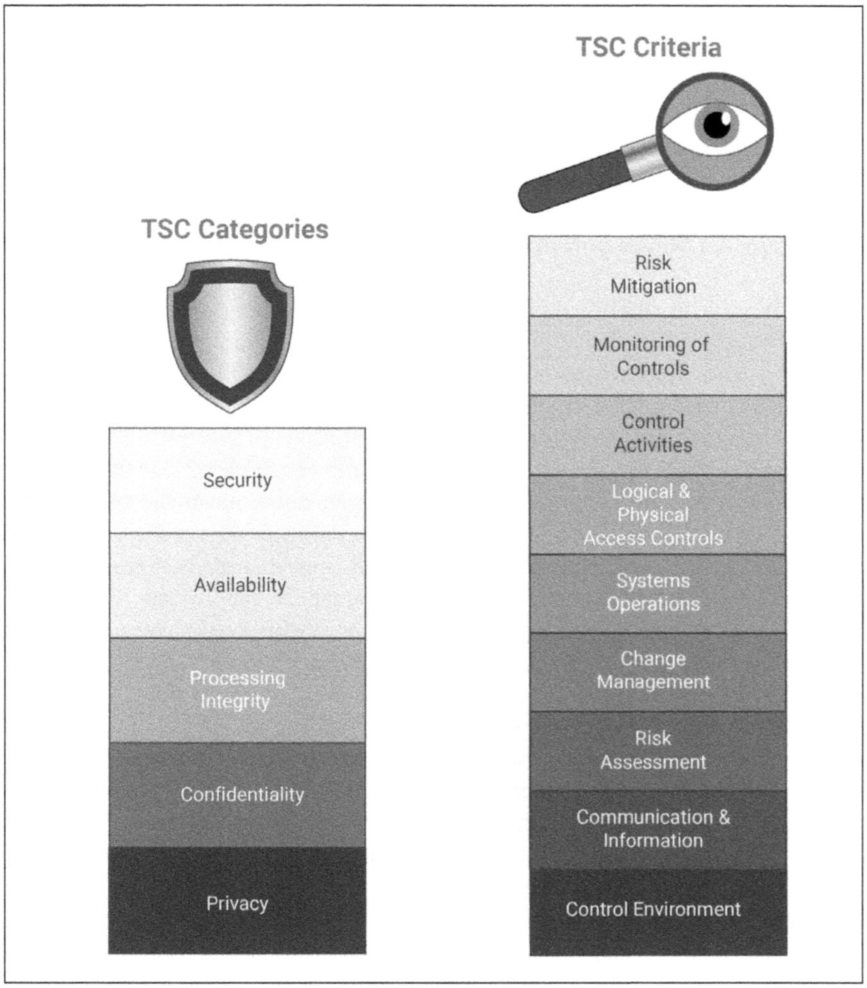

Figure 8-3. A Summary of AICPA Trust Services Criteria

System and Organization Controls (SOC) Reports

As noted earlier, the current suite of SOC reports that CPAs can issue include the following:

- SOC 1
- SOC 2
- SOC 2+
- SOC 3
- SOC for Cybersecurity
- SOC for Supply Chain

These reports are System and Organization Controls (SOC) **attestation** reports, and most rely on the AICPA Trust Services Criteria to provide a level of assurance for business process controls and information security. Organizations that provide the audit services typically refer to this work as **SOC Engagements**, but we will focus on the reports themselves instead of the audit procedures that make up an engagement. The reports use the criteria to evaluate and document an entity's nonfinancial controls over CIA, plus security and privacy.

SOC (System and Organization Controls) Engagement A third-party audit that a CPA or accounting firm performs to evaluate and report on the controls of a service organization—typically related to security, availability, processing integrity, confidentiality, or privacy of systems and data.

Service Auditor An independent auditor who evaluates and reports on the controls of a service organization.

Opinion (Auditor) The formal conclusion that an independent auditor provides after evaluating a company's financial statements and related disclosures.

Subsequent Event An event that occurs after the formal end of the audit testing period but before the audit opinion is issued that may require disclosure or adjustment of the management statements.

Three parties are associated with the report: the service organization, the user entity (report recipient), and the service auditor. The user entity requests the report from the service organization. A **service auditor** (who must be a CPA) issues SOC reports on behalf of the service organization. These reports were originally created for outsourced service providers, such as data centers, but over time they have become increasingly popular for all types of organizations. Now these reports may be requested by a user entity of any third party that will be maintaining or processing their data, and they are no longer restricted to service organizations. For example, a potential customer may request a SOC report from a company that will be processing their orders through an online portal.

The main deliverable of all SOC reports is the Independent Service Auditor's Report, in which an independent auditor gives their **opinion** as to the fairness of the presentation of the system and the suitability and effectiveness of presented controls. An *unqualified opinion* indicates that the organization has passed its audit. The service auditor has tested controls and found them to be appropriately designed or, depending on the type of SOC report, operating as intended during the period covered by the audit. An unqualified opinion does *not* mean that there is no room for improvement, merely that, in all material respects, the organization meets the minimum standards.

A *qualified opinion* might be considered a partial failure (or partial success!). The service auditor has found that one or more controls being assessed were materially ineffective in design or implementation. However, the exceptions noted in the qualified opinion are not pervasive to the extent that the controls are completely ineffective. Whether exceptions warrant a qualified opinion depends on their number, scope, and the severity of the impact on internal control systems. Materiality is a matter of professional judgment. In financial audits, auditors may make ballpark decisions based on percentages of revenue, for example. In a SOC 2 audit where the service auditor is rendering an opinion, the judgment concerns the appropriateness of management's information system and the ability of controls to meet control objectives. The auditor considers both quantitative and qualitative variables. The engagement report will include all exceptions, either in the auditor's opinion, or in the section of the report concerning control design and/or effectiveness.

> **EXAMPLE**
>
> **From a hypothetical engagement report, "ABC Company did not have sufficient logical access controls in place during the period to prevent unauthorized access to information assets, including customer data. Specifically, we identified a lack of a requirement for multifactor authentication for remote access to the internal network. We also found multiple instances where terminated employees still had access to customer data."**

A *disclaimer of opinion* is no opinion. There may be different reasons behind this lack of decision. The organization may not have provided enough information for the auditor to make an assessment about the controls, or perhaps there were limitations to the scope of the engagement. Another cause could be that the service auditor believed that there were material and pervasive misstatements that they were unable to detect.

An *adverse opinion* signifies that the controls have been improperly designed or are not operating effectively, to the extent that they are unreliable in providing security for the confidentiality, integrity, availability, security, and/or privacy of data. The material exceptions in this instance are pervasive.

It is possible that a **subsequent event** requiring disclosure may occur following the preparation of the service auditor's report. The omission of such an event could cause the report to be unreliable. The service auditor will need to discuss the event with the service organization management and determine whether the report and/or the auditor's decision requires revision. It is also possible that subsequent events may require the service auditor to withdraw from the engagement. Subsequently discovered facts are new, relevant information that come to light following the date of the service auditor report. The auditor needs to determine if these facts would have impacted the report content and auditor decision. This is a professional judgment decision that considers whether someone relying on the report would draw a different conclusion with the addition of these new facts. The service auditor may need to perform more procedures, and a revised report may be required.

The concept of materiality is a bit different with respect to SOC reporting versus financial statement reports. *Materiality*, simply put, means that "it matters." In financial reporting, what matters are misstatements that would impact the decisions of stakeholders, including lenders and investors. In SOC reporting, materiality concerns material misstatements with respect to the suitability of the control design, the system description, the operating effectiveness of controls, and reporting of all of these items. The auditor will consider qualitative factors in judging the materiality of misrepresentations in management's description of the service organization's system and design of controls. They may also have to consider both qualitative and quantitative risk factors in assessing whether the controls are operating effectively.

The period that a SOC report covers might differ from a service organization's calendar or fiscal year. In that case, a **bridge letter** might be issued to cover the time gap. The letter would assure the user entity that controls are continuing to operate effectively and will also note any significant changes that occurred between reporting periods.

Bridge Letter A written statement that a service organization (or its auditor) provides to bridge the gap between the end date of the latest SOC report and the user entity's financial reporting period or audit timeline.

The following subsections describe SOC 1, SOC 3, SOC for Cybersecurity, and SOC for Supply Chain reports. A separate section discusses SOC 2 and SOC 2+ reports in greater detail.

SOC 1 Reports

SOC 1 reports have been in a few forms since the early 1990s. Originally, they were called SAS 70 reports, but their name was formally changed to SOC 1 reports in 2011 with the implementation of SSAE 16. These reports are for outsourced service entities that have the potential to impact the contracting company's financial statements. Examples of service organizations providing these reports are payroll processors and financial institutions. The SOC 1 attestation provides a user entity with some assurance that the service organization has appropriate controls in place for both its business processes and its IT. In turn, it provides a reasonable assurance that the financial reporting is accurate. The reports specifically address **internal control over financial reporting (ICFR)**. The focus is on controls for transactions and processing. SOC 1 reports have restricted use; they are not public documents. Their recipients are the user entities and their auditors.

Internal Control Over Financial Reporting (ICFR) The policies and procedures a company puts in place to reasonably ensure that its financial statements are accurate, complete, reliable, and prepared in accordance with the applicable accounting standards.

SOC 1 reports may be Type 1 or Type 2. A SOC 1 Type 1 report focuses on the design and implementation of controls at a point in time. The report includes the following parts:

- Management's description of the service organization's system
- Management's assertion that their description of the service organization is a fair representation of the system and that the control objectives in the system description were suitably designed to achieve the objectives, both at a specified date
- Independent service auditor's report on the suitability and existence of the controls at a specified date

The SOC 1 Type 2 report evaluates both the design and operating effectiveness of the controls over a period, generally six months or longer. It includes the same parts as a Type 1 report, except that it covers a specific period versus a point in time, and the service auditor report includes a description of the tests of the controls the auditor performed and the results of those tests. SOC 1 reports provide some assurance to external stakeholders that they may have confidence in financial reporting, with Type 2 reports providing a higher level of assurance.

SOC 1 reports may also complement compliance with laws and regulations, including the Sarbanes-Oxley Act of 2002 (SOX), reducing or eliminating duplication of effort. SOX was enacted in the wake of the Enron, WorldCom, and other accounting scandals to protect stakeholders from financial reporting errors and fraud. **SOX Section 404** requires public companies to have a system of internal controls over financial reporting. The company must document, test, and maintain these

SOX Section 404 A provision of the Sarbanes-Oxley Act of 2002 (SOX) that requires public companies in the United States or companies that trade in the United States to establish, assess, and report on the effectiveness of their internal control over financial reporting (ICFR).

controls. SOX compliance reporting is somewhat like SOC 1 attestation and reporting. In both cases, an auditor tests internal controls and processes. However, SOX is a mandatory regulatory requirement, while SOC reports are voluntary. SOC 1 reports are also narrower in scope, limited to the controls for specific services impacting the entity's internal control over financial reporting.

SOC 3 Reports

Like SOC 2 (discussed in the next section), a SOC 3 engagement provides assurance for internal controls protecting the confidentiality, integrity, availability, security, and privacy of data. However, while SOC 2 reports are prepared for specific stakeholders and have restricted use, a SOC 3 report is tailored for public consumption and is a general use report. A service organization might engage a CPA to prepare a SOC 3 attestation report about its internal controls that it may provide to current and prospective third parties. The user may publish the report on its website to be transparent about how its controls and processes protect third-party data. Because these reports are more of a marketing tool than an assurance, they are less detailed about the specific controls, systems, and processes tested by the service auditor. There are no Type 1 and Type 2 SOC 3 reports, as these reports provide more of a summary.

> **CASE IN POINT**
>
> **Amazon Web Services (AWS) offers cloud computing services to more than a million customers of all sizes and across all industries. In the 2024 SOC 3 report, Ernst & Young states that, "in our opinion, AWS' controls over the system were effective throughout the period April 1, 2023, to March 31, 2024, to provide reasonable assurance that its service commitments and system requirements were achieved based on the applicable trust services criteria." AWS also has SOC 1 and SOC 2 reports that are restricted use and available only to AWS customers.***

SOC for Cybersecurity

In response to the growing threat posed by cyberattacks, the AICPA created SOC for Cybersecurity in 2017. These engagements are suitable for any type of company that maintains or processes sensitive data. They are broader in scope than SOC 2 reports and attest to an organization's enterprise-wide cybersecurity risk management program in general, although management may choose to focus on just a portion of their cyber risk program. Somewhat like a SOC 3 report, a SOC for Cybersecurity is intended for use by a broad audience that includes any stakeholders with an interest in trusting that an entity's risk management program is comprehensive and appropriately designed. SOC for Cybersecurity may use the AICPA Trust Services Criteria for control evaluation, but flexibility allows for the use of other frameworks, such as NIST CSF 2.0 or COBIT.

The report includes the following parts:

- Management's description of its cybersecurity risk management program
- Management's assertion that their risk management program is effective
- Independent service auditor's report

Management's description of its cybersecurity risk management program is a narrative description that is based on the AICPA Description Criteria. These criteria provide guidance to management in describing their risk management program and to auditors in evaluating the program. The following are the AICPA description criteria for risk management programs:

- Nature of business operations
- Nature of information at risk
- Cybersecurity risk management program objectives

* https://aws.amazon.com/compliance/soc-faqs/ provides links to the SOC 3 AWS white paper.

- Factors that have a significant effect on inherent cybersecurity risks
- Cybersecurity risk governance structure
- Cybersecurity risk assessment process
- Monitoring of the cybersecurity program

In addition to the detailed criteria, the AICPA Description Criteria (DC) includes implementation guidance and notes the need for judgment in describing and evaluating a risk management program. The following example details one of the criteria and the accompanying implementation guidance.

> **EXAMPLE**
>
> **DC8: The process for board oversight of the entity's cybersecurity risk management program. Implementation Guidance: When making judgments about the nature and extent of disclosures to include about this criterion, consider the following:**
>
> - **The extent of the board of directors' cybersecurity and IT expertise or access to external cybersecurity and IT expertise, or both**
> - **Identification of the board committee designated with oversight of the entity's cybersecurity risk management program, if any**
> - **The frequency and detail with which the board or committee reviews or provides input into cybersecurity-related matters, including board oversight of security incidents (AICPA, 2017)**

The CPA or firm provides an opinion that the description is presented consistently with the AICPA Description Criteria, as well as an opinion on control effectiveness.

SOC for Supply Chain

A weakness in any part of the link in our **supply chain** threatens the entirety of the chain. The COVID pandemic highlighted this vulnerability to a level such that organizations, investors, and regulators began to pay close attention to the entire process for delivering goods and services to customers. Cyberattacks on any part of the supply chain can create disruption like the one described in the following case.

Supply Chain The network of individuals, organizations, resources, activities, and technologies involved in the creation and distribution of a product or service—from raw materials to final delivery to the customer.

> **CASE IN POINT**
>
> **When Clorox suffered a data breach in 2023, supply chain delays in processing orders and producing goods led to reduced sales and profitability. Clorox makes many household products, including bleach, Pine Sol cleaner, and disinfecting wipes. The breach demonstrated how hackers can exploit vulnerabilities in a supply chain of multiple partners, including manufacturers, distributors, and retailers.**

In response to this heightened attention, the AICPA released SOC for Supply Chain in 2020. This new SOC report provides producers, manufacturers, and distributors with a risk management tool and reporting framework. The SOC for Supply Chain is an attestation on the controls for Security, Availability, Processing Integrity, Confidentiality, and/or Privacy. The SOC is flexible, and users may select among the Trust Services categories, though the Security category is required. The objective of this latest SOC service is to give, to the customers and business partners of third parties that include manufacturers and distributors, assurance that appropriately designed cybersecurity controls are in place and operating effectively to manage supply chain risks.

SOC 2 Reports

Although businesses may choose among the suite of SOC engagements when opting to obtain some assurance from a CPA for their information security controls, to date, the SOC 2 report has been the most popular choice. While originally intended only for service organizations, SOC 2 has grown in usage to be a choice for any company wishing to provide evidence that it has designed its controls related to the AICPA's five Trust Services Criteria appropriately, and potentially that these controls are also operating effectively.

> **CASE IN POINT**
> **Many accounting firms offer SOC 2 attestation services. They include the Big Four public accounting firms and smaller entities that specialize in SOC reporting. An example of the former is PricewaterhouseCoopers (PwC), a Big Four firm that offers SOC and other attestation reporting under its Audit and Assurance service line. PwC can help clients with SOC readiness assessment to identify any gaps or any improvements needed before a SOC examination. It can also prepare SOC reports and SOC 2+ reports for clients to share with user entities and other auditors (PricewaterhouseCoopers, 2025).**

As with SOC 1, SOC 2 reports have two different types. A Type 1 report concerns the design and implementation of controls, and a Type 2 assessment provides all of the parts of a Type 1 report but tests and evaluates the effectiveness of controls as they function over time. Type 1 assessments are as of a specific date and can take just a few weeks for the service auditor to complete. A SOC 2 Type 2 evaluation may take several months to a year, as it is a test of the existing controls over a stated period. The Type 2 report includes descriptions of the auditor's tests of the controls and the results of those tests.

SOC 2 engagements are based on the AICPA Trust Services Criteria, specifically the Common Criteria. As previously explained, each of the Common Criteria includes points of focus that expand and detail necessary controls under the broad topic. CC1 to CC5 encompass the 17 COSO Principles, while CC6 to CC9 encompass the historic tests of general controls and risk mitigation.

> **EXAMPLE**
> **Under Trust Services Supplemental Criteria, Logical and Physical Access Controls is CC6.4, which concerns restricting physical access. One point of focus is that processes exist to remove physical access to facilities and protected information assets when, for example, an employee no longer requires that access. In a Type 1 SOC 2 assessment, the service auditor would check to see if a policy and a procedure were in place for this control. The Type 2 attestation would require the service auditor to observe that the control was working as designed. They might do this by, for example, examining access logs and comparing them to employee listings.**

SOC 2 Readiness Assessment

SOC 2 Readiness Assessment A pre-audit evaluation designed to help an organization prepare for a SOC 2 examination. It identifies gaps in its security, privacy, and internal control practices before the formal audit begins.

Organizations may decide to undergo a **SOC 2 Readiness Assessment** prior to beginning the SOC 2 assessment process. This readiness exercise is a preliminary evaluation to identify any gap corrections or improvements in policies, procedures, or controls that would create better alignment with the AICPA Trust Services Criteria. This process can be done internally by the information security leaders and staff or with the help of a service auditor. It will include a review of the organization's documented policies and procedures related to CIA and privacy and security. The assessment process should end with remediation of any identified gaps or implementation of any enhancements. Figure 8-4 summarizes the steps in a SOC 2 Readiness Assessment.

> **CASE IN POINT**
> **Scott McAuliffe, a Risk Advisory Services Partner at Keiter CPAs, recommends several steps to prepare for a SOC 2 engagement. One recommendation is to examine a SOC 2 report of a similar entity. He also recommends performing risk assessments and ensuring**

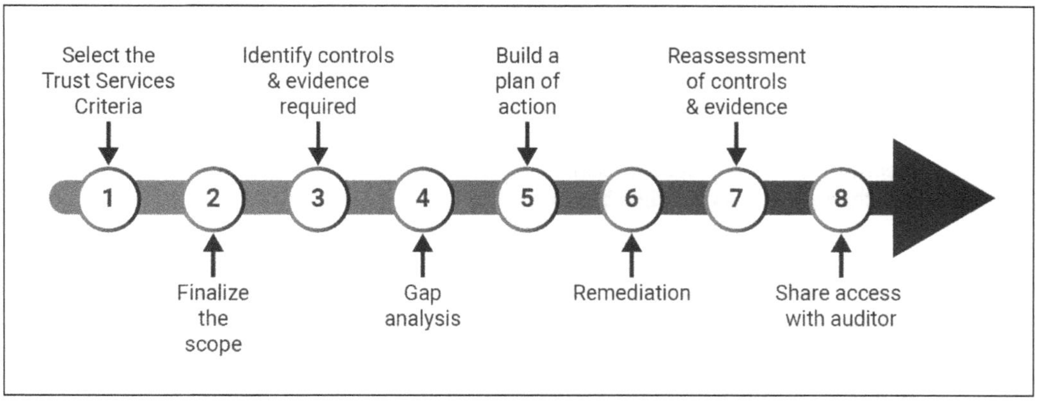

Figure 8-4. Steps in a SOC 2 Readiness Assessment

> **that critical entity-level controls have been implemented, including HR and IT policies. Policies and procedures should be documented to be easily communicated to the service auditor during the engagement, and procedures should be in place to monitor key vendors and subservice providers (McAuliffe, 2020).**

SOC 2 Report Sections

SOC 2 reports have four required sections, with an optional fifth that contains any additional information management wishes to provide.

Management Assertions

The Management Assertion section of a SOC 2 report is, as the name implies, a statement by management that they have prepared a description of their information system in accordance with the requirement for a SOC 2 report. This section is a written representation by management that acknowledges their responsibilities and affirmations about control objectives being suitably designed to meet Trust Services Criteria and, in the case of a Type 2 report, operating effectively.

Subservice Organizations

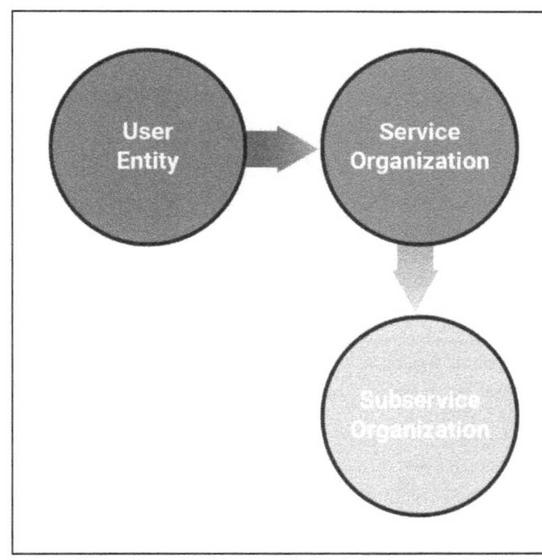

Figure 8-5. Parties in a SOC 2 engagement

This part of the SOC 2 report provides information about the use of any **subservice organizations**. Subservice organizations are vendors that the service organization has engaged to perform some of its services as they relate to the processes the SOC 2 engagement covers. Figure 8-5 shows the three entities that might be parties involved in an SOC 2 engagement.

A subservice organization may be included in the SOC 2 report or covered by the **carve-out method** of the report. If the subservice entity's services to the service organization are extensive and the subservice organization does not have its own SOC 2 report or similar assurance, it may be included in the engagement. Under the **inclusive method**, the subservice organization's control objectives and related controls must be disclosed. In combination with the service organization's controls, they provide the assurance that the SOC 2 Trust Services Criteria cover the service commitments (promises the service organization makes to its stakeholders) and the system requirements for delivering their services.

Subservice Organization A service organization used by another service organization to perform some of the services provided to user entities.

Carve-Out Method An approach used in SOC 1 or SOC 2 audits when a service organization relies on a subservice organization but does not include that organization's controls in the scope of its own SOC report.

Inclusive Method An approach used in SOC 1 or SOC 2 audits when a service organization includes the controls of a subservice organization within the scope of its own SOC examination. The auditor(s) will therefore evaluate and test the subservice organization's controls as if they were part of the primary organization's system.

> **EXAMPLE**
>
> **A payment processing company (service organization) uses a cloud provider (subservice organization) to store its data. To meet the AICPA Trust Services Security category, the service organization would describe its access and monitoring controls, including data encryption. The subservice organization would describe its network's logical and physical security measures. In combination, these two sets of actions provide protection against unauthorized access and data breaches.**

If the subservice entity is included in the service organization's SOC 2, that organization's management will need to provide an assertion in addition to the originating company's assertions, and the subservice organization's controls will be included and tested where applicable.

Although a SOC 2 report must always describe the services performed and the types of controls of a subservice organization, the specific controls and control objectives may be excluded, or carved-out, if the subservice entity either is not relied on or has its own SOC 2 or similar attestation regarding its controls. In any event, the service organization assumes that **complementary subservice organization controls (CSOCs)** are in place with the subservice organization. The main service organization will describe the subservice organization's responsibility for these controls and note that it can only achieve its control objectives and meet relevant Trust Services Criteria if these controls are in place and, in a Type 2 examination, operating effectively. These controls are included in the System Description, discussed later in this section.

Complementary Subservice Organization Controls (CSOCs) Specific controls that a subservice organization is expected to implement and operate effectively to support the overall system of controls described in a SOC 1 or SOC 2 report. These controls are excluded from testing by auditors using the carve-out method but are included in testing using the inclusive method.

SOC 2 Description Criteria The required elements that a service organization must include when describing its system in a SOC 2 report.

Management Representation Letter A formal, written statement provided by an organization's management to their external auditor at the end of an audit engagement. It confirms that management has fulfilled its responsibilities and provided the auditor with accurate, complete, and truthful information necessary for the audit.

Parts of the Management Assertion

The Management Assertion section of a SOC 2 report concludes with statements of assurance from management that their system description has been prepared in accordance with the AICPA 2018 **SOC 2 Description Criteria**, that the controls have been suitably designed for the period, and, in the case of a Type 2 report, that they are operating effectively in accordance with the Trust Services Criteria. The AICPA Description Criteria are benchmarks that provide guidance for describing a system in terms of its inputs, processing, and outputs.

> **EXAMPLE**
>
> **DC1 requires that the system description include the type of services provided. Implementation guidance provides examples that include customer support and managed services. The guidance notes that the system description needs to clearly identify the services that are the focus of the SOC 2 report.**

As with an SOC 1 report, in a SOC 2 engagement, the service auditor must obtain a **management representation letter** acknowledging responsibility for the services provided and the controls over the system. The representation letter will also affirm that the system description is appropriate and control objectives are designed to meet the Trust Services Criteria. In a Type 2 report, management will also assert that the controls are operating effectively. Finally, the management assertion letter confirms that management has provided the service auditor with all relevant information.

Now that we have examined the Management Assertions section, we will look more closely at the rest of the SOC 2 report.

Independent Service Auditor Report

Section 2 of the SOC 2 report is the independent service auditor's report. The information presented in this section includes a description of the scope of the assessment, the relative responsibilities of the service organization and the service auditor, and the auditor's opinion (unqualified, qualified, adverse, disclaimer) (Figure 8-6).

The service auditor is responsible for examining the service organization's system in accordance with AICPA attestation standards. These standards require an examination that will provide the

Figure 8-6. Auditor Opinions

service auditor with reasonable assurance of two things: 1) that management's system description is consistent with description criteria and 2) that the identified controls are designed (Type 1) and operating effectively (Type 2) to ensure that the service organization's service commitments and system requirements are achieved via the applicable Trust Services Criteria. A SOC 2 engagement does not need to be based on all Trust Services categories. While all SOC 2 engagements include Security, the scope may also include any or all of the following:

- Confidentiality
- Integrity
- Availability
- Privacy

The service auditor report will detail the steps taken in performing the examination of the client's controls. An example is that the service auditor garnered an understanding of the system and its requirements, and they performed procedures to determine that the description presented matched the description criteria. This section of the report also includes any inherent limitations, such as scope limitations or future changes, that may impair the effectiveness of controls. The final section of the service auditor report notes its restricted use. SOC 2 reports (unlike SOC 3 reports) are not intended for the general public. They are restricted to third-party organizations that rely on the organization's controls for their own business.

System Description

The service organization's management team prepares this section of the report, which includes a description of the types of services provided, the principal service commitments and system requirements, the components of the system providing the services, identified significant system incidents, the applicable Trust Services Criteria and related controls, **complementary user entity controls (CUECs)**, subservice organizations, and changes to the processing system. We will discuss each of these items briefly here.

Complementary User Entity Controls (CUECs) Controls that a service organization assumes its customers (user entities) will implement for the service organization's system of controls to operate effectively.

Provided Services

This section of the report briefly describes the service organization's general business and the specific processing system being assessed. The report notes that the policies, procedures, and controls evaluated are limited to the specific processing system identified.

Principal Service Commitments and System Requirements

Service commitments are obligations identified in contracts with stakeholders. They explain the service being offered and provide some assurance about the system's security and availability. They also define how the service organization will protect stakeholder data. Service contracts describe how the organization invoices customers and handles payments. The report may discuss any cybersecurity frameworks that have been adopted (e.g., NIST CSF 2.0) and provide additional detailed information about specific policies and procedures in place to protect data.

System Components Used to Provide Services

This section tends to be the most extensive part of the System Description, as it details all the elements of IT infrastructure, software, people, and procedures. In describing IT infrastructure, the service organization will again identify any subservice organizations, such as those providing cloud hosting services. The software component includes all programs and applications supporting the system, including security software such as antimalware and patching software/policies. The people identified here are those who have roles supporting the processing system being evaluated. They include managers, engineers, security personnel, HR personnel, and software product managers. Policies and procedures are an important and complex system component that contributes to the operation of an information processing system. They include those related to cybersecurity, personnel, data handling, risk assessment and management, logical and physical access, and incident response and recovery.

Applicable Trust Services and Controls

As mentioned, a SOC 2 assessment does not have to include all of the Trust Services Criteria, but each engagement must include security and may also include one or more of the other categories. This section of the report identifies the included categories and their related controls. For example, if availability criteria are included, the report will state that the controls have been designed to ensure that information and information systems are available to meet the service organization's objectives. In the case of a Type 2 engagement, the report will also state that the controls operate effectively for this purpose and will show the results over the testing period.

Complementary User Entity Controls (CUECs)

As the word *complementary* implies, these controls work in conjunction with the controls of the service organization to provide assurance that the information and systems are designed, exist, and, in the case of Type 2 audits, are operating effectively to achieve the service organization's objectives and commitments relative to the selected Trust Services Criteria category. The user entity is relying on the service organization's controls, but the same must be true with respect to the service organization's reliance on the user entity. The following example illustrates this relationship.

> **EXAMPLE**
>
> **AAA Company is a payroll processing company obtaining a SOC 2 report for specific clients. These clients share their payroll data with AAA. AAA is the service organization, and the clients are user entities. AAA needs to provide assurance that it is protecting client data in its systems, but the clients must also demonstrate that they perform control activities, such as using encryption when transmitting data to AAA.**

System Incidents

Management will catalogue any significant cybersecurity incidents that resulted from insufficiently designed or ineffective controls, or that prevented the organization from achieving its service commitments or requirements. These incidents could be a disruption of service (lack of availability), failed or successful attempts to gain unauthorized data access, or a data privacy protection failure. Virtually all organizations are continuously bombarded with phishing and other malware, but this section of the report describes only those incidents that are deemed to be significant in that they resulted in a failure of some type or indicated a failure of controls. Users should not conflate this section with the **SEC cybersecurity breach disclosure rules** described in Chapter 9.

SEC Cybersecurity Breach Disclosure Rules SEC guidelines requiring public companies to disclose their cybersecurity risk management, strategy, governance, and incidents.

Subservice Organizations

The Management Assertion portion of the SOC 2 report describes any subservice organizations that the service organization uses. As noted in the Management Assertion description, these entities may be included in the review or excluded via a carve-out. The System Description will describe any complementary subservice organization controls (CSOCs) that the subservice organization is expected to have that ensure it achieves its service commitments and system objectives. The service organization will describe how it ensures that the CSOCs are in place and how those controls are verified at the subservice organization. For example, the service organization may require independent third-party reviews of these subservice entities.

Changes to the Processing System

This final section of the System Description briefly notes any changes to the system during the period under audit that would impact the ability to comply with service commitments and meet system objectives.

Trust Services Criteria, Related Controls, Tests of Controls, and Test Results

This final required section of the SOC 2 report may be considered the meat or the heart of the report. The service auditor completes this section, which is typically presented in a table format, as shown in Table 8-5. Regardless of the engagement, the report will include the auditor's review of the design of controls and test for existence and, in the case of a Type 2 report, the tests of the controls' effectiveness and test findings. This section of the report is likely the lengthiest and can run to dozens of pages. The **Testing Matrix** may show that the control is in place and working effectively or note any exceptions. The test results will determine the type of opinion the service auditor will give. Table 8-5 also serves as a good example of the Testing Matrix for a SOC 2 report.

Testing Matrix A structured tool used in audits, risk assessments, or compliance reviews to organize and track the testing of internal controls. It helps auditors or control testers clearly map what is being tested, how it's being tested, and what the results are.

Table 8-5. Sample Testing Matrix Section of a SOC 2 Report

Trust Services Criteria	Control Activity	Service Auditor Tests	Test Results
CONTROL ENVIRONMENT			
CC1.1 The entity demonstrates a commitment to integrity and ethical values.	The company has a code of conduct.	The auditor examines the standards of conduct and tests to ensure they are understood.	No exceptions noted.
CC1.2 The Board of Directors demonstrates independence from management and exercises oversight of the development and performance of internal controls.	The Board of Directors has among them the necessary expertise to be able to operate effectively independent of management.	Obtain the resumes of board members and evaluate levels of expertise in functions such as finance, IT, and marketing.	One new board member's résumé has not been reviewed. Since discovery, management has reviewed this résumé and formally noted their expertise level in cybersecurity.
INFORMATION & COMMUNICATION			
CC2.2 The entity internally communicates information, including objectives and responsibilities for internal control, necessary to support the functioning of internal control.	The service organization provides annual cybersecurity training for all employees.	Evaluate training content and check employee participation in training.	No exceptions noted with respect to training content. 95% of employees completed annual training.

Other Information Provided by Management

This final section of the SOC 2 report is discretionary. The service organization's management may use it to respond to any exceptions the service auditor has noted in their review of system control design and effectiveness. For example, in the report example provided in Table 8-5, management may wish to explain why not all employees received annual **cybersecurity awareness training** and how they might mitigate this exception going forward. Management may also use this section of the report to describe any future plans relevant to system organization controls, changes planned

Cybersecurity Awareness Training Education programs designed to teach employees and users how to protect themselves and the organization from cyber threats.

for policies and procedures, or plans to implement new software or hardware, along with controls over those new systems.

SOC 2+ Engagements

In 2024, the AICPA introduced a SOC 2+ framework. This new service offering extended the SOC 2 report to be useful for assessed entities wishing to avoid reporting redundancy in meeting regulatory requirements. The goal of the framework is to reduce an organization's compliance burden. SOC 2+ reports are basically the same as a SOC 2 report, except they include additional control criteria beyond those in the AICPA Trust Services Criteria. The control criteria within SOC 2+ align with those of several specialized industry requirements, such as the NIST CSF or SP 800-53, the HITRUST common security framework (used primarily for healthcare entities), and the EU's GDPR requirements.

> **EXAMPLE**
>
> **To map SOC 2+ to NIST SP 800-53, the organization needs to identify controls that are common to AICPA Trust Services Criteria and the NIST guidance. For example, SOC 2 Privacy criteria map to the Privacy NIST control family. Overlaps between the two frameworks can be used to reduce redundancy. A mapping document will show the links and overlaps and demonstrate compliance in a SOC 2+ engagement.**

Using SOC 2 Reports

There are three perspectives to a SOC 2 engagement. A service organization or a third-party organization that maintains, stores, or handles customer data will prepare a report with the help of a service auditor. The organizations that utilize the service organization will receive the report, and the CISO or other information security analyst should evaluate it carefully to see that it meets their organization's needs. Figure 8-7 shows the typical steps in this review process.

The first thing a SOC 2 report user should do is verify that the auditor preparing the report is a CPA. You can do this by checking the www.CPAverify.org database, maintained by the National Association of State Boards of Accountancy. As with any product, the quality of SOC reports may vary, so the user might also check on the CPA firm's reputation if possible.

As noted earlier, a SOC 2 report does not need to include every Trust Services Criteria category, so the user will next check to see which additional categories beyond Security the report

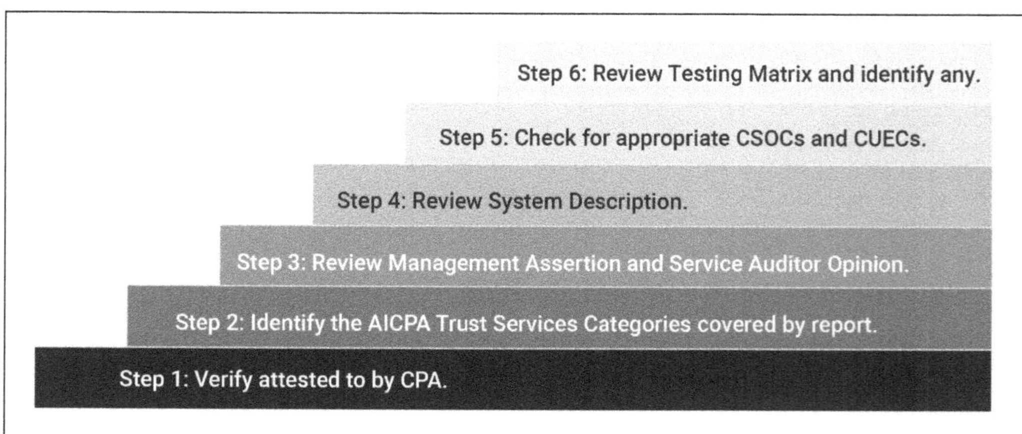

Figure 8-7. Steps in Evaluating a SOC 2 Report

covers. If the third party is holding and processing sensitive data, such as customer PII, the privacy and confidentiality categories will likely be of most concern.

The next step is a review of the Management Assertion section of the report. The user will check to see if the statements contained in this section vary from typical boilerplate language with respect to their assertions. The user will also check the Service Auditor Opinion to be sure it is unqualified. Qualified or adverse opinions from the auditor may require that the organization review its exposure from this vendor.

Reviewing the System Description will be one of the most time-consuming parts of the user report review. Security staff who review many SOC 2 reports are experienced in knowing how to spot red flags (such as system interface links to other parties that are not covered in the report) or any items in the description that do not align with expectations; for example, some interface links that are not using encryption. If the user encounters any discrepancies or uncertainties in the System Description, they may ask additional questions of the Service Auditor.

The user needs to identify all subservice organizations and make sure that all expected controls, such as maintaining cybersecurity insurance, and CSOCs are in place. A weakness in a subservice organization can be just as catastrophic as weak controls in the entity obtaining the report. At this time, the user will also review their own CUECs to be sure they meet control objectives. For example, the user entity might be expected to have a certain level of role-based access control. In some respects, the CUECs provide an opportunity for the user entity to review its own cybersecurity controls and risk management.

Step 6 in the SOC 2 report process is to review the Testing Matrix and look for any exceptions noted by the auditors. In the case of exceptions, there should be a management response noting that the exception has been resolved.

Finally, management has the discretion to add any information, such as planned security enhancements. Once the user is satisfied with the SOC 2 report, it becomes assurance evidence for the user organization.

Summary

This chapter explained COSO, which provides the foundation for AICPA Trust Services Criteria. The evolution of Trust Services and SOC reporting began with SAS 70, SSAE 16, and SSAE 18, which provide authoritative standards from the AICPA for attestation engagements. The next section of the chapter discussed the relationship between COSO and AICPA Trust Services Criteria, which are the foundation for SOC 2 engagements. The AICPA Trust Services Criteria help organizations to ensure the confidentiality, integrity, availability, privacy, and security of the data they maintain and process. Although the chapter's focus is primarily on SOC 2 (and SOC 2+) reports, we also discuss the full suite of SOC reports, including SOC 1, SOC 3, SOC for Cybersecurity, and SOC for Supply Chain. The chapter goes into detail in discussing SOC 2 engagements and each section of a SOC 2 report. While CPAs prepare SOC reports, users of the reports come from a variety of backgrounds. The final section of the chapter explains how to peruse a SOC 2 report.

Chapter Review

Key Terms

Assurance
Attestation
Bridge Letter
Carve-Out Method
CIA Triad
Complementary User Entity Controls (CUECs)
Complementary Subservice Organization Controls (CSOCs)
Control Criteria
Cybersecurity Awareness Training
Enterprise Risk Management
Financial Reporting
Inclusive Method
Internal Controls

Internal Control Over Financial Reporting (ICFR)
International Federation of Accountants (IFA)
Management Representation Letter
Opinion (Auditor)
Sarbanes-Oxley Act (SOX)
SEC Cybersecurity Breach Disclosure Rules
Service Auditor
SOC (System and Organization Controls) Engagement
SOC 2 Description Criteria
SOC 2 Readiness Assessment
Software as a Service (SaaS)
SOX Section 404
Statement on Auditing Standards (SAS)
Subsequent Event
Subservice Organization
Supply Chain
Testing Matrix

Review Questions

1. What are SOC reports? Why can they only be issued by a CPA?
2. How do COSO and the AICPA Trust Services Criteria relate to SOC reporting?
3. Compare and contrast SOC 1, SOC 2, and SOC 3 reports.
4. Explain the difference between Type 1 and Type 2 reports.
5. Why did the AICPA add SOC for Cybersecurity and SOC for Supply Chain reports to the original three reports?
6. How do the SOC for Cybersecurity and SOC for Supply Chain reports differ from an SOC 2 report?
7. List and describe the main sections of a SOC 2 report.
8. What does an auditor or business professional need to know about SOC reports and their impact on business and business decisions?
9. What are CSOCs and CUECs, and how do they work together with a service organization's controls?

Multiple Choice

Identify the choice that best completes the statement or answers the question.

1. Which of the following SOC reports does not primarily relate to cybersecurity?
   a. SOC 3 Report
   b. SOC 1 Report
   c. SOC for Cybersecurity
   d. SOC 2 Report
   e. SOC for Supply Chain
2. What is a qualified opinion in a SOC report?
   a. Indicates that all controls exceed the required standards
   b. An opinion given without any reservations
   c. Indicates total failure of control systems
   d. Indicates partial failure with some controls materially ineffective
   e. None of the above
3. What does an adverse opinion in a SOC report signify?
   a. The controls are both improperly designed and ineffective to the point of unreliability.
   b. The controls are properly implemented but poorly designed.
   c. The controls are functioning perfectly with no need for improvement.
   d. The organization has provided all necessary information satisfactorily.
   e. The controls are designed well but not properly implemented.
4. What is the main difference between SOC 2 and SOC 3 reports?
   a. SOC 2 reports are prepared for the public, while SOC 3 reports are restricted.
   b. SOC 3 reports require more frequent auditing than SOC 2 reports.
   c. SOC 2 reports focus on financial controls, while SOC 3 reports focus on operational controls.
   d. SOC 2 reports are restricted to specific stakeholders, whereas SOC 3 reports are for general public use.
   e. SOC 3 reports are more detailed than SOC 2 reports.
5. Which of the following is **not** included as a supplemental criterion under the AICPA Trust Services Criteria?
   a. System operations
   b. Risk mitigation
   c. Marketing strategies
   d. Change management
   e. Logical and physical access
6. What does the COSO framework primarily focus on?
   a. Environmental impact assessments
   b. Corporate governance and shareholder relations
   c. Data privacy laws and regulations
   d. International financial reporting standards
   e. Internal control systems and their effectiveness
7. What does an unqualified opinion in a SOC report indicate?
   a. The controls are designed and operating as intended, but improvements are possible.
   b. The service auditor has found minor issues that do not affect the overall reliability of the controls.
   c. The service auditor has found pervasive and material misstatements.
   d. The organization has major flaws in its audit.
   e. None of the above.

8. What is the main difference between SOC 2 Type 1 and Type 2 reports?
   a. Type 1 reports take longer to complete than Type 2 reports.
   b. Type 1 reports assess design and implementation as of a specific date, whereas Type 2 reports evaluate control effectiveness over time.
   c. Type 2 reports are less detailed than Type 1 reports.
   d. Type 1 reports are cheaper than Type 2 reports.
   e. Type 1 reports are only for service organizations, while Type 2 reports are for all types of companies.

Case Studies

Moore Printing—SOC 2 Engagement—Part I

Natalie Magee has now been working at Herron Consulting LLP for two years, and she was recently promoted to Senior Associate. She passed the CPA exam during her first few months at Herron and is considering pursuing other credentials.

Natlie knows that SOC engagements are a significant consulting service line at Herron, and she is eager to work on a SOC 2 project. Her boss, Chris Mason, thinks this is a good idea for her and has decided to pair her with a manager, Kara Johnson, on a new SOC 2 engagement for Moore Printing.

Moore Printing is a small Software as a Service (SaaS) company that offers clients printing and photo services. The company has been in business for just two years but has quickly grown to four offices and now has several large industrial clients. Some of its clients use Moore for outsourced printing of customer proposals and other proprietary documents. Moore hopes to continue its expansion but recognizes that its customers may have some concerns about having Moore host their data and photographs without some assurance that it has appropriate cybersecurity controls in place to safeguard their data. The company has engaged Herron Consulting to provide it with a SOC 2 Type 1 report.

Kara and Natalie are having their first meeting to discuss this engagement.

Case Requirements

1. Is a SOC 2 Type 1 report the best option for Moore Printing? Why or why not? Explain your answer.
2. As they prepare to meet with their client, what should Kara and Natalie tell the client with respect to the division of responsibilities between management and the service auditors?
3. What data/information will Kara and Natalie need to collect from their client as they conduct their engagement?

Moore Printing—SOC 2 Engagement—Part II

Assume that it is one year following the first SOC 2 Type 1 report engagement that Natalie worked on with Kara. Natalie now has completed several SOC engagements, including a SOC for Cybersecurity and a SOC for Supply Chain project. Moore Printing believes it is now ready to contract for a SOC 2 Type 2 report and will include Security and Confidentiality as part of the engagement.

Case Requirements

1. How will the SOC report change in this new engagement?
2. Referencing AICPA Trust Services Criteria (available for download at the AICPA/CIMA website with registration), describe three possible auditor tests Natalie could perform to demonstrate that Moore has effective controls in place for "CC1.1 The entity demonstrates a commitment to integrity and ethical values."
3. Referencing AICPA Trust Services Criteria (available for download at the AICPA/CIMA website with registration), what are the additional criteria to be tested for Confidentiality? Why might they be important for Moore's customers?

Exercises

8-1. Both SOC 1 and SOC 2 reports have been used to provide assurance regarding third parties for many years. Required:
   1. Compare and contrast four differences between these two reports (e.g., purpose and audience).
   2. Where in a business school curriculum would business or accounting school students typically learn about each of these two types of reports?
   3. Identify the roles within an organization that would typically interact with each of these reports (e.g., internal auditor or cybersecurity analyst).

8-2. Using online resources, find an example of a SOC 3 report that has been made publicly available. Required:
   1. Which categories of the AICPA Trust Services Criteria does the report address? Why do you think they were chosen?
   2. Does the company have any subservice organizations and complementary subservice organization controls (CSOCs)? If so, describe three of these controls and the category of Trust Services Criteria they address.
   3. Does the report include any complementary user entity controls (CUECs)? If so, describe three of them.

8-3. Visit the AICPA/CIMA website and download a copy of *AICPA 2017 Trust Services Criteria for Security, Availability, Processing Integrity, Confidentiality, and Privacy (with Revised Points of Focus—2022)*. Required:

1. Find CC6.1 in the document under Logical and Physical Access Controls. Choose two Ponts of Focus and describe how a service organization or vendor would address each point with two or more specific controls. (e.g., use of MFA to authenticate users).
2. Within the Trust Services Criteria, you will find Additional Points of Focus for specific categories. Perusing the document, identify two COSO Principles (CC) where there are Additional Points of Focus for Privacy. Describe them. Why do you think they are important points for addressing Privacy?

References

American Institute of Certified Public Accountants (AICPA). (1992). Internal Control-Integrated Framework.

American Institute of Certified Public Accountants (AICPA). (2017, April 15). *Description criteria for management's description of the entity's cybersecurity risk management program.* https://assets.ctfassets.net/rb9cd-njh59cm/5Xsf3QaCNRgklTyA3FJPQt/157c8046f90c-97f6ee0dff458db04a21/description-criteria.pdf

American Institute of Certified Public Accountants (AICPA). (2022). *2017 Trust Services Criteria for security, availability, processing integrity, confidentiality, and privacy (with revised points of focus—2022).* https://www.aicpa-cima.com/resources/download/2017-trust-services-criteria-with-revised-points-of-focus-2022

Coca-Cola Company. (2024). *FORM 10-K.* US Securities and Exchange Commission. https://www.sec.gov/Archives/edgar/data/21344/000002134424000009/ko-20231231.htm

Committee of Sponsoring Organizations of the Treadway Commission. (2023). *Internal control—Integrated framework.* Retrieved September 26, 2025, from https://www.coso.org/internal-control

McAuliffe, S. M. (2020, May 28). 8 steps to prepare for a System and Organization Controls (SOC) report. *Insights.* https://keitercpa.com/blog/prepare-system-and-organization-control-report/

PricewaterhouseCoopers. (2025). *SOC reporting services.* Retrieved September 26, 2025, from https://www.pwc.com/us/en/services/audit-assurance/digital-assurance-transparency/soc-reporting.html

Chapter 9

Cybersecurity Legal and Regulatory Environment

Chapter Outline

US Federal Laws and Regulations
Cybersecurity and US State Laws
International Cooperation
International Cybersecurity Legislation
Prosecuting Cybercrimes

Learning Objectives (LO)

After reading this chapter, you should be able to:

- LO1. Understand the complexity of cybersecurity legislation and regulation.
- LO2. Be familiar with US federal laws against cybercrime.
- LO3. Know about the Cybersecurity Infrastructure and Security Agency (CISA).
- LO4. Be familiar with US state laws against cybercrime.
- LO5. Be familiar with the Budapest Convention and international cybersecurity regulation in select countries.
- LO6. Understand why cybercrimes are difficult to prosecute successfully.

Introduction

The global average cost of a data breach reached $4.44 million in 2024 (IBM, 2025). The aggregate cost to the world may be in the trillions of dollars. Most of the chapters in this book are concerned with preventing or responding to cybercrimes. In this chapter, we look at the legal and regulatory environment. Cybersecurity has an interesting tension with the legal framework in most countries. Citizens expect laws to be enacted to safeguard sensitive data, privacy, and the integrity of computer systems, but the variety of crimes involving digital devices makes it difficult to ensure this protection.

Cybersecurity is a global issue, and cybercrime occurs everywhere in the world. In this chapter, we briefly touch on various international laws, but the main emphasis is on the US legal and regulatory environment. This environment includes a diverse range of federal and state laws, regulations, and industry standards, all working in tandem to protect individuals, organizations, and critical infrastructure from the ever-evolving landscape of cyber threats. Chapter 11 covers privacy legislation and regulation.

US Federal Laws and Regulations

As computers became more and more ubiquitous in organizations, the current US criminal code was found lacking when crimes involved computers. Courts found laws that work well with physical crimes to be inapplicable to computer crimes and thus technically outside the law, possibly because of the lack of any actual physical threat. Over the past 40 years, the US Congress has enacted a variety of legislation to deal with this inadequacy. These laws are found within Title 18 of the United States Code, the US federal criminal code (Figure 9-1).

Part I: Crimes
Part II: Criminal Procedure
Part III: Prisons and Prisoners
Part IV: Correction of Youthful Offenders
Part V: Immunity of Witnesses

Figure 9-1. Title 18 of the US Code

States have enacted their own laws, some more rigorous than federal laws, that make navigating the legal framework difficult for cybersecurity professionals and litigators. Although not all of these laws were originally written with cybersecurity in mind, the cybersecurity laws have been updated, and prosecutors have used them to charge cybercriminals. Legislation directed at privacy protection is covered separately in Chapter 11.

Today, at the US federal level, the Computer Fraud and Abuse Act (CFAA) and the Cybersecurity Information Sharing Act (CISA) (note that this is different from the Cybersecurity and Infrastructure Security Agency [CISA], discussed later) form the backbone of legal measures aimed at curbing cybercrimes and facilitating information sharing. Other laws not specifically targeting computer crimes, such as wire fraud, can also be applied. Wire fraud is often cited in cybercrime cases due to its established legal precedents and familiarity among prosecutors, who may be less certain about newer cybercrime laws. Additionally, state data breach notification laws and sector-specific regulations support the federal legal framework. We will outline here descriptions of the most common US federal laws used to prosecute cybercrimes in chronological order.

Wire Fraud (18 U.S.C. § 1343)

Wire Fraud A criminal act involving the use of telecommunications or the Internet to defraud someone of money or property.

The law governing wire fraud was originally enacted in 1952 and has been updated and amended over the years to incorporate new technologies and techniques. **Wire fraud** is like mail fraud, except that it involves wired and wireless communications that range from telephones to Internet connections. The wire fraud law has been one of the primary tools to prosecute crimes and fraudulent activities that have been conducted using digital devices. Wire fraud is a felony and may include a sentence of up to 20 years in federal prison and a fine that could reach $250,000. Wire fraud against financial institutions may result in longer prison sentences and steeper fines.

Wire fraud has several components. There must be a scheme and criminal intent to defraud, there must be a material misrepresentation or omission, and electronic (wired or wireless) communications must be used to perpetrate fraud. The criminal intent in wire fraud is to extract monetary resources from a victim.

Many crimes today can fall under wire fraud, as the use of phones, email and text messages, social media, and computer networks constitute electronic communication.

CASE IN POINT

Wire fraud prosecutions have a broad range. The 2022 trial of Elizabeth Holmes, founder of the now defunct blood analysis company Theranos, included charges of both conspiracy and fraud. She was convicted and sentenced to more than 11 years in prison. Her conviction included three counts of fraud on individual investors, which used more than $140 million worth of wire transfers, so prosecutors used the wire fraud statute. Another example in which wire fraud was used to prosecute cybercrime is *United States v. Parks, III*. The defendant engaged in a cryptojacking scheme, in which he obtained unauthorized access to cloud computing service providers and used their computer resources to mine

cryptocurrency. Parks was charged with several crimes and ultimately pled guilty to wire fraud in 2024.

Racketeer Influenced and Corrupt Organizations Act (RICO) (18 U.S.C. §§ 1961–1968)

The **Racketeer Influenced and Corrupt Organizations Act (RICO)** is part of the Organized Crime and Control Act of 1970 and is used primarily to combat organized crime and enterprises, individuals, or organizations that engage in a pattern (two or more activities over a 10-year period) of racketeering activities. Although its original target was organized crime enterprises, prosecutors have used the RICO framework to prosecute cybercriminal organizations and individuals engaged in cybercrime. Racketeering related to cybercrime is defined as specific criminal activities by individuals or organizations that engage in computer fraud, wire fraud, identity theft, hacking, and other activities using cyber technologies.

Racketeer Influenced and Corrupt Organizations Act (RICO) A US federal law designed to combat organized crime by allowing prosecution of all individuals involved in a corrupt organization.

To be applied, RICO requires a financial motive and makes it illegal to invest any gains from those illegal activities or to acquire or maintain any interest in organizations that engage in the pattern of racketeering activities. A conviction for violation of the RICO Act requires forfeiture of any ill-gotten gains and potential fines of up to $25,000 and 20 years in prison for *each* count of racketeering.

CASE IN POINT

In 2013, David Ray Camez became the first cybercriminal convicted under the RICO Act. *United States of America v. David Ray Camez* was upheld on appeal in 2016. Camez was one of a group of alleged criminals who operated a website called carder.su. (The su domain extension is for the Soviet Union.) The website was an online marketplace that specialized in identity theft and sales of credit card data. Eventually, several dozen individuals were charged and prosecuted under a variety of laws as part of a criminal alliance that operated the site.

Computer Fraud and Abuse Act (CFAA) (18 U.S.C. § 1030(a))

The **Computer Fraud and Abuse Act (CFAA)** became US law in 1986. The primary purpose of the legislation was to address crimes related to computers and to protect systems, data, and networks from unauthorized use or access. The law has been updated and amended several times (in 1988, 1994, 1996, 2001, and 2008) to reflect new technologies and evolving cyber threats. Prosecutors have used the CFAA to prosecute a wide range of cybercrime activities, including hacking, unauthorized access to computer systems or networks, identity theft, and other computer-related crimes. The following two cases illustrate the act's broad application.

Computer Fraud and Abuse Act (CFAA) A US federal law that criminalizes unauthorized access to computer systems and related offenses.

CASE IN POINT

An early case prosecuted under the CFAA was *US v. Morris*. Robert Morris, a student at Cornell University, created worm malware. It may be that the damage from the worm was more extensive than intended, but thousands of compromised computers were infected multiple times, and they were inoperable until they could be disinfected. Although Morris's defense asserted that he had authorized access to the computers that he infected and he did not intend for the worm to cause so much damage, he was ultimately convicted, and the appeals court upheld the decision.

CASE IN POINT

In 2013, computer programmer and hacktivist Aaron Swartz tragically took his own life after being charged with multiple felony counts, including wire fraud and violation of the CFAA. In 2011, Swartz downloaded more than four million academic papers from the online JSTOR database, which constituted about 80% of JSTOR's library. His motive was

> his belief that these data should be available openly and for free. The prosecution of Aaron Swartz has been controversial, with many feeling that the government went too far in pursuing the case.

Criminal penalties under the CFAA typically range from one- to ten-year prison sentences. The CFAA is mostly about access to digital devices. This includes unauthorized access, exceeding authorized access, access to compromised computers, and unauthorized damage to computers. In addition, the law covers trafficking (selling) passwords. The law is not without controversy, as it does not clearly define some of the terms incorporated in its language. Many of the amendments to the legislation have been aimed at strengthening it by expanding coverage of felonious acts and guilty parties. In some cases, adjudication may narrow the law's reach.

> **CASE IN POINT**
>
> The Supreme Court case *Van Buren v. United States* illustrates the evolution of the CFAA. In this case, a policeman used his patrol car computer to access a database and obtain license plate data that he could sell. He was charged with a violation of the CFAA for exceeding authorized access to the data. The Supreme Court found that he did not exceed his access but rather that he had an inappropriate motive for obtaining this information. This 2021 case served to narrow the scope of the CFAA in prosecuting cases of unauthorized access.

Identity Theft and Assumption Deterrence Act (ITADA)

The Federal Trade Commission enacted ITADA in 1998. The law makes it illegal for someone to transfer or use someone else's personal identification to break the law. The ITADA is not limited to digital identity theft; the law covers any unauthorized use of an individual's identifying information for fraudulent or illegal purposes.

Violation of the identity theft law is a felony and may be punishable by up to 15 years in prison and a financial penalty. Many cybercrimes involve theft of identity, so this law can be used effectively to prosecute digital crimes.

> **CASE IN POINT**
>
> Evgeniy Bogachev has spent many years on the FBI's Most Wanted List for a variety of cybercrimes dating back to 2011. He is alleged to have installed malicious software named "Zeus" on his victims' computers, which allowed him to capture personal information leading to financial fraud. In 2012, Bogachev was indicted by a federal grand jury in Nebraska for racketeering, bank fraud, violation of the CFAA, and conspiracy to violate the ITADA. Two years later, Bogachev was charged in a Pennsylvania federal court on charges that included wire and computer fraud and money laundering.

Gramm-Leach-Bliley Act (GLBA) (15 U.S.C., various sections)

The GLBA is a federal law enacted in 1999 and is also known as the Financial Services Modernization Act. The law addresses the privacy and security of data and financial information held by financial institutions. The purpose of this law was to modernize and deregulate the financial services industry and to provide consumer protection. Table 9-1 describes the law's three provisions. Note that the Privacy Rule is discussed in more detail in Chapter 11.

Enforcement actions under the GLBA typically involve the failure to protect customer PII against cyber threats.

> **CASE IN POINT**
>
> In 2014, Morgan Stanley learned that proprietary customer data that it held was available for sale online. Galen Marsh, a Morgan Stanley financial adviser who was subsequently convicted for the breach, uploaded data from more than 730,000 of Morgan Stanley's

clients to his personal computer. In 2016, Morgan Stanley was fined $1 million by the SEC under the GLBA Safeguard Rule for failing to protect the clients' PII.

Table 9-1. Provisions of the GLBA

GLBA Provisions	Description
Pretexting Prohibition	Financial institutions and others are prohibited from obtaining customer information under false pretenses from financial institutions.
Privacy Rule	Financial institutions must give customers the right to opt out of having their information shared with third parties and must also notify them about how they collect, use, and share nonpublic PII.
Safeguards Rule	Financial institutions must develop, implement, and maintain information risk management to protect customer information from unauthorized use, access, or disclosure.

Prosecutions under GLBA are likely to involve other government agencies, notably the Federal Trade Commission (FTC) and the Securities and Exchange Commission (SEC), as illustrated in the following case.

CASE IN POINT

In 2015, Lifelock, Inc., an identity theft protection company, was required to pay a $100 million financial settlement for failing to comply with a 2010 order to establish a security program to protect the data the company collects from customers. The case was not a direct GLBA violation prosecution, but the settlement's data security provisions indirectly related to GLBA safeguard protections. The data security practices in GLBA align with the FTC's mandate for data security.

In 2021, the Federal Trade Commission made significant updates to the Safeguards Rule, expanding its coverage to a broader range of financial institutions, including nonbanking financial institutions, such as real estate appraisers and credit rating agencies. The rule requires those who the legislation covers to have a cybersecurity program that protects customer information. The program must be documented and appropriate to the nature of the activities of the organization, its size and complexity, and the sensitive nature of the data it handles. In addition, other requirements of the Safeguards Rule include access controls, data inventories, customer information encryption, multifactor authentication, risk assessments, a log of authorized user activity, evaluation of system changes, and secure information disposal.

The updates also require reporting certain data breaches and cyber incidents to the FTC within 30 days if unencrypted information for 500 or more customers has been exfiltrated without authorization. Finally, the amendments noted the importance of ensuring appropriate information security for service providers, including that their contracts with third parties include safeguards. Financial institutions must periodically assess the service providers to ensure that these safeguards are in place and operating.

Homeland Security Act and the Cybersecurity Infrastructure and Security Act

The **Homeland Security Act of 2002** was enacted in response to the terrorist attacks of September 11, 2001. The primary goal of this act was to enhance the security and protection of the United States. The act created the Department of Homeland Security (DHS) and a multitude of rules and regulations that reorganized federal agencies to streamline and improve homeland security efforts. This law relates to cybersecurity by establishing a framework for addressing cybersecurity threats as part of the legal enhancement of homeland security. The Homeland Security Act explicitly recognizes that cybersecurity is a critical aspect of national security and gives legislative authority for cybersecurity enforcement.

Homeland Security Act of 2002 The act that created the Department of Homeland Security (DHS) by consolidating 22 diverse organizations to help prevent terrorist attacks in the United States, reduce the vulnerability of the United States to terrorist attacks, and minimize damage and assist in recovery from attacks that do occur.

Cybersecurity and Infrastructure Security Agency (CISA) A US federal agency responsible for protecting the nation's critical infrastructure from physical and cyber threats.

In 2018, President Donald Trump signed into law the Cybersecurity and Infrastructure Security Agency Act, which created the **Cybersecurity and Infrastructure Security Agency (CISA)** under the Department of Homeland Security.

Among other mandates, the act made CISA responsible for "leading cybersecurity and critical infrastructure security programs, operations, and associated policy." Since its founding, CISA has acted as a liaison between the US government, state government, and private sector entities to protect critical infrastructure, such as the electric grid and transportation. CISA also develops policies and provides information to the private sector regarding cybersecurity risks, vulnerabilities, and incident response. The agency's website (www.cisa.gov) has many programs and resources to keep individuals, organizations, and our world safe from cyber incidents. The site also allows for cyber incident reporting.

> **EXAMPLE**
>
> **Shields Up is a CISA program. The program provides guidance for everyone to be cyber aware and protect themselves online. Shields Up offers guidance and resources for individuals, organizations, and leadership to stay safe in cyberspace. The guidance includes recommended steps for organizations to reduce the likelihood of a cyber intrusion, detect potential intrusions, respond to an incident if it occurs, and be resilient (CISA, n.d.).**

> **CASE IN POINT**
>
> **Leading up to the 2024 US Presidential election, CISA created a program called #Protect2024. The program's mission was to assist election officials and other related parties in protecting against cyber, physical, and operation security risks that might threaten election infrastructure. CISA notes that election hacking is a serious threat posed by nation states and cybercriminals. The CISA website provides information and advice to help election stakeholders ensure a safe and fair election. One example is a web page, available at the time of the election, named Rumor vs. Reality, which sought to help dispel election disinformation narratives.**

Cyber Incident Reporting for Critical Infrastructure Act of 2022 (CIRCIA) A US law that requires critical infrastructure entities to report cyber incidents to the federal government.

As cyber threats to national infrastructure continue to grow and evolve, CISA has gained more authority and responsibilities. In 2022, President Joseph Biden signed into law the **Cyber Incident Reporting for Critical Infrastructure Act of 2022 (CIRCIA)**. This act required CISA to develop and implement regulations that would require certain entities to notify CISA of cyber incidents and any ransomware payments. This new regulatory authority requires several stages before taking effect as a Final Rule. Until that time, CISA is encouraging voluntary reporting of unusual cyber activity or incidents.

SEC Public Company Cybersecurity Risk Management, Strategy, Governance, and Incident Disclosure SEC guidelines requiring public companies to disclose their cybersecurity risk management, strategy, governance, and incidents.

US SEC Disclosure Rules

In July 2023, the US SEC issued a much-anticipated cybersecurity ruling, the **SEC Public Company Cybersecurity Risk Management, Strategy, Governance, and Incident Disclosure**. The SEC's foray into cybersecurity was not completely new, as in 2011 its Division of Corporation Finance had issued interpretive guidance on cybersecurity risk and incident disclosures. Since that time, the commission has issued further amendments, but the 2023 ruling provided more extensive requirements and guidance. The new ruling was needed in part due to the increasing significance of cybersecurity incidents, underreporting of cybersecurity events, and inconsistency in public company disclosure content and timing.

8-K Form Disclosure

The SEC cybersecurity disclosure ruling has two primary provisions. One is a requirement for public companies to file publicly available Form 8-K disclosure of material cybersecurity incidents within four business days of materiality determination. (Chapter 10 discusses materiality in some detail, but, as we have described in other chapters, it means that "it matters.") The rule defines *incidents* as an "unauthorized occurrence" or a series of occurrences that pose a threat to the CIA of the public company's information systems or information within those systems. The disclosure should describe what has happened and the perceived impact on the company's business. The SEC provided some flexibility on the disclosure content, recognizing that the entity may not know much about the incident at such an early date and should avoid giving excessive information that could prove incorrect later.

CASE IN POINT

VF Corporation issued a Form 8-K on December 15, 2023, describing an intrusion it detected on December 13. The form noted that the company had detected unauthorized occurrences and had taken steps to assess and remediate the incident. It also noted that the adversary encrypted some IT systems and stole data. The form recorded that the company was working to bring its systems back online and create workarounds to minimize disruptions. VF was still able to take most orders, but fulfillment was impacted by the incident. The form stated that the "full scope, nature, and impact of the incident are not yet known," and VF could not determine whether it would prove to have material impact on operations.

EXAMPLE

Krispy Kreme Form 8-K Filing—December 11, 2024

In this report, Item 1.05 describes a cybersecurity incident that the company learned about on November 29, 2024. There was unauthorized activity in its IT systems, and after learning of it, the company took steps to address it. The 8-K filing noted that the incident was impacting operations, such as online ordering, but the company also reported that its store deliveries were continuing without delays. The 8-K was filed because the company believed that the incident was reasonably likely to have a material impact on its financial condition in at least the short run, although Krispy Kreme did hold cybersecurity insurance.

A company submitting an initial Form 8-K may need to file subsequent reports as more information becomes available. The affected entity will also need to provide updates and disclosures on its quarterly 10-Q report and annual 10-K report. These disclosures should include material updates, changes, and additions to previous filings. Additionally, these reports need to disclose current and potential material effects on operations and financial condition, remediation progress, and changes to the SEC registrant's cyber policies and procedures resulting from the incident.

CASE IN POINT

In 2024, Debevoise & Plimpton, a large law firm, began tracking Form 8-K cybersecurity incident disclosures. The tracker is a database of the SEC filings and provides links to the SEC's EDGAR filing, along with the date of the filing. It also includes any amendment filings, which are not uncommon, given that the disclosure requirement window is short and the incident victims are continuing to acquire information. Companies that filed a Form 8-K cybersecurity incident disclosure in 2024 included UnitedHealth Group, Inc., Prudential Financial, Inc., Microsoft Corporation, and Bassett Furniture Industries, Inc. (Gesser et al., 2024).

Annual 10-K Disclosure

The second primary requirement of the SEC cybersecurity disclosure ruling is for annual 10-K reporting about cybersecurity risk management, strategy, and cybersecurity governance. The risk management and strategy disclosures in the 10-K report should describe the company's processes in place to assess, identify, and manage material risks from cybersecurity threats such that a reasonable investor could understand them. The risk management disclosure should also describe whether and how processes are integrated into an overall risk management system or processes, as well as any risks that have or likely may have a material impact on strategy, results of operations, or financial condition. Finally, the filer should disclose the use of third parties to identify and manage third-party risk.

> **EXAMPLE**
>
> **NIKE 10-K Filing—May 31, 2024**
>
> **Item 1C. Cybersecurity. This portion of the annual report states: "At NIKE, cybersecurity risk management is an important part of our overall risk management efforts. We have cybersecurity processes, technologies and controls in place to aid in our efforts to assess, identify and manage material risks associated with cybersecurity threats. We assess cybersecurity risk at both the board and management levels." Item 1C includes information about management's role in managing risk, board oversight, and risks from cybersecurity threats. The 10-K notes that no material incidents have occurred to date but that they and third-party relationships may experience one in the future.**

At one time, the SEC had considered requiring disclosure about whether any Board of Directors member had cybersecurity expertise. Instead, the 10-K governance disclosure needs to describe how the SEC registrant's board oversees risks from cyber threats, as well as management's role and expertise in both assessing and managing the material risks from cyber threats. This includes identifying the board committee or subcommittee responsible for oversight, as well as management committees or positions responsible for assessing and managing cyber risks. Cybersecurity expertise of people in these positions should also be disclosed, as well as the processes in place to inform and monitor cyber incidents and their reporting lines.

Cybersecurity and US State Laws

While many cyber incidents cross US state borders, each US state and several US territories have their own laws addressing cybercrime. These laws largely concern breach notification, but they may also protect against computer fraud, election fraud, identity theft, unauthorized access, cyberbullying, and other crimes that involve digital devices. As cyber threats evolve, states frequently enact new legislation to proactively guard against them. An example is that several states created legislation prohibiting specific social media applications (e.g., TikTok) from being downloaded on government-issued devices.

> **CASE IN POINT**
>
> **In 2023, Montana became the first US state to impose a ban on the social media and video app Tik Tok, based on data privacy and national security concerns. Senate Bill 419 was passed, signed by the governor, and then blocked by a federal judge. The law was meant to block mobile application stores from offering the app to Montana residents. The law included possible fines of $10,000 for each violation—that is, every time a user in Montana accessed the app—with an additional $10,000 each day for as long as the violation continued.**

In 1978, Florida became the first state to pass specific cybersecurity legislation, Stat.815.01, the Florida Computer Crimes Act. This law prohibited unauthorized computer access to delete or modify data and damaging computer hardware. The act carried a penalty of up to five years of incarceration and a monetary fine of up to $5,000 or more, depending on the proceeds from the crime. Today, the Florida statutes cover several different types of digital crimes, including cyberstalking, extortion, transmission of pornography, criminal use of PII, and sexting.

Another Florida bill related to cybersecurity is 2024 FL H 275, Offenses Involving Critical

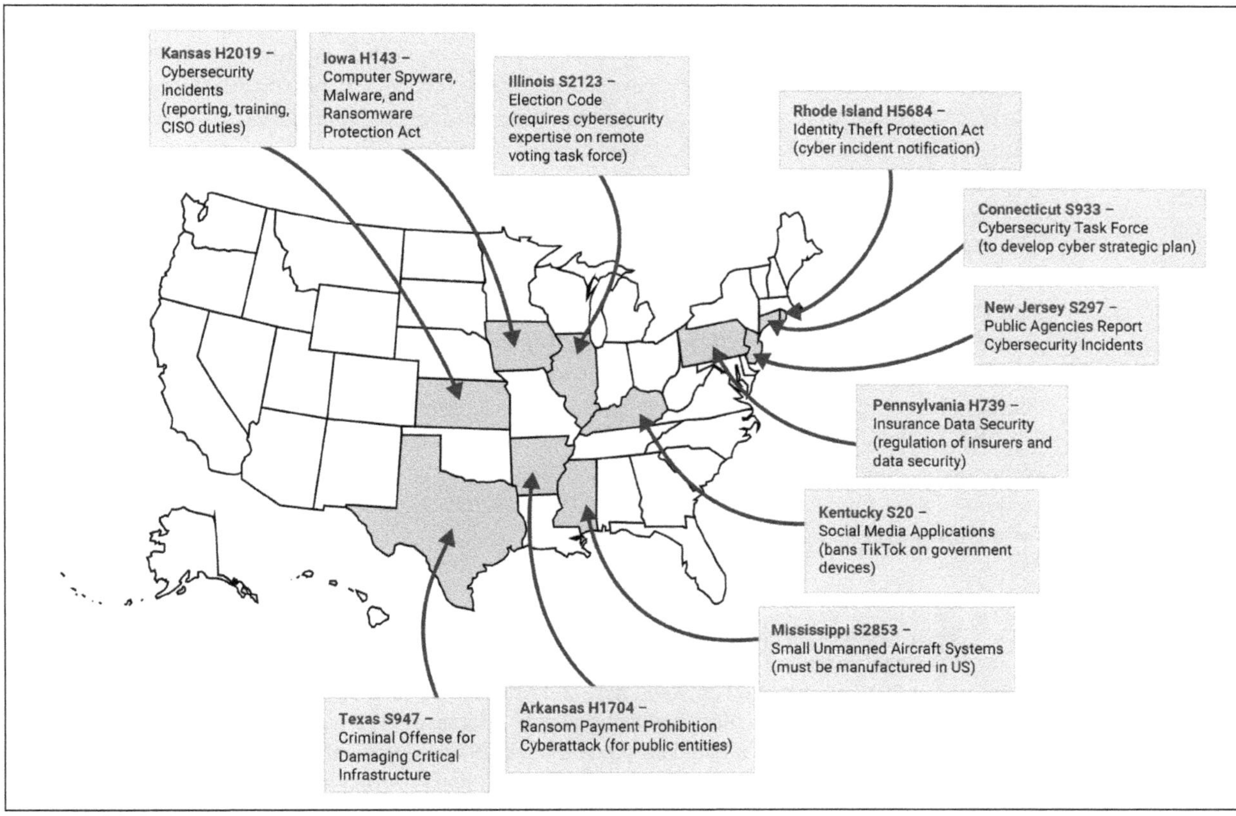

Figure 9-2. Sample of 2023 State Cybersecurity-Related Laws

Infrastructure. This is an example of legislation that is not limited to the use of digital devices to cause harm but encompasses both physical and cyber incidents that target critical infrastructure. The bill provides criminal and civil ramifications for improper tampering with critical infrastructure, and it includes provisions against unauthorized access to electronic devices or networks of critical infrastructure. The bill was signed into law in July 2024.

California is often recognized as a state with substantial legislation protecting against computer crime. Cybercrimes in California include deleting or changing digital data or inserting a computer virus. California laws also make unauthorized taking or copying of computerized data a crime. Other cyber activities prohibited in California are cyberbullying, using a service without permission, and disrupting or denying access to computer services, as with a denial of service (DoS) attack.

Figure 9-2 shows a sampling of state laws enacted in 2023, demonstrating the variety of cybersecurity legislation and its protections or mandated proactive measures.

International Cooperation

Cybercrime frequently crosses international borders. The following are two initiatives aimed at international cooperation in tracking and prosecuting threat actors.

The Budapest Convention

In 2001, the Council of Europe on Cybercrime met in Budapest to create the first international cybercrime treaty. The primary objectives of the participants were to develop a common policy and adopt legislation that would protect member countries from crimes that included

Budapest Convention The first international treaty aimed at combating cybercrime through the coordination of national laws, investigative techniques, and international cooperation.

child pornography, hate crimes, and computer fraud. The **Budapest Convention**, as it is referenced now, was intended to ensure international cooperation so that computer crimes that crossed country borders could be prosecuted effectively in all the countries impacted. As of early 2025, more than 70 countries were parties (ratifiers) of the convention, including Brazil, Japan, Ukraine, and the United States. Neither China nor Russia have ratified the Budapest Convention as of this writing.

As digital technologies and electronic communications have evolved, member parties have updated the Budapest Convention to reflect new challenges. The Second Additional Protocol became open for signature in 2022. The new protocol incorporated updates that increase cooperation in light of the difficulties in obtaining electronic evidence and enforcing laws in foreign jurisdictions. The update included provisions regarding respect of privacy. The new tools included in the protocol include direct cooperation with registrars and service providers and give a legal basis for requesting domain name registration data in foreign territories.

One of the most significant results of the Budapest Convention is its influence on the development of domestic cybercrime legislation in member countries and beyond. The convention requires its parties to make illegal many of the criminal offenses contained within the treaty, such as unauthorized access to computer systems.

CASE IN POINT

The Council of Europe Cybercrime Convention Committee provides many cases of international cooperation resulting from the Budapest Convention treaty. One such case describes a 2013 cooperation between Italy and Norway. Criminal authorities in Norway discovered that a criminal suspect had two Italian IP addresses. The suspect's mobile phone was geolocated in Rome, and Italian officers were able to identify and arrest the suspect, who was subsequently extradited to Norway ("Convention on Cybercrime—Budapest Convention," 2001).

United Nations Convention Against Cybercrime

UN Convention Against Cybercrime A multilateral treaty currently under negotiation with the United Nations. It is intended to create a globally agreed legal framework for combating cybercrime and promoting international cooperation.

The General Assembly of the United Nations (UN) adopted the **UN Convention Against Cybercrime** in December 2024. This convention is a global treaty providing member parties with measures to prevent and fight cybercrime. It also strengthens international cooperation in sharing digital evidence of such crimes.

The convention to date presents a resolution that is open for signature in 2025. While embracing many of the provisions of the convention, the United States noted that they would be unlikely to ratify the convention without more protections of human rights by the other signatories.

Figure 9-3. The General Assembly discusses the UN Convention Against Cybercrime

International Cybersecurity Legislation

Almost every country and national jurisdiction has a law addressing some form of cybercrime. These laws and regulations may be prescriptive, establishing frameworks or accountability for cybersecurity and protections, or they may be more punitive, with fines and other penalties.

> **CASE IN POINT**
> **In 2024, the Ukrainian cyber police arrested three people who were suspected of stealing more than 100 million email and Instagram accounts from users in multiple countries. If found guilty, they could face up to 15 years in prison. The hackers used brute-force attacks to guess user IDs and passwords, which they attempted to sell on the dark web. Because of geopolitical conflicts, Ukraine is continuing to evolve its cybersecurity laws and regulations. Ukraine's current regulation regarding computer crime is the Law of Ukraine on Main Principles of Maintaining Cybersecurity of Ukraine, which was enacted in 2017.**

The following sections describe a few different countries' legislation against cybercrime.

China

China enacted the **Cybersecurity Law of the People's Republic of China** in 2016. The intention of the 2016 law was to both increase data protection and strengthen national security. Chinese data is required to be stored on digital devices within China. The law prohibits unauthorized access to networks and network data. The Cybersecurity Law outlines the requirements for collecting, using, and protecting PII and critical information infrastructure. Penalties for violating the law include suspension of business activities and fines of up to 1 million yuan (~$140,000). Serious violations may result in shutting down a business or suspending a business license.

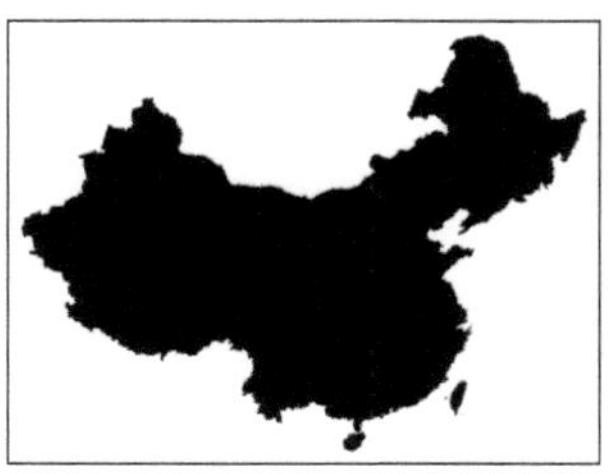

Cybersecurity Law of the People's Republic of China China's foundational law for regulating cybersecurity, data protection, and Internet governance enacted in 2017.

Two additional cybersecurity laws became effective in China in 2021: The Data Security Law and the Personal Information Protection Law. The **Data Security Law** provided guidelines for governing sensitive data classified as national core data and important data. The law imposes significant fines for mishandling national *core* data, with somewhat less severe fines for doing the same with *important* data. Definitions of what constitutes important data and infractions are not clear within the law and may be difficult to follow. The Data Security Law expands on the 2016 law regarding data location and transfer. As with the earlier law, data generated and stored within China needs to remain there, requiring approval for transfer abroad.

China's **Personal Information Protection Law** is a privacy law that covers all Chinese citizens' PII, regardless of where it is collected. Company-collected data require individual consent, except where the data are needed for the public interest. Companies violating the law face significant financial penalties, and there may be ramifications with respect to obtaining financing. The law includes provisions for penalties to individuals. The Personal Information Protection Law also makes it difficult to transfer personal data outside of China, and companies that have large volumes of PII may be required to undergo a security review by the Cyberspace Administration of China prior to transferring any data outside the country.

Data Security Law (China) A Chinese law to regulate data processing activities (including collection, storage, use, processing, transmission, provision, and disclosure of data) with the goals of protecting national security and the public interest, strengthening data security governance, and promoting the development and use of data in a lawful and safe manner.

Personal Information Protection Law (China) China's first comprehensive law specifically focused on protecting personal information.

> **CASE IN POINT**
> **Soon after China's Cybersecurity Law took effect, action was taken against multiple entities deemed to be in violation. One example was an investigation by China's Cyberspace Administration Offices against Internet companies WeChat, Weibo, and Teiga. The companies were penalized and required to remove select users because these users allegedly distributed misleading or inappropriate information as well as information that could threaten national security (Jones Day, 2017).**

European Union (EU)

General Data Protection Regulation (GDPR) (EU) A regulation in EU law governing data protection and privacy for all individuals within the European Union and the European Economic Area. It also addresses the transfer of personal data outside the EU and EEA.

Network and Information Security (NIS2) Directive (EU) A European Union directive aimed at improving the cybersecurity of network and information systems across the EU.

Cybersecurity Act (EU) A law that strengthens the EU Agency for Cybersecurity (ENISA) and establishes a cybersecurity certification framework for products and services.

Digital Operational Resilience Act (DORA) (EU) A law that provides resources to organizations in the financial sector of the EU, specifically in the area of risk management and communications technologies.

The European Union (EU) is well known for its privacy regulation, the **General Data Protection Regulation (GDPR)**, which impacts all countries doing business with member jurisdictions. (GDPR is covered in Chapter 11.) The EU also has specific legislation regarding cybercrime. The European Parliament updated its cyber law in 2022 with the **Network and Information Security (NIS2)** Directive. The NIS2 directive updates an earlier 2016 version and strengthens cybersecurity measures with respect to risk management, corporate accountability, reporting, and business continuity. Elements of the directive vary depending on whether an entity is considered *essential* or *important*. Essential entities are those in which a cyber incident causing system disruption would threaten the country's economy or society in total. Examples are public utilities providing water or energy. Under the directive, entities in critical sectors must implement basic security measures that include risk assessment, use of MFA, and cyber awareness training. Noncompliance with NIS2 can carry both criminal penalties for management and fines. Each member state has flexibility in determining penalties, but the directive requires member states to impose a maximum fine on essential entities of the greater of €10 million or 2% of yearly global revenue. For important companies, the maximum fines are set at the greater of €7 million or 1.4% of yearly global revenue.

In 2019, the EU introduced the **Cybersecurity Act** to help member states protect against and respond to cyberattacks. The act allocated resources to the European Union Agency for Cybersecurity (ENISA) and provided it with more power and a permanent mandate. With the new act, ENISA serves as the core within a single framework for cybersecurity that creates consistency in cybersecurity certifications of products and services and provides a common approach for member states in coping with cyber incidents. For example, the act includes best practices for cybersecurity and steps to take against ransomware attacks.

In 2023, the EU enacted the **Digital Operational Resilience Act (DORA)**, which provides entities in the financial sector with a risk management framework for information and communication technologies (ICT). The framework is mandatory for many types of financial institutions, including banks, insurance companies, investment entities, and trading platforms. These companies must use DORA to strengthen their operational resilience against attacks on their ICT. Requirements within DORA include benchmarks and cybersecurity measures such as penetration and vulnerability testing. As the word "resilience" connotes, DORA emphasizes business continuity, ensuring that financial institutions can quickly respond to and recover from a cyberattack. Reporting, intelligence sharing, and collaboration are also components of the act.

Israel

Protection of Privacy Law (Israel) An Israeli law that regulates the collection, processing, and dissemination of personal data to protect individual privacy.

Israeli Computers Law of 1995 A law that criminalizes unlawful computer-related misconduct, hacking, or making malicious use of computers and computer programs.

Israel is a small country that takes its national security very seriously. They would be expected to have robust cybersecurity laws and regulations, and they do. Israel's **Protection of Privacy Law** dates to 1981 and concerns general privacy as well as privacy over digitized data. Privacy regulations have been updated since that time. The 2017 Protection of Privacy Regulations stipulated the rules for securing computerized databases and expanded on privacy protection for companies with operations in Israel.

The **Israeli Computers Law of 1995** is the country's primary legislation covering computer crimes. The law criminalizes many activities involving computers, including hacking, illegal access to computer material to commit a crime, unauthorized system disruption, data interference, and malicious computer usage. Depending on the nature of the offense, the law carries a prison sentence of three to five years. For example, under Section 3 of the law, a five-year incarceration punishment applies for anyone who passes or stores false information or output or uses software that creates false output.

Consistent with the evolution of computer crime and emerging threats, in 2012 the Israeli Parliament passed an amendment to the 1995 law. The amendment aligned the law with the Budapest Convention on cybercrime and expanded prohibitions on computer-related actions, strengthening its ability to address computer software crimes.

> **CASE IN POINT**
> **Israel has long battled attacks on infrastructure, including an attack on the postal service and a cyberattack against the water controllers for irrigation systems, which occurred in 2023 prior to the onset of the Israeli-Gaza war. In the wake of the war, cyberattacks have increased significantly. Many of the cyberattacks on Israel stem from Iran. One such attack was an AI-generated fake news video that interrupted streaming television programming (Watts, 2024).**

Japan

Japan's 2014 **Basic Act on Cybersecurity** created a general framework for the country's cybersecurity principles, strategy, and responsibilities. The law created a Cybersecurity Strategic Headquarters and a National Center of Incident Readiness and Strategy for Cybersecurity that could respond to cyber emergencies. The law emphasized public-private collaboration and international cooperation that includes sharing information with other countries.

In 2025, Japan passed the **Active Cyber Defense Law**, which amended and extended its prior cybersecurity legislation. The law allows the government to defend against foreign Internet traffic coming into the country across its own IT infrastructure, thus providing a proactive defense against possible cyber incidents. In addition, the new legislation required critical infrastructure operators to report cyber incidents to governmental authorities.

In addition to general cybersecurity laws, Japan has other legislation that can prosecute computer crime: the 1999 **Japanese Act on Prohibition of Unauthorized Access (UCAL)** and the more general 1907 Penal Code. These laws carry penalties for anyone committing a computer-related crime that impacts Japan, regardless of where the crime originates. UCAL, as the name implies, specifically prohibits unauthorized computer access. Hacking and phishing are both examples of unauthorized access, and penalties include up to three years of imprisonment or a fine not to exceed 1 million yen (approximately $7,000).

As with most countries, Japan has had a general penal code that addresses a variety of criminal activities and has evolved over time to deal with computer crimes. For example, the Japanese Penal Code references giving improper commands to a computer that would cause the device to take an action not intended by the user. An example would be a computer virus or being part of a botnet. The Penal Code carries prison terms or a fine. The Penal Code's provisions as they relate to computer crime concern electromagnetic records, which are defined as records used in computer systems. Article 161 of the Penal Code concerns the illegal production and use of these records, and Article 163 is for payment card records. The Penal Code also carries penalties for computer fraud.

Basic Act on Cybersecurity (Japan) Japan's foundational cybersecurity law, enacted in November 2014. It provides the legal and policy framework for protecting Japan's national security, economy, and society against cyber threats.

Active Cyber Defense Law (Japan) An amendment to existing cybersecurity legislation that shifts from caution and restraint to proactive cyber defense. It empowers government agencies to take anticipatory actions against foreign cyber threats while instituting oversight mechanisms to uphold civil liberties and constitutional limits.

Japanese Act on Prohibition of Unauthorized Access (UCAL) A Japanese law that criminalizes unauthorized access to computer systems and networks.

> **CASE IN POINT**
> **A 2020 cyberbullying case gave impetus to an amendment to the Penal Code that strengthened the penalties for this type of crime. The amendment, passed in 2022, provides punishment of up to one year in jail or a fine of 300,000 yen ($2,000) for "online insults." In 2020, Kimura Hana, a professional wrestler and reality TV star, committed suicide after receiving abusive messages on social media. At that time, the Penal Code penalty for cyberbullying was a maximum of 30 days of incarceration and a fine not to exceed 10,000 yen ($70). Many perceived this as inadequate punishment.**

United Kingdom

The United Kingdom (UK) was a member of the European Union (EU) from 1973 until 2020. During that time, data security and cybersecurity were largely governed by EU laws and regulations. Following Brexit, the UK has made some changes to its cyber laws while also maintaining alignment with EU legislation, given the close business relationships and trade with EU countries.

Computer Misuse Act (UK) A law in the United Kingdom that makes unauthorized access to computer systems a criminal offense.

Apart from data protection regulation, which closely aligns with the EU's GDPR, the UK has two primary cybersecurity laws. The first is the 1990 **Computer Misuse Act**. This legislation covers multiple offenses. Unauthorized access to computers, which includes hacking, carries a fine of up to £5,000 ($6,600) and a prison term not to exceed six months. Unauthorized access with criminal intention, such as fraud or theft, has more severe punitive damages of up to five years in prison and no limit on the level of fine. Another criminal offense this law covers is unauthorized modification of computer material, including software or data. The scope of these acts includes computer viruses. The penalty for violation is an unlimited amount of fine and up to five years of incarceration. The law also covers acts that may impair the operation of a computer system.

CASE IN POINT

The UK Computer Misuse Act has been amended many times since its enactment in 1990. It is frequently used to prosecute hackers. According to *The Register*, despite hundreds of successful prosecutions under the Computer Misuse Act, almost all of those convicted did not serve a prison sentence. Fines imposed were typically between about £300 and £500 ($400–$660). A majority of those who did receive prison terms spent only six to nine months in jail (Corfield, 2019).

Network and Information Security Regulations (UK) UK regulations designed to enhance cybersecurity by requiring organizations to improve their network and information systems.

The second government authority over cybercrime is the 2018 **Network and Information Security Regulations**. The Brexit Referendum was approved in 2016, which meant that the EU's Network and Information Security directives no longer applied. The UK undertook to mandate similar regulations through this set of requirements. These regulations impose various computer security requirements for companies including cloud service providers and those offering essential services. Compliance with these regulations is mandatory for the service providers that they cover. Requirements include maintaining an incident response plan, reporting cyber incidents, regular auditing, and risk assessments. Financial penalties can be as much as the greater of £17 million ($22.5 million) or 4% of annual global turnover.

Comparing International Laws

The examples provided describe cybercrime legislation in China, the EU, Israel, Japan, and the UK. Table 9-2 summarizes the key aspects of these countries' cybersecurity laws.

Table 9-2. Summary of Sample Countries' Major Cybercrime Legislation

	China	EU	Israel	Japan	UK
Legislation	Cybersecurity Law of the People's Republic of China (2017); Data Security Law (2021); Personal Information Protection Law (2021)	General Data Protection Regulation (2016); Network and Information Security Directive (NIS2) (2016 and 2022); Cybersecurity Act (2019); Digital Operational Resilience Act (DORA) (2023)	Protection of Privacy Law (1981); Computers Law (1995)	Japanese Act on Prohibition of Unauthorized Access (1999); Basic Act on Cybersecurity (2014); Active Cyber Defense Law (2025)	Computer Misuse Act (1990); Network and Information Security Regulations (2018); EU legislation prior to 2020

Prosecuting Cybercrimes

Prosecution for breaking the law is usually straightforward: a person violates a law (civil or criminal), they are identified and matched with the law they have broken, and they are either required to be imprisoned or pay damages for the crime they have committed, depending on the type of crime. Prosecution under US federal cybersecurity law is different from prosecution under most other laws regarding damages to organizations and individuals in the following ways:

- Cybersecurity laws are part of the US criminal code (with imprisonment being the consequence), whereas typical damage lawsuits are part of civil law (in which the lawbreaker will have to pay fines/damages to whomever they have committed the crime against).
- Unlike most physical attacks, the perpetrator of the attack is typically (and may always be) unknown. The interconnectedness that makes modern computing possible allows for anonymous attacks that span legal and national jurisdictions.
- The perpetrator may be either an individual or a group of organized individuals with either criminal or foreign nation backing.
- In loss litigation, the penalty for violating these laws is typically jail time, in addition to fines large enough that the perpetrator will (probably) not have the means to provide restitution to the victims.

It is not possible to know how many cybercrimes are successfully prosecuted, in part because they are difficult to define and are frequently underreported. Businesses and other entities prefer not to report them because their reputation may be damaged and they may lose third parties' trust. Individuals might not report a computer crime either because they are not sure about where they would do so, or they feel foolish for having been taken advantage of online. This underreporting is part of the impetus behind recent breach disclosure legislation.

It is much easier to catch a bank robber whose image is captured on a closed caption television (CCTV) than it is to track down a sophisticated criminal operating anonymously on the dark web. Hackers have many methods available to hide their identity. Since perpetrators are decentralized,

Figure 9-4. The Difficulty of Identifying Cybercriminals

even when a gang takes credit for a crime, it can be challenging to identify the individual criminals. Criminal gangs may also disappear and then rebrand and relaunch under a new name.

As noted by the Budapest Convention and the UN Convention on Cybercrime, much of computer crime occurs across jurisdictional borders and requires international cooperation to track the criminals and capture them. Even when they are identified, some countries may not wish to extradite them and may even consider them political pawns for negotiating issues with other countries.

Yet another reason why computer crimes are difficult to prosecute is a lack of an integrated legal framework dedicated specifically to these crimes. Digital offenses range broadly from cyberbullying to ransomware attacks to data breaches. Countries have sometimes relied on outdated laws to prosecute these crimes; for example, in the United States, laws against wire fraud, and in the UK, the Computer Misuse Act. Computer crime is constantly evolving, and many illegal activities today involve the use of digital devices in one way or another.

CASE IN POINT

Examples of successful prosecutions of computer crimes resulting in prison time are difficult to find, particularly when the offense crosses international borders. One example of such an outcome is the case of Roman Seleznev, a.k.a. Track2. In 2016, he was sentenced to 27 years in prison for committing credit card theft. In a separate case, in 2017, Mr. Seleznev was sentenced to 14 years of imprisonment for his participation in an organized cybercriminal gang that committed identity theft and bank fraud. He was escaping to Moscow when he was apprehended in the Maldives and extradited to the United States. The criminal counts resulting in Mr. Seleznev's convictions included bank fraud, wire fraud, identity theft, and intentional damage to a protected computer.

CASE IN POINT

In 2024, in an example of how international cooperation against cybercrime can work, agencies in 10 countries worked together in Operation Cronos to arrest and indict members of the Lockbit ransomware gang. This gang is allegedly responsible for extorting more than $120 million globally in ransom payments. Britain's National Crime Agency (NCA) and the FBI led the operation, which led to multiple US arrests, hundreds of frozen cryptocurrency accounts, and two arrests in Poland and Ukraine. Operation Cronos in effect hacked the hackers by infiltrating and controlling Lockbit's dark web site. The takedown also enabled law enforcement to obtain decryption keys that can be used to help victims recover their data.

Summary

This chapter described the complex legal and regulatory structure under which computer crimes are prosecuted. The chapter began with descriptions of laws in the United States and then identified and explained several dedicated US state laws. The United States, like so many other jurisdictions, does not have a single integrated country-wide legal framework covering cybercrime. The chapter also described the legal and regulatory environment in several other countries and two examples of international cooperation, the Budapest Convention and the UN Convention Against Cybercrime. Some countries are known for their robust or evolving legal frameworks. In most cases, the explosion in hacker activity globally is spurring new laws and regulations or amendments to old ones. We discussed several countries/jurisdictions where this is happening, including China, the European Union, Israel, Japan, and the United Kingdom. The chapter concludes with an explanation of the factors that make it difficult to prosecute and convict cybercriminals.

Chapter Review

Key Terms

Active Cyber Defense Law (Japan)
Basic Act on Cybersecurity (Japan)
Budapest Convention
Computer Fraud and Abuse Act (CFAA)
Computer Misuse Act (UK)
Cyber Incident Reporting for Critical Infrastructure Act of 2022 (CIRCIA)
Cybersecurity Act (EU)
Cybersecurity and Infrastructure Security Agency (CISA)
Cybersecurity Law of the People's Republic of China
Data Security Law (China)
Digital Operational Resilience Act (DORA) (EU)
General Data Protection Regulation (GDPR) (EU)
Homeland Security Act of 2002
Israeli Computers Law of 1995
Japanese Act on Prohibition of Unauthorized Access (UCAL)
Network and Information Security (NIS2) Directive (EU)
Network and Information Security Regulations (UK)
Personal Information Protection Law (China)
Protection of Privacy Law (Israel)
Racketeer Influenced and Corrupt Organizations Act (RICO)
SEC Public Company Cybersecurity Risk Management, Strategy, Governance, and Incident Disclosure
UN Convention Against Cybercrime
Wire Fraud

Review Questions

1. What are the challenges in enforcing cybersecurity laws?
2. Why did legislators create new laws for cybercrime?
3. Which of the US cybercrime laws are most commonly used in prosecutions?
4. Describe the US wire fraud law.
5. Describe the RICO Act.
6. Describe the Computer Fraud and Abuse Act.
7. Describe the Gramm-Leach-Bliley Act.
8. Describe the cybercrime portions of the Homeland Security Act of 2002.
9. Describe the 2023 SEC cyberattack disclosure rule.
10. How is the 2023 SEC disclosure rule controversial?
11. Describe the efforts of individual states' cybersecurity laws.
12. What was the Budapest Convention?
13. Describe China's cybersecurity laws.
14. Describe the EU's cybersecurity laws.
15. Describe Japan's cybersecurity laws.
16. Describe Israel's cybersecurity laws.
17. Describe the United Kingdom's cybersecurity laws.
18. What are the issues regarding prosecuting individuals and/or groups related to cybersecurity?

Multiple Choice

Identify the choice that best completes the statement or answers the question.

1. What is a key difference between prosecution under US federal cybersecurity law and commercial damages or breach of contract laws?
   a. Cybersecurity laws do not involve the US criminal code.
   b. Commercial damages are only addressed in international courts.
   c. Cybersecurity laws are part of the US criminal code.
   d. Breach of contract laws provide for harsher penalties than cybersecurity laws.
   e. Commercial damages typically result in the perpetrator being unknown.
2. What complicates the prosecution of cybercrime compared to most physical attacks?
   a. The perpetrator of the attack is typically known and easily traceable.
   b. Cyberattacks cannot cross national or legal jurisdictions.
   c. The interconnectedness of modern computing allows for anonymous attacks across jurisdictions.
   d. Cybercriminals are usually required to provide restitution to the victims.
   e. All cyberattacks are conducted by individuals without any group or national backing.
3. What does the Computer Fraud and Abuse Act (CFAA) primarily address?
   a. Physical theft of computer hardware
   b. Unauthorized access to and use of computer systems and networks
   c. Copyright infringement on software products
   d. Sales of computer components without a license
   e. Installation of legal software on multiple devices
4. What is the main purpose of the Gramm-Leach-Bliley Act (GLBA)?
   a. To solely increase the profits of financial institutions
   b. To deregulate the financial services industry and provide consumer protection
   c. To enforce strict government control over bank transactions
   d. To eliminate all privacy rules within financial institutions
   e. To prosecute all financial advisors who access client data

5. What information must be included in the initial Form 8-K disclosure about a cybersecurity incident, according to the 2023 SEC ruling?
   a. Detailed financial losses from the incident
   b. Names of the hackers involved
   c. Specific technical details about the security breach
   d. Description of the incident and its perceived impact on the business
   e. A complete list of stolen data and compromised systems
6. What is one reason why businesses underreport computer crimes?
   a. The cost of upgrading cybersecurity measures
   b. Fear of reputational damage
   c. High success rate of prosecuting these crimes
   d. Lack of impact on their operations
   e. Legal immunity provided to businesses

Case Studies

Harper Construction—Ransomware Attack

One of Herron Consulting's clients, Harper Construction, recently suffered a ransomware attack. The company did not pay the ransom and immediately reported the crime to the FBI. The perpetrators were traced to a hacker group called the Robbin Spiders, which operates primarily in the United States and the United Kingdom.

Case Requirements

1. Which US laws could be used to prosecute the individual hackers in the Robbin Spiders?
2. Conduct research to find an example of a successful US ransomware prosecution. Describe the case facts.

Capital One—Criminal Prosecution

In 2019, Capital One suffered a data breach in which a threat actor gained unauthorized access and stole PII of customers and credit card applicants. The data stolen included payment history, contact information, credit scores, Social Security numbers, and bank account numbers. More than 100,000 customers were impacted, and at the time the breach was considered to be one of the largest in US history. Thanks to an anonymous email to Capital One, the FBI arrested Paige Thompson a few months later.

Case Requirements

1. How did Paige Thompson access the data?
2. Under which laws was the threat actor prosecuted? What was the result of the criminal case?
3. Do you agree with the result of the court case and the penalty imposed? Provide a rationale for your answer.

Martin Pharmaceuticals, Inc.—Ransomware Attack (Part 1)

Martin Pharmaceuticals, Inc., is a Fortune 500 drug company with about $60 billion in annual revenue. Its primary source of revenue is patents on specialized vaccines, but the company also has several other drug patents, including for drugs related to diabetes and heart health. The company's CEO, Peggy Lynn, has been encouraging the company to develop new revenue streams, and the research and development team has been working on developing a weight loss drug that appears promising.

In 2024, Martin suffered a ransomware attack that paralyzed operations for 10 days while the Incident Response team worked to restore backups following a decision **not** to pay the ransom. Martin's stock price fell 10% in the days following public disclosure of the attack and estimates that supply chain and operations complications cost the company about $20 million daily. Customer and employee data, including Social Security numbers, were exfiltrated, but to date none of this PII has appeared on the dark web. It has also been unclear as to whether IP was stolen, including research on the new weight loss drug. Since the attack, the company has spent $100 million on information system upgrades and cybersecurity governance. Martin has multiple cybersecurity insurance policies, covering extortion, stakeholder notification, and business interruption. The company expects that these policies will cover about 75% of the financial loss from the attack.

As soon as the breach was discovered, Martin contracted with Deloitte, a Big Four public accounting and consulting firm with one of the largest cybersecurity practices globally. The consultants were able to trace the breach to a third-party phishing email that led to stolen credentials of a high-ranking Martin administrator in the IT department. It is believed that the breach went unnoticed for about two months, until the computer systems were frozen, and the attackers demanded $20 million in ransom.

Martin worked closely with the FBI and other law enforcement following the attack. In August 2023, Anastasiya Bagryanov, a Bulgarian national living in the United States, was arrested. Bagryanov is believed to be a member of the hacker group Robbin Spiders, which was responsible for several of the largest ransomware attacks in 2022 and 2023.

Case Requirements

1. Draft a Form 8-K SEC cybersecurity breach disclosure for this event.
2. What law or laws will be used to prosecute Anastasiya, and what punishment do you expect will be levied?

Exercises

9-1. This chapter discussed the SEC Public Company Cybersecurity Risk Management, Strategy, Governance, and Incident Disclosure. Required:
1. While the SEC is considering a ruling, it invites comment letters. The SEC received many comments on the SEC Public Company Cybersecurity Risk Management, Strategy, Governance, and Incident Disclosure Ruling. Search the SEC website for these comment letters. You will note that some of these letters come from individuals, and others are written by representatives of companies, including the Big Four public accounting firms. Find letters from two of the Big Four firms. What are the pros and cons of the rulings they cite in their letters?
2. Conduct research to find a Form 8-K disclosure of an attack and any subsequent 8-K reports updating the original report. What information is included in the report? Do you believe the incident was ultimately material in having an impact on investor decision making?
3. Conduct research to find a public company's cybersecurity risk management and risk strategy and governance in their 10-K report. What information is included for the company you chose? How do you think this information might be helpful to investors and creditors?

9-2. Who Did What? Required:
1. Either individually or in small groups, complete the table to the right by researching and providing information about each of the cybercriminals listed. What did they do? Were they prosecuted, and, if so, under what law(s)?

9-3. The United States has many federal and state laws governing cybercrime and cyber disclosures. Required:
1. There are arguments in favor of a unified US federal law to address cybercrime, though some believe that state laws are needed. Develop a case for a unified law and a case for maintaining state laws. Cite at least three reasons for each position. (Note to Instructors: You many organize this exercise as a classroom debate.)

Cybercriminal	What?	Prosecution?
David Ray Camez		
Albert Gonzalez		
Gummo		
Kimura Hana		
Elizabeth Holmes		
Adrian Lamo		
Galen Marsh		
Gary McKinnon		
Robert Morris		
Kevin Poulsen		
Roman Seleznev		
Steve Wozniak		

References

Convention on Cybercrime—Budapest Convention, January 7, 2001. https://www.coe.int/en/web/conventions/full-list?module=treaty-detail&treatynum=185

Corfield, G. (2019, May 29). Guilty of hacking in the UK? Worry not: Stats show prison is unlikely. *The Register*. https://www.theregister.com/2019/05/29/computer_misuse_act_prosecutions_analysis/

Cybersecurity and Infrastructure Security Agency (CISA). (n.d.). *Shields up: Guidance for organizations*. Retrieved September 26, 2025, from https://www.cisa.gov/shields-guidance-or
ganizations

Gesser, A., Pedersen, B. R., Jacob, J., & Lorch, T. (2024, March 6). Announcing the Debevoise tracker for cybersecurity incident disclosure on Form 8-K. *Data Blog*. https://www.debevoisedatablog.com/2024/03/06/cybersecurity-form-8-k-tracker/

IBM. (2025). *Cost of a Data Breach Report 2025*. https://www.ibm.com/reports/data-breach

Jones Day. (2017, October 12). *China's new cybersecurity law brings enforcement crackdown*. https://www.jonesday.com/en/insights/2017/10/chinas-new-cybersecurity-law-brings-enforcement-crackdown

Watts, C. (2024, February 6). Iran accelerates cyber ops against Israel from chaotic start. *Microsoft on the Issues*. https://blogs.microsoft.com/on-the-issues/2024/02/06/iran-accelerates-cyber-ops-against-israel/

Chapter 10

Accounting for Cybersecurity Damages

Chapter Outline

The Need for Loss Estimates
Classifying Costs Associated with Losses
Direct and Indirect Losses
Tools for Estimating Losses
Measuring Losses Versus Controls with Annualized Loss Expectancy (ALE)
Materiality
Reporting Cyber Losses on Financial Statements

Learning Objectives (LO)

After reading this chapter, you should be able to:

- LO1. Know why it is important to estimate losses from cyber incidents.
- LO2. Distinguish between tangible and intangible costs from cyber incidents.
- LO3. Distinguish between direct and indirect costs from cyber incidents.
- LO4. Know the losses associated with operational, compliance, and reputational cyber risk.
- LO5. Be familiar with tools for estimating cyber losses.
- LO6. Be able to use Annualized Loss Estimation to weigh costs and benefits of risk management.
- LO7. Understand the importance of materiality and the challenges in determining it.
- LO8. Know how cyber losses are presented in financial reporting.

Introduction

Oh no! We have had a cyberattack! How much did we lose? What is this going to cost our company? These are questions not just for the accountants to answer—everyone in an organization needs to assist in estimating the losses. Marketing can help figure out the cost associated with lost sales and harm to reputation. Production can contribute by calculating losses from downtime and supply chain disruptions. Human resources personnel can assist in calculating employee overtime and reallocation costs.

Calculating losses from a cyberattack is an exercise that should take place before an incident occurs so you can allocate resources to cybersecurity. Cybersecurity is expensive, but cyberattacks can bankrupt a company. Determining losses and how much to spend (i.e., cost versus benefit) to mitigate cyber risk is necessary but can be challenging. For example, how much should a company spend to protect against a ransomware attack? Often, the total costs from such an attack are not known until a year or more has passed.

This chapter explains why it is so important to develop loss estimates, including recognizing materiality thresholds. The primary focus of the chapter is on categorizing and identifying various costs associated with cybersecurity intrusions. The more easily identifiable and estimated costs are direct, while indirect costs can be particularly difficult to assess. We discuss these losses using the same three categories used in assessing cyber risk: operational, compliance, and reputational costs. Tools and techniques are available that can be useful in estimating losses, and we discuss a few of these qualitative and quantitative approaches. The chapter concludes with discussion of the financial statement disclosures for cybersecurity events.

The Need for Loss Estimates

Incident An event that compromises the confidentiality, integrity, or availability of an information system or the data it processes.

Breach An incident where unauthorized access to data, applications, services, networks, or devices occurs.

Cyberattacks may be categorized as **incidents** and **breaches**. An incident is an event that impacts the core elements of information security: data confidentiality, integrity, and/or availability. Events are simply activities or occurrences. They are not necessarily bad. Incidents are security events impacting the CIA triad. A breach is an event where unauthorized access is made to data, applications, services, networks, or devices, during which data may or may not be exfiltrated.

For almost two decades, IBM has issued a Data Breach Report, conducted by the Ponemon Institute, detailing the data collected regarding cybersecurity damages. The 2024 communication reports $4.88 million as the global average cost of a data breach, but costs in the United States were much higher, averaging $9.36 million. These losses lead to increased prices for consumers, reduction in market capitalization, possible bankruptcy, and even physical harm. Data in the 2024 IBM report point to an increasing need to invest in cybersecurity and demonstrate how various cybersecurity investments, including in AI, can reduce the cost of an attack (IBM, 2024).

> **CASE IN POINT**
>
> **Many of the increasing cybersecurity breaches in the healthcare industry are ransomware attacks. These attacks cause significant financial loss, but they can also lead to physical harm. In 2023, Ardent Health Services was forced to shut down many of its services that relied on computers because of a cyberattack. The ransomware attack forced hospitals in multiple states to reroute some patients in hospital emergency rooms to other locations. Fortunately, no deaths resulted from the attack, but this case does highlight that ransomware and other malicious cyberattacks can cause physical injury and pose a threat to life (Collier, 2023).**

With the increased cost to businesses and the economy, the need to account for cyber incident costs is paramount and continues to grow. Understanding the overall cost structure of an attack is critical for risk assessment, management, mitigation, and transfer. Being able to show the costs of a breach and any subsequent savings by implementing preventive, detective, and corrective controls regarding cybersecurity allows the organization to justify cybersecurity spending and show due diligence to its stakeholders, including stockholders, management, boards, vendors, customers, governments, and investors.

New demands for transparency with respect to cybersecurity risk management and incident reporting also provide impetus for threat and damage valuations. For example, SEC disclosure laws require reporting of *material* cybersecurity incidents. Determining materiality requires damage estimates, sometimes far into the future. Additionally, organizations need to be able to quantify their losses to involve law enforcement and to report to insurers to file a loss claim. There is no minimum threshold of financial loss that would automatically trigger involvement of federal law enforcement, but federal authorities typically prioritize investigation into attacks impacting critical infrastructure, national security, or substantial financial or data loss.

Organizations holding insurance policies against cybercrime attacks will need to provide loss information to qualify for loss coverage. The details on which types of losses as well as the amount covered would be detailed in the insurance policy. Often organizations will need to get overlapping coverage, as most insurers are unable or unwilling to cover all losses from any specific cybercrime attack, and the losses can be high enough that the insurance company would not be able to cover them. Organizations also need to estimate losses from cyber incidents so they can report them on their balance sheets and income statements.

Classifying Costs Associated with Losses

There are two common ways to classify losses: direct versus indirect or tangible versus intangible. Costs may be easily quantified or difficult to estimate, in which case judgment and perhaps ranges are needed.

Direct costs are directly related to a specific cyber incident. They include the immediate costs of recovery, including consulting and investigation, legal fees and fines, and reporting costs. **Indirect costs**, which are often difficult to quantify, can include reputational damage, decline in stock price, and/or employee turnover. Tangible losses are like direct costs in that they are relatively easy to measure in financial terms. These losses associated with cyberattacks include infrastructure damage, consulting fees, and additional IT employee costs. Intangible losses are similar to indirect costs and are similarly difficult to quantify; for example, the loss of customer loyalty or lower employee morale. Both methods are similar but have subtle differences. As we identify costs associated with cyber incidents, we will use the direct and indirect classification approach in this chapter.

Direct Costs A price that can be directly tied to the production of specific goods or services. A direct cost can be traced to the cost object, which can be a service, product, or department.

Indirect Costs Any cost not directly identified with a single, final cost objective, but identified with two or more final cost objectives or an intermediate cost objective.

Types of Losses

Regardless of whether costs are direct or indirect, they can be related to each of the risk categories identified in Chapter 5 on Risk Assessment. These categories of risk are operational, reputational, and compliance. Cyber losses can and should have a price attached to them, though those that are difficult to quantify may have only a best guess, given limited information. Some of these losses can be directly attributed to a specific cyberattack, but others may be caused indirectly by the same cybersecurity attack. For example, how do you quantify the cost of grant or investment money not obtained because of reputational damage from a cyberattack when there could be other reasons for the failure to obtain the funds?

As stated at the beginning of this chapter, there are several reasons for estimating losses from cyber incidents. This presents an interesting dilemma with the conflicting objectives described in Table 10-1.

Table 10-1. Conflicting Objectives in Estimating Cyber Costs

Conflicting Objective	Description
Reporting	Organizations will want to minimize the cost of a cybersecurity incident either to avoid any legal or regulatory disclosure requirements or to minimize the total material cost to reduce volatility in the overall valuation and reputation of the organization.
Law enforcement	Organizations will want to show higher cost estimates to be able to engage law enforcement to investigate the crime.
Insurance claims	Organizations will want to maximize the cost of the cybersecurity disruption to be able to collect the maximum amount from their insurers.
Resource allocation	Departments within organizations may have conflicting views. Management would like to minimize cybersecurity spending, while the IT/Cybersecurity Department would like to maximize investment to more adequately secure the organization.

Although it would be convenient if all costs were both easily quantifiable and insurable, the reality is that cybersecurity incident costs often rely heavily on estimates and need to be justified with hard data to meet the objectives in Table 10-1. The direct costs of a cybersecurity incident are more easily identifiable, while indirect costs are challenging to determine and estimate and may be disputed_by those paying for them. Table 10-2 is an example of cost classification associated with each type of cybersecurity risk.

Table 10-2. Direct vs. Indirect Cost Classification

	Direct	Indirect
Operational	Lost sales from customers due to supply chain disruption	Website unavailable
Compliance	Fines	Lawsuits from third parties who had data stolen
Reputation	Lost sales from current customers	Low employee morale

The following case demonstrates the complexity of losses stemming from a cybersecurity incident.

CASE IN POINT

In 2017, malware compromised tax software used in Ukraine and caused major disruption to global shipping company Maersk, which has about 800 ships at 76 international ports. Financial losses to the company have been estimated at about $300 million, primarily in lost revenue. The malware infiltrated every part of the IT infrastructure, and employees and contractors spent 10 days rebuilding the network of servers, PCs, and about 2,500 applications. Normal IT operations took several more months. Losses stemmed from idle trucks, spoiled goods, and packages lost in transit. Interestingly, while Maersk suffered short-term reputational damage due to media publicity, its quick response to the attack may have enhanced its reputation in the long run, making it difficult to identify any costs associated with this risk.

Direct and Indirect Losses

In this section, we will describe the direct and indirect costs associated with each cybersecurity risk factor. In some cases, the categorization is ambiguous and may depend on the particular context and circumstances around a specific incident. However, this list does include most of the potential losses associated with an array of attacks, ranging from IP theft to data breach to ransomware attacks.

Operational Losses

Operational Loss A financial loss that arises from the failure of internal processes, people, or systems, or from external events, rather than from market movements or credit risk.

Core business functions may be impaired or disrupted so that the organization is not able to carry out its normal operating activities, such as serving customers or paying suppliers, and that is an **operational loss**. Most losses from a cybersecurity incident are operational. Financial impacts from these risks include expenses associated with getting core business systems and functions back up and running, and any costs associated with protecting the organization from future operational risks. Examples of operating losses from cyber incidents include cyber risk insurance, downtime, shoring up infrastructure, theft of intellectual property, lost revenues, theft of monetary assets, professional fees, ransom payments, and system restoration. We will discuss each type in some detail here.

Cyber Risk Insurance A type of commercial coverage that protects businesses from financial losses caused by data breaches and other types of cyberattacks. It covers computer-related risks that are not covered by your general liability or commercial property insurance.

Cyber Risk Insurance.

As discussed in Chapter 6, one way to manage risk is to transfer it by purchasing **cyber risk insurance**. In the event of a cyber incident, auditors and others will look to the existing cyber insurance policies to determine how much they might mitigate the damage. In that sense, they are a deduction from losses rather than a contributor. They are mentioned here as an operational loss because the cost of cyber risk insurance is likely to rise following a cyberattack, and the victim company is also likely to purchase more coverage. The victim business and others may see premiums go up after an attack, and an attack within an industry is likely to increase demand and cost in that industry.

> **CASE IN POINT**
>
> **Following the Colonial Pipeline ransomware attack, energy companies scrambled to purchase cybersecurity insurance. At the time of the incident, an insurance executive noted that only about half of US pipeline companies had protection. Insurance companies were planning to increase premiums significantly because of the growth in the attacks, but the energy industry was expected to have particularly high costs. The Colonial Pipeline had cyber risk insurance coverage and filed a claim for the ransom payment (Sanicola, 2021).**

Cyber risk insurance is a dynamic product. It has a short history and has evolved along with the threats from which it protects companies. As the cyber threat landscape has grown, insurance prices have escalated, and they may become unaffordable for some of the most vulnerable organizations. A feature of cyber risk insurance that can affect its price and availability is that the buyer must demonstrate that they have some level of security in place to obtain the product. This is akin to a fire insurance policy demanding that you have sprinklers in your home before issuing you a policy. If the insurance purveyor views an entity as a poor risk, they will decline coverage.

There are two types of basic cyber insurance. First-party coverage are policies that cover direct expenses associated with the cyber breach. **Third-party coverage** is for damages or losses claimed by a third party based on the actions of the insured. This insurance is important if a data breach results in privacy violations. Cyber risk insurance policies may cover the direct cost of a breach, including notification costs, data recovery, and consulting fees. Specific policies may pay for ransom demands and business interruption costs. Most policies do not cover intentional acts committed by insiders, future lost profits, or nation-state terrorist cybersecurity attacks.

Third-Party Coverage A type of insurance policy that the insured (first party) purchases from the insurance company (second party) for protection against the claims of another (third party).

Downtime

The most significant loss from downtime may be foregone revenues (discussed later in the chapter), but there are other costs as well. Employees will continue to be paid, regardless of their inability to perform their job functions. If digital devices and/or data are inaccessible, employees cannot do much of their work. Additionally, many workers, particularly in IT, will redeploy their time to work on the cybersecurity incident. To calculate the cost of downtime, you will need to compute the salaried workers' lost time and add to that any hourly workers' wages lost. These computations should also include a percentage cost for fringe benefits, such as health insurance, retirement plan contributions, and withholding taxes. The amount of cost associated with downtime will vary by industry. In manufacturing, if the production line is brought to a halt, the costs can be much higher than for a business that earns revenue by providing services.

Infrastructure

Infrastructure losses from cyberattacks include costs to repair computer hardware and software. These costs focus primarily on direct labor costs, overtime costs, and applied overhead. There can be physical damage or damage that renders devices inoperable for any period. The Stuxnet virus provides an example of physical damage. In that attack, centrifuges were made to spin out of control. The resulting vibrations, if left to continue, would destroy the equipment. Many cyberattacks result in at least temporary inoperability of infrastructure connected to the Internet. Cyber threats to hospitals and utility providers are especially concerning in this respect.

Physical damage to hardware and software can create direct and indirect losses. Direct financial losses can stem from downtime and repair costs. Indirect losses can include physical harm to humans or animals; for example, when ambulances need to divert patients from a hospital undergoing a cyberattack or when patients cannot receive treatment because of a power grid or medical device failure. According to Deloitte, infrastructure costs might include equipment repairs, building temporary infrastructure to handle business while repairs are made, and/or switching operations from one incapacitated infrastructure component to another (Deloitte, 2017).

Following a cyberattack, organizations often upgrade their computer systems to protect against future attacks. The financial costs associated with these investments are another example of losses associated with operational risks.

CASE IN POINT

Following the Equifax cyber breach incident in 2017, in which more than 140 million people had their PII stolen, the company invested heavily in upgrading its information security. While this is typically an investment decision a company may make, in the case of Equifax, these enhancements were partially mandated by its settlement with the Federal Trade Commission. In the FTC settlement, the company was required to implement a comprehensive data security program that included annual risk assessments. According to Equifax's 2019 annual report, the company spent $401.2 million pretax in expenses related to the cyber breach. These costs included expenses for transforming their IT infrastructure and information security.

Intellectual Property Theft

A motive in some cyberattacks is the theft of intellectual property (IP). IP is an intangible asset in that you cannot physically touch it. Examples are copyrights, trademarks, and patents (Figure 10-1). For many businesses, these intangible assets are their most valuable property. Hackers, including nation-states and competitors, may seek to steal these assets through a cyberattack. Valuing the loss from these thefts is difficult because these assets are often carried on the books of the owners at a much lower value than their actual value or the costs incurred in their development. For example, a company that develops a patent will charge research and development costs to expense and may only capitalize (recognize as an asset) the legal fees incurred in obtaining the patent.

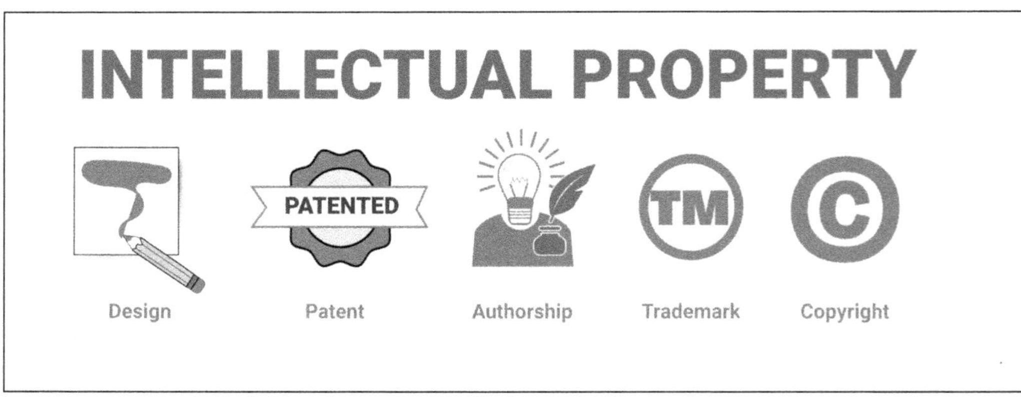

Figure 10-1. Examples of Intellectual Property

With-and-Without Method Also called the "incremental cash flow" or "differential income" method, a common approach in intellectual property (IP) valuation, particularly for patents, trademarks, and technology assets.

Judgment is necessary in valuing the loss from IP theft. Consider the case of a company that developed a COVID vaccine early in the pandemic. The value of having that patent stolen depends on whether competitors develop similarly useful vaccines and if the virus will mutate, making the vaccine less effective. These are difficult scenarios to predict. Methodologies are available that can help accountants in making these loss estimates. One approach is the **with-and-without method** (Valentiam Group, 2021). Using this analytical tool, you would estimate the value of the enterprise with and without the asset in question. What is the enterprise's profitability and net worth with the IP and without it?

Lost Revenue and Profits

The average cost of post-breach activities and business lost from a data breach in the United States in 2023 was estimated to be about $2.8 million. This includes operational disruption, lost customers, and post-breach responses and fines (IBM, 2023). Revenue is lost when disruptions in the supply chain, ordering and billing dysfunction, and website outages occur. The following two cases illustrate disruption impact.

> **CASE IN POINT**
>
> **In August 2023, Clorox filed an 8-K report with the SEC, disclosing that the company had identified unauthorized activity in its IT systems. The cyber incident led to disruption in order fulfillment, which caused product shortages. The SEC disclosure is required when there is a likely material impact stemming from a cyberattack, and in the case of Clorox, the company's November 1, 2023, 10-Q disclosed that "the effects of the cyberattack are expected to negatively impact fiscal year 2024 results" (Clorox Company, 2023 p. 31).**

> **CASE IN POINT**
>
> **MGM Resorts International suffered a cyberattack in September 2023 that disrupted casino and hotel computer systems. Customers shared stories online about not being able to access their hotel rooms and reservation systems being inaccessible online. The disruption lasted for more than a week and was expected to cost the company as much as $100 million, including lost sales revenue. One industry analyst put the daily losses from lost revenue at approximately $8.4 million, with about $1 million less in cash flow each day (Braithwaite, 2023).**

Calculating the amount of revenue lost because of supply chain disruption or website outages can be relatively straightforward as orders received but not fulfilled. You can also estimate the amount of lost revenue by comparing same year-to-year sales and adjusting for any expected increases or decreases due to environmental factors and changes in estimated demand or pricing. It is not easy to determine any losses in revenue from contracts that do not materialize because of the cyber incident. Judgment, assumptions, experience, and economic knowledge are useful in making this valuation. Other financial indicators can help with the valuation, such as unusual patterns or trends in sales to current customers.

Lost profits can stem from reduced revenues and/or increased expenses. Forensic accounting provides several methods that can be used individually or in combination to assist in estimating these losses, based on the circumstances and the judgment of the accountant. Table 10-3 describes four of these, which may be combined if necessary. Choosing an approach depends on factors such as industry and available data.

Table 10-3. Methods for Estimating Lost Profits

Estimation Method	Description
Before and after	Compare the actual profits before the attack to those after the attack.
Yardstick (benchmark)	Compare company profitability to similar companies or to industry averages.
Sales projection	Calculate lost profits based on relevant projections (forecasts) with future growth assumptions and reasonable expense assumptions.
Market share	Compare the market share before the attack to the market share after the attack.

Total lost profits includes both financial and nonfinancial data, like customer visits in person and online.

EXAMPLE

An online retailer experiences a cyber incident. Determining lost sales requires understanding changes in web traffic. The organization will need to maintain and analyze web traffic logs on a regular basis. Accountants can look at the history to calculate the average profit per visit from first time and repeat customers, as well as the probability of selling additional merchandise to new and repeat customers. These profits are lost during downtime. Calculating the profits lost permanently requires comparisons of past annual website visits and purchases with those in a future period. The most challenging calculation is estimating potential new business that is not acquired.

Wire Transfer Scams Scams where attackers trick individuals or businesses into sending money through wire transfers to accounts controlled by the fraudsters.

Monetary Assets

Hackers have many motives in staging cyberattacks. Political activism, revenge, and ego are examples. However, many cyberattacks are about money. The payout might be from a ransom or from the sale of proprietary data, or it could be a redirection of funds. **Wire transfer scams** are a common way for hackers to steal monetary assets.

EXAMPLE

Company XYZ's customer receives an email saying that the company has changed its bank account, and the outstanding account payable should be directed to the new account number. The customer pays their bill, but XYZ does not receive the funds.

In terms of loss valuation, the direct costs of monetary asset theft are easy to calculate. Indirect costs also ensue, however. These include reputational damage, time spent in investigating the incident, and increased employee training and awareness campaigns, along with other increased cybersecurity measures. The following case is an extreme example of this type of cyber incident.

CASE IN POINT

In 2024, deepfake technology persuaded a finance employee at a multinational firm in Hong Kong to pay more than $25 million to scammers. The employee had attended a fake video conference call that was purported to include the company's Chief Financial Officer in Hong Kong. The group call included individuals who appeared to be the employee's colleagues. Deep fake scams are getting increasingly sophisticated as they become AI enabled. This fraud was discovered when the employee checked with the corporate office . . . *after* remitting the payment (Chen & Magramo, 2024).

Professional Fees

During and in the aftermath of a cybersecurity incident, management likely needs the services of professionals, including auditors, consultants, and lawyers. Both external and internal auditors will be involved in assessing losses from the incident, ensuring that appropriate disclosures are made, ensuring compliance with laws and regulations, and perhaps helping the victim company in risk assessment and enhanced internal controls going forward. Losses will impact the financial statements. Not only are these losses included as expenses or loss from the incident on the income statement, but the balance sheet may also need to include contingent or possible losses related to the attack. Audit fees tend to rise following a cyberattack. A 2023 article in *Harvard Business Review* reported that audit fees on average rose about 13.5% following a data breach (Huang et al., 2023).

Many consulting firms specialize in helping clients cope with the impact of a cyberattack. IT specialists and forensic accountants may be called on to identify the source and extent of the intrusion, as well as calculate the damage. Marketing and public relations staff might help calculate the cost of notifications. In-house counsel and outside attorneys are necessary for their input on potential regulatory fines and potential lawsuit costs. Production staff have the best knowledge about disruptions in manufacturing and the supply chain. Even if the victim company has an internal Incident Response Team, it will most likely need the services of incident response consultants to manage the intrusion and disinfect systems. These specialists can also assist in recovering stolen or encrypted data and helping to manage restoring from backup systems. They may even negotiate with the threat actors if they are known.

> **CASE IN POINT**
> **CyberSteward is a cybersecurity company that can provide assistance in the wake of a cyberattack. The services it offers include responding to the incident and managing the breach. In addition, it has expertise in negotiations and reaching settlement agreements with the threat actors. Its website notes: "We engage directly with the cybercriminals, leveraging our advanced threat intelligence and extensive experience and understanding of their tactics to inform and implement outcome-driven strategies and action plans"** (https://cybersteward.com/).

The affected victim firm may hire digital forensic experts to trace the source of the attack and its progression to shore up defenses against future intrusions. Once this is done, consultants can help with communications and public relations for the sake of transparency and to mitigate reputational damage.

Consultants are needed both during an attack and afterward to prevent future cyber incidents. They can analyze threats and vulnerabilities to develop a comprehensive cybersecurity strategy. This includes implementing employee awareness training, yet another cost of a cybersecurity attack. Fees for consultants are hundreds of US dollars per hour, with charges varying with the size of the business, the industry, the nature of the incident, and specific services and expertise required. Consultants may be hired after an incident, but another approach is to pay a retainer or have a subscription to ensure availability of help when needed.

In addition to hiring outside consultants, one of the first actions victims of cyberattacks must take is to contact legal counsel. In-house counsel may consult with lawyers specializing in helping clients comply with regulatory and legal requirements, as well as determine responsibility to customers and other third parties impacted by the event. Lawsuits may ensue from a cyberattack, making the victim company liable to stockholders if a data privacy violation occurs. There also may be litigation associated with broken contracts, such as if they result in unfulfilled orders. If the victim is insured, lawyers can be helpful in filing claims. Legal assistance will also be useful in helping consultants and management in communicating with stakeholders and in managing public relations during and following the incident.

Managing and recovering from a cyberattack is a team effort. Many members of that team are in-house, but management will not have the specialized expertise of professionals who deal with cyberattacks as part of their practice. Consulting with these professionals is costly but necessary.

> **CASE IN POINT**
> **Alston & Bird is a law firm with specialization in cybersecurity services. It has knowledge of global data protection laws and regulations and understands the landscape of cyber threats. It can assist in running tabletop exercises to simulate investigations and responses to threats. The firm advertises that it maintains a Global Privacy and Security Network with attorneys around the world, which is particularly important, given that many cyber incidents stem from outside an organization's headquarters country. It also notes that it has relationships with forensic consultants and can leverage these partnerships for incident response** (https://www.alston.com/).

As with consulting fees, lawyers charge in the hundreds of dollars per hour. These charges depend on the scope and complexity of the incident, the reputation (prestige) of the law firm, and the services needed. For example, outside legal counsel may need to be retained to help with regulatory requirements or in the case of lawsuits stemming from a cyber incident.

Ransom

Most likely the easiest direct cost to calculate resulting from a cybersecurity attack is the cost of any ransom paid. In 2023, the average ransom payment increased from $812,380 to $1,542,333. The median in 2023 was $400,000 (Sophos, 2024). Not all victim enterprises pay the ransom demanded, and many negotiate the payment downward from the original ask. Additionally, some cybersecurity insurance policies mandate negotiations with cybercriminals. Paying ransom creates a **moral hazard** because if no one paid, criminals would stop doing it. However, if a company lacks complete backups of its data, or there are extenuating circumstances that make payment the best option for the company, it will choose to pay. Those circumstances may be the threat of **double extortion**. This occurs when the hackers not only encrypt the data, making them inaccessible to the owners, but also threaten to sell them or make them public. First learning of a ransomware attack can be quite frightening, as Figure 10-2 demonstrates with the computer screen that appeared in a NotPetya ransomware attack.

Moral Hazard The risk that a party insulated from risk may behave differently from how they would if they were fully exposed to the risk.

Double Extortion A ransomware attack where attackers not only encrypt the victim's data but also threaten to release it publicly unless a ransom is paid.

Cryptocurrency A digital or virtual currency that uses cryptography for security and operates independently of a central authority.

Bitcoin A decentralized digital currency that operates without a central bank, based on blockchain technology.

Ransom is paid in **cryptocurrency**, most often in **Bitcoin**. Cryptocurrency, as a pseudonymous, decentralized payment system, has enabled ransomware attacks and cyber ransom demands, which have contributed to the growth of Bitcoin. Thieves are challenged with hiding the assets they take and then spending them. Bitcoin allows cybercriminals to be paid under a fictitious name, and the cryptocurrency can easily transcend international borders. Ransom paid with Bitcoin or other cryptocurrencies will fluctuate with the market value of the currency, but transparent market prices make the direct payment loss valuation easy. Cyber insurance or recovery of a portion of the payment can reduce the loss.

Ransom payments are but one of the costs associated with ransomware attacks. The other direct and indirect costs detailed in this chapter accompany these types of malware intrusions. These include supply chain interruption, lost revenues, and consulting fees. In addition, when ransom is paid, it can negatively impact the organization's reputation. The good news is that ransom can

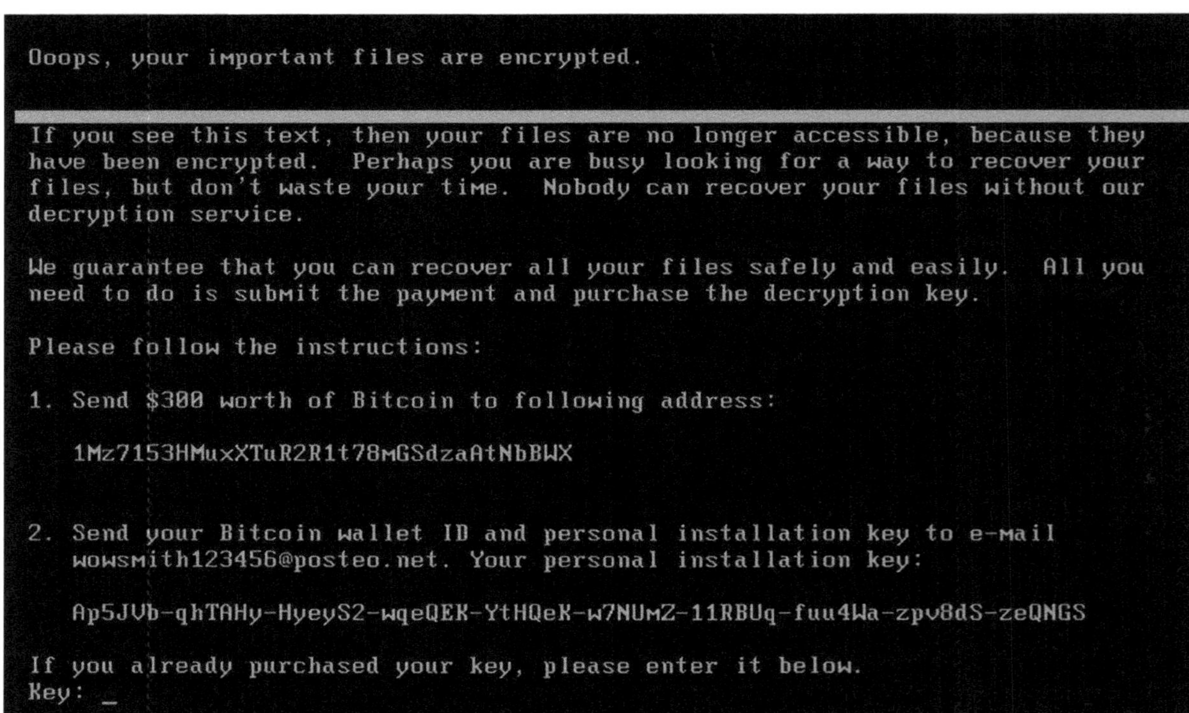

Figure 10-2. Screenshot from a NotPetya Ransomware Attack

be recovered in some cases. Bitcoin transactions, for example, are recorded permanently on an immutable blockchain (discussed in Chapter 2). The true identity of the individual making the transaction may not be tied to the Bitcoin address, but the IP address can be logged, and your transactions are visible on the chain.

> **CASE IN POINT**
> **The Colonial Pipeline ransomware attack is one of the more well-known cyber incidents in which the FBI was able to recover part of the ransom payment by tracing the bitcoin payment. They did this by obtaining a private key to access the Bitcoin account that held the funds. It is not known how they obtained the private key, though it may have been through a compromised VPN password.**

System Restoration

System restoration costs are all costs associated with diagnosing, restoring, checking, testing, data restoration, and so on. This includes the direct expense of hiring a "tiger team" and any internal efforts to restore the system to its original state. Once malware has been detected in a computer network, the infected devices need to be isolated. The steps needed to restore network operations depend on the type of malware. For example, did the malware encrypt files or prevent network access? The most effective way to remove the infection and restore service is to wipe all devices and reinstall all software and data from scratch. This includes reformatting hard disks, patching vulnerabilities, and accessing backup data. The success of the restoration will depend on the comprehensiveness of backups. It is not unusual for a company to find out that backups are inadequate only when a cyberattack occurs.

Several direct costs are associated with system restoration. These include IT staff overtime and consultant fees. System upgrades and additional safeguards to protect against another attack are other direct costs. In addition to the direct costs incurred in fixing the network, indirect costs are incurred as well, including downtime and its effect on operations.

Compliance Costs

Entities that suffer a cybersecurity incident must satisfy many legal and regulatory requirements, which may include national, federal, state, or industry compliance requirements. In the United States, complying with notification and data protection regulations is challenging because there is a lack of federal regulation. Many companies operate in multiple jurisdictions, adding to compliance complexity. The following are some potential **compliance costs**.

Compliance Costs Expenses that a company incurs to adhere to industry regulations and government legislation.

Credit Monitoring/Identity Theft Protection

Credit monitoring and **identity theft protection** are other customer remediation costs when a data breach occurs. Stolen PII and PHI may give hackers the ability to assume a victim individual's identity and transact in their name. Hackers could use the stolen personal data to open credit cards, apply for government benefits, take out loans, or simply assume your persona as an alias. Entities that are hit with a data breach may need to offer those affected free services to protect them from these threats. Some companies offer these services as a public relations measure, while others offer these services as part of a legal case settlement. Hackers can use the stolen data at any time, though credit monitoring and identity theft protection is generally offered for a limited period, such as one year. To calculate the value of the loss, you would look at the cost of these services per individual, per year, and adjust for the number of periods of protection. Affected stakeholders might be advised to freeze their credit and purchase identity theft protection. The protection plans vary in price, with most purveyors offering a range of prices and services.

Credit Monitoring A service that tracks changes in your credit report to alert you of potential fraud or identity theft.

Identity Theft Protection Services and tools that help protect individuals from identity theft, such as credit monitoring and identity recovery assistance.

LifeLock by norton

EXAMPLE

In 2024, LifeLock offered three levels of protection, with costs ranging from about $89 to several hundred dollars annually (https://lifelock.norton.com/).

Fines and Lawsuits

A hacked entity may acquire legal counsel for many different services following a cyberattack. As discussed earlier in the chapter, some of these services may be operational, or necessary to determine what is needed to keep the lights on. Legal counsel specializing in cybersecurity may offer expertise to help their client victims determine priorities and compliance with national, regional, or industry requirements for notifications and disclosures. Legal costs may also stem from use of counsel in negotiations with either the attackers or insurance providers. Legal counsel may also be needed if the cyber incident results in lawsuits, where the victim is either suing or being sued. Hackers can be sued, as the US Computer Fraud and Abuse Act allows for private lawsuits. That said, the identity of the threat actors may not be known, or they may reside in an unfriendly jurisdiction, making it difficult to obtain restitution.

According to IBM's *Cost of a Data Breach Report 2024*, about one-third of organizations were fined following a data breach. The report found an increase in fines from the prior year, with those paying more than $50,000 up 22.7% and those paying in excess of $100,000 increased year over year by 19.5% (IBM, 2024). These fines may be levied for failure to notify third parties or regulators about the breach, or because the victim entity failed to protect the privacy of those whose data was compromised.

CASE IN POINT

A 2018 investigation by a privacy watchdog in the European Union resulted in an $887 million fine imposed on Amazon for alleged violation of the Global Data Protection Regulation (GDPR). The company unsuccessfully appealed the fine, arguing that there was no breach or unlawful data exposure. In a separate case, Amazon did agree to pay $25 million in civil penalties to the US FTC for violating children's privacy regulations by maintaining records of children's voice recordings made on its Alexa digital assistant.

Victims of cyberattacks may sue for damages when their data have been stolen in a breach. The amount of loss per individual's PII stolen depends on the nature of the data stolen. IBM's *Cost of a Data Breach Report 2024* identified customer data as the most common record compromised, with the cost per customer PII record at about $179. Employee PII cost per record was even more expensive, at $189 per record compromised (IBM, 2023).

The outcome and final loss amount in data breach class action lawsuits can be difficult to predict. In many cases, data are stolen, but no harm comes to the individual whose data have been compromised, making it difficult for the plaintiff to prove injury. In other cases, the defendant (entity breached) may be able to show that a third party caused the breach or that there were unusual circumstances despite appropriate security measures. Increasingly, large data breaches result in class action suits, and where harm can be shown, the cost to settle the suit has the potential to be the most expensive cost associated with the breach.

CASE IN POINT

T-Mobile has suffered multiple data breaches. A 2021 breach resulted in a class action settlement of $350 million. The breach exposed the names, addresses, and Social Security numbers of about 76 million customers. The settlement, with payments beginning in April 2025, included cash payments, identity protection, and credit monitoring for individuals who were customers when the breach occurred. An additional aspect of the breach was a requirement for T-Mobile to make significant investment in upgrading cybersecurity.

Notification and Customer Remediation

When a cyber breach occurs, the victim is responsible for notifying key stakeholders. Public companies must file a Form 8-K report with the Securities and Exchange Commission within four days of determining that the breach was material. Federal credit unions must notify the National Credit Union Administration within 72 hours. Many US states require notification of any party whose data has been compromised, often within a period specified as "without unreasonable delay." The cost of these notifications is the hourly wages of each of the individuals involved in the notification process, including a percentage for fringe benefits. (A salaried worker's hourly wage can be calculated from their total cost.) Digital communications are relatively costless, including social media postings, but mailed notifications will include the supplies and postage costs. The compromised enterprise may need to create a website or a call center to handle inquiries and communicate with affected stakeholders.

> **CASE IN POINT**
>
> **Target and Equifax both added capacity to call centers to handle customer inquiries in the aftermath of a cyber data breach. In both cases, many customers had their data compromised by hackers, and the call centers were part of an overall strategy to remediate the crisis. Unfortunately, in both cases, the companies struggled to handle the large volume of calls, and customers were frustrated by long wait times.**

Reputational Loss/Cost

Do you associate luxury with Prada or Porsche? Band-Aid, Toyota, USAA Insurance—all of these are brands people trust. A brand name may be an organization's most valued asset, yet it does not appear on the balance sheet unless it has been purchased, and then it is capitalized as goodwill. It can be exceptionally difficult to value the indirect, intangible loss from damaging an asset that may not even exist in financial reports. Many companies have seen at least a temporary erosion of brand in the wake of a cyber breach. This is especially the case when the brand is one that is trusted to keep data safe. We may decide not to shop at a retailer when we learn that our credit card data have been stolen, or not to trust a cloud computing company that has failed to secure our PII.

There are many consequences to brand damage that have financial implications. These include loss of customer trust, low employee morale, third-party mistrust, reduction in credit rating, and lost opportunities, such as merger and acquisition activity.

> **CASE IN POINT**
>
> **Verizon was set to acquire Yahoo in 2016. In 2017, the planned merger was restructured, with Yahoo receiving a reduced purchase price because of two cyber incidents. The data breaches, which were disclosed in 2016, occurred in 2013 and 2014 and impacted billions of customers. Verizon renegotiated the deal, referencing the breaches as significant adverse events, ultimately paying $350 million less for Yahoo.**

Customer mistrust leads to lost sales from current customers and customer turnover, which has associated customer acquisition costs in addition to lost profits. Financial consequences from low employee morale include employee turnover, employee mistakes, and loss in employee productivity, among others. Third-party mistrust can cause suppliers to no longer wish to work with a company for fear their own data could be compromised. Credit rating reduction stemming from a cyber incident, at a minimum, results in higher borrowing costs. Lost opportunities, such as mergers and acquisitions, can be calculated if the activity is planned, such as in the case of the Verizon–Yahoo acquisition.

CASE IN POINT

In 2019, Moody's reduced Equifax's credit rating from "stable" to "negative." The financial consequences of Equifax's data breach were significant and long-lasting, as class action suits threatened and legal expenses mounted. Moody's also placed SolarWinds' credit ratings on review for downgrading following its 2020 supply chain attack, and S&P reduced their rating for the company from a B+ to a B. These examples demonstrate how weak cybersecurity can lead to increased borrowing costs and may endanger an organization's ability to secure financing.

Decline in Stock Price

Public companies that experience a cybersecurity incident may see at least a temporary fall in their stock price. A *Harvard Business Review* report stated in 2023 that stock prices of public companies fell 7.5% on average following a breach. These declines are typically temporary, but the same report asserts that it takes an average of about 46 days for stocks to return to their original value prior to the cyber incident. The time to recover and whether the stock price recovers depend on many factors, including investor perception of how the organization responded to the attack (Huang et al., 2023). Another variable is the industry in which the affected company operates. Businesses in the tech industry, especially those that are cloud or security entities, may have their market capitalization significantly damaged. For example, a 2023 cyber breach at Okta, a cloud-based identity and access management company, resulted in a market cap decline of more than $2 billion in the immediate aftermath of the attack. However, often stock prices rebound in the long run if the company's earnings stabilize or grow.

CASE IN POINT

MGM Resorts International discovered a cyber breach on September 11, 2023. At that time, several of the company's operations were impaired, and there were estimates that losses would be several million dollars per day. The stock price volatility for MGM demonstrated a not atypical pattern following a data breach. On September 11, the stock traded at $42.70 per share. October 5 and October 27 were the low points, with the price at $34.79 and $34.23, respectively. By December 12, 2023, the stock had fully recovered and was trading at $43.37 per share (Velotta, 2023).

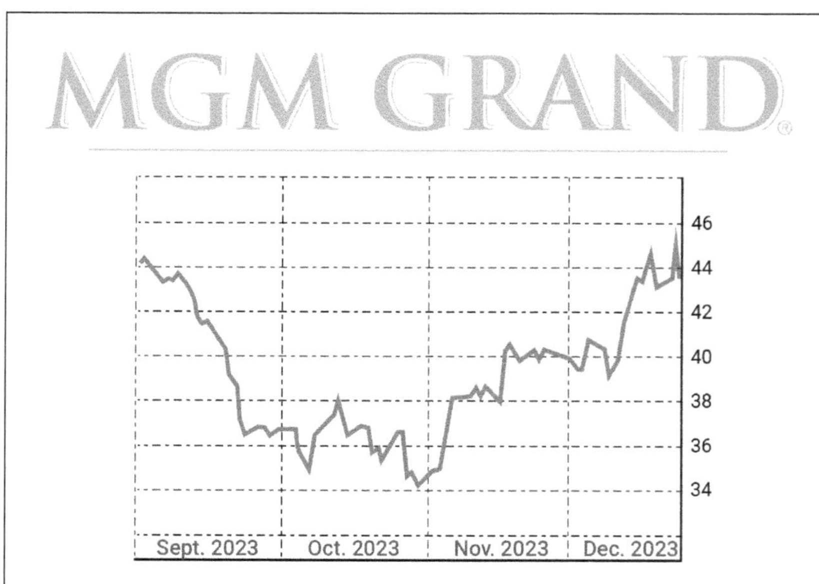

Figure 10-3. MGM Resorts' Stock Price, 9/1/2023–12/31/2023

Public Relations

One way to mitigate damage from a cybersecurity incident is to engage in a **public relations (PR)** campaign. A compromised entity most likely needs to contract with a PR agency to manage communications. In-house marketing and PR specialists may lack the expertise for crisis communications.

The cost of PR in defending a brand and managing crisis communications is sometimes overlooked, though the money is well spent. Communication with many types of stakeholders, including the media, customers, and regulatory authorities, is necessary. Significant reputational damage can ensue when a company is viewed as not being transparent or as lacking in empathy. PR agents specializing in crisis management may charge several hundred dollars an hour for their expertise, and the need for assistance is not likely to end with the containment of the threat. PR professionals will need to monitor public media and online groups to analyze public perception of the breach impact. Estimating the anticipated cost depends on the level of in-house expertise, type of incident, industry, and timing. Contracting with an agency after the incident will likely cost more than if a firm is on retainer.

Public Relations (PR) The set of techniques and strategies related to managing how information about an individual or company is disseminated to the public, and especially the media.

> **EXAMPLE**
> **Matter (www.matternow.com) is a communications agency offering services in crisis communications and reputation management. They advertise that they can help in creating proactive communications following security breaches. Their services include active incident management and brand audits.**

Tools for Estimating Losses

Cyber risk quantification is the process of calculating the loss associated with a cyber incident. Given all the direct, indirect, and opportunity costs that can be associated with a cyber intrusion, this is a difficult task, but tools are available to help. They include quantitative methods and breach calculators, which are built on assumptions and assigned risk probabilities.

Cyber Risk Quantification The process of measuring the financial impact of cyber risks on an organization.

Quantitative Models

Many quantitative models can be used to assist in calculating anticipated cybersecurity losses. Some are proprietary and are offered by cybersecurity consulting firms. Two comprehensive quantitative techniques are ABC Costing and FAIR, which we will describe here.

Activity-Based Costing

Accounting professionals are likely familiar with **activity-based costing (ABC)**, as it is commonly used in manufacturing and other industries to determine costs and pricing. IBM's *Cost of a Data Breach Report 2024* applies this method to calculate the cost of a data breach. ABC assigns costs to the specific activities that drive them. As applied to data breaches, the IBM report identifies four such activities: detection and escalation, notification, post-data breach response, and lost business (IBM, 2024). The total cost of the breach is calculated by estimating the number of individuals involved in the subactivities in each activity and their average cost for the time involved. Costs for professional services require estimating the number of hours and average fees charged needed under a range of scenarios.

Activity-Based Costing (ABC) A costing method that assigns overhead and indirect costs to related products and services. This cost accounting method recognizes the relationship between costs, overhead activities, and manufactured products, assigning indirect costs to products less arbitrarily than traditional costing methods.

Factor Analysis of Information Risk

Chapter 5 discussed **Factor Analysis of Information Risk (FAIR)** quantitative modeling's use in risk assessment. Similarly, the model can be used to estimate the loss from a current or prospective cyber incident. This requires estimating several attributes of the specific intrusion, including its severity and prospects for mitigation. The FAIR model is complex and requires significant data

Factor Analysis of Information Risk (FAIR) A risk management framework used to understand, analyze, and quantify information risk in financial terms.

gathering and probability assignments. As a result, it is likely best used in risk assessment rather than calculating the loss from a current incident.

Cyber Breach Calculators

Online cyber breach calculators can assist in approximating the loss from a cybersecurity incident. The questions they pose in making their calculations are helpful in identifying the factors that impact the severity of loss. For instance, a question related to a data breach is whether the breached data included employee PII, customer PII, or both. Other questions concern the number of records breached and whether the breach was accidental, stemmed from a hack, or was the consequence of a stolen device.

> **EXAMPLE**
>
> **At-Bay, an insurance and security company, offers a free calculator (https://www.at-bay.com/cyber-risk-calculators/) that can quantify dollar losses from either a ransomware attack or a data breach. The calculator asks 10 questions in making a cost determination.**

Measuring Losses Versus Controls with Annualized Loss Expectancy (ALE)

Annualized Loss Expectancy (ALE) A quantitative risk management metric used to estimate the expected monetary loss from a specific risk over the course of a year.

Annualized loss expectancy (ALE) is an approach you can use to calculate potential cyber losses associated with cyber threats and vulnerabilities. This loss estimation technique is valuable in risk management, as accountants and IT specialists can use it to weigh the cost versus the benefit of a control to mitigate specific risks. Here is the formula for calculating ALE:

In this formula, SLE is an estimate of the cost of a single risk event occurring, and it includes both direct and indirect costs. For example, you might estimate that you might incur specific direct and indirect costs in a cyber incident. To calculate SLE, you need to estimate each of those losses. The ARO is an estimate of the number of times a risk event might occur within one year. Many external sources of data are available to help in making this calculation, including historical data, industry averages, consultants and experts, and surveys (e.g., by Gartner or IBM).

> **EXAMPLE**
>
> **Suppose a company estimates that a ransomware attack could cost it $3,000,000 in lost revenue, reputational damage, compliance, and mitigation. This is the SLE. If a ransomware attack would likely occur once within 10 years, the ARO is 1/10, or 0.1. The ALE would be $300,000. In managing the risk of a ransomware attack, an organization should be willing to spend no more than $300,000 annually on risk mitigation.**

Materiality

Materiality The term that determines whether a cyber incident is significant enough that it must be reported to regulators, investors, or stakeholders. Just like in financial reporting, it is not only about the size of the financial loss but also about the nature, timing, and context of the incident.

The primary concern of financial auditors is to ensure that financial statements fairly represent the underlying financial condition of an entity. Accounting involves estimates, and the final number on an income statement more closely resembles a painting than a photograph, but that painting should be representative. Any activity or misstatement that would have a **material** impact on financial condition should be, at a minimum, disclosed. A simple way of thinking about materiality is that an item is material if it *matters*. Of course, what matters to one individual or with respect to one reporting entity may be quite different from another. A dollar amount of $1,000,000 is material to individuals and many small companies, but not so much to a Fortune 500 company. While there is no hard and fast quantitative threshold for materiality, a general rule of thumb auditors may consider is the SEC guidance that misstatements or omissions under 5% of a financial statement item are usually not considered material. The SEC's primary concern with financial statement reporting is whether an amount would affect a reasonable person's decisions (US Securities and Exchange Commission, 1999). However, expressing a materiality threshold as a numerical value is not always appropriate. Investors and creditors may view losses from financial fraud as mattering more than a simple mistake.

Materiality is important in the context of cybersecurity because cyber incidents can have material impact on financial statements. Many of the losses discussed in this chapter can reach a materiality threshold. Materiality has become more important in the wake of the 2023 SEC Cybersecurity Rules, which require that a public company disclose on Form 8-K any cybersecurity incident determined to be material. The rules stipulate that the disclosure should describe the material aspects with respect to nature, scope, and timing and the material or likely material impact. Chapter 9 discussed this disclosure in some detail. Again, many factors go into deciding whether a cyber incident could impact stakeholder decisions about the financial condition of an organization.

CASE IN POINT

The PwC Cyber & Privacy Innovation Institute offers a summary of materiality that includes the harm to reputation, impact of business disruption, vendor and customer relationship damage, competitive harm, possibility of litigation, and actual and expected losses associated with the event. Also, according to the PwC Institute, the type of attack, industry in which the entity operates, type of bad actor, and response to date are all variables that can impact these factors (PricewaterhouseCoopers, 2023).

Making the materiality call requires a significant amount of judgment and experience. It also takes a team. Accountants will work with legal counsel, IT security professionals, communications experts, and finance executives in valuing qualitative and quantitative factors that determine the estimated loss amount.

Reporting Cyber Losses on Financial Statements

Losses that have a material impact on financial statements should be reported on those statements—but how? And what should accountants do if they do not really know if a loss is material? Estimated and actual cyber losses are reported on multiple financial statements, including Form 8-K , Form 10-Q, and the annual Form 10-K, as we explain here.

Form 8-K Report

Cybersecurity losses—realized and potential—appear in financial reporting in several places and a variety of ways. A **Form 8-K** report is often the first financial document that recognizes a cybersecurity loss. These reports are used to communicate unscheduled material changes that impact financial statements. As explained in Chapter 9, this report is required to be filed with the SEC if an unanticipated material cybersecurity loss occurs. The amount of financial loss may not be known, but the report is filed with the SEC to make the public aware of the cybersecurity event.

Form 8-K A report that publicly traded companies must file with the SEC to announce major cybersecurity events that shareholders should know about.

Form 10-Q

The reporting period following a cybersecurity incident may be a quarter of a year or year-end. **Form 10-Q** is an unaudited, comprehensive quarterly report of financial activity. It includes financial statements, management discussion and analysis, disclosures, and internal controls. Cybersecurity losses are disclosed in the Management Discussion and Analysis section. There may also be some discussion of a cybersecurity event in the section on Internal Controls. The Income Statement reports actual cybersecurity losses realized most often in Operating Expenses as "Other Expenses," or the losses may be consolidated into one or more expense line items.

Form 10-Q A quarterly financial report filed by publicly traded US companies, providing a snapshot of their financial performance.

Form 10-K

Form 10-K A comprehensive annual report filed by publicly traded companies in the United States that details financial performance and other relevant data.

Form 10-K is likely to contain information about a cybersecurity incident in many sections, and the event may be discussed for many years following the event. As with the 10-Q form, the 10-K report will include any realized expenses on the Income Statement. The Balance Sheet may include cybersecurity losses in the Commitments and Contingencies line item, with an explanation of the event and its likely impact in the Notes to the Financial Statements. Figure 10-4 provides an excerpt from Capital One's 2020 Annual Report.

Cybersecurity Incident (page 200) Capital One

As a result of the Cybersecurity Incident announced on July 29, 2019, we are subject to numerous legal proceedings and other inquiries and could be the subject of additional proceedings and inquiries in the future. Consumer class actions. We were named as a defendant in approximately 73 putative consumer class action cases (61 in U.S. federal courts and 12 in Canadian courts) alleging harm from the Cybersecurity Incident and seeking various remedies, including monetary and injunction relief. The lawsuits allege breach of contract, negligence, violations of various privacy laws and a variety of other legal causes of action. The U.S. consumer class actions have been consolidated for pretrial proceedings before a multi-district litigation ("MDL") panel in the U.S. District Court for the Eastern District of Virginia, Alexandria Division, where the remaining 29 consumer class actions are currently pending. In the third quarter of 2020, the MDL court denied in part and granted in part Capital One's motion to dismtiss and permitted pretrial discovery to continue. Securities class action. The Company and certain officers have also been named as defendants in a putative class action pending in the MDL alleging violations of certain federal securities laws in connection with statements and alleged omissions in securities filings relating to our information security standards and practices. The complaint seeks certification of a class of all persons who purchased or otherwise acquired Capital One securities from July 23, 2015 to July 29, 2019, as well as unspecified monetary damages, costs and other relief. Governmental inquiries. We have received inquiries and requests for information relating to the Cybersecurity Incident from Congress, federal regulators, relevant Canadian regulators, the Department of Justice, and the offices of approximately fourteen state Attorney General. We are cooperating with these offices and responding to their inquiries. In August 2020, we entered into consent orders with the Federal Reserve and the OCC resulting from regulatory reviews of the Cybersecurity Incident and relating to ongoing enhancements of our cybersecurity and operation risk management processes. We paid an $80 million penalty to the U.S. Treasury as part of the OCC agreement. The Federal Reserve agreement did not contain a monetary penalty.

Figure 10-4. Part of the 2020 Capital One Annual Report

The Risk Factors section of the 10-K will discuss cybersecurity risk as a threat and will contain information about a current or past incident. The Management Discussion and Analysis will have information as well. As with the 10-Q form, the 10-K may also discuss cybersecurity in the section on internal controls.

CASE IN POINT

As mentioned in an earlier case in this chapter, MGM Resorts International discovered a cybersecurity breach on September 11, 2023. It subsequently disclosed the event in two Form 8-K reports. Its September 30, 2023, 10-Q report mentioned the cybersecurity

incident in several places. The Notes to Financial Statements identified Cybersecurity Litigation in the Note on Commitments and Contingencies. The cyber breach was also described in Management's Discussion and Analysis of Financial Conditions and Results of Operations. This description mentioned the possibility of class action lawsuits and explained how the incident impacted Adjusted Property EBITDAR (earnings before taxes, depreciation, amortization, and restructuring or rent costs) for Las Vegas Strip Resorts and Regional Operations during the month of September. The discussion also identified technology consulting services, legal fees, and other third-party fees as nonmaterial expenses for the quarter.

Summary

Calculating the loss from a cyber incident is no easy task. This chapter began with the need for estimating losses and how such losses might be classified. It identified direct and indirect costs associated with operating, compliance, and reputational risks. Each type of loss was described in some detail. The chapter also discussed the concept of materiality. This concept is important for accountants to consider, as it pertains to how losses are disclosed in financial reporting, and we described how cyber incidents are typically presented in financial reports. The next section of the chapter described a few of the tools that may be used in accounting for cyber damages and how Annualized Loss Estimates may be used to weigh the cost versus benefit of controlling cyber risk.

Chapter Review

Key Terms

Activity-Based Costing (ABC)
Annualized Loss Expectancy (ALE)
Bitcoin
Breach
Compliance Costs
Credit Monitoring
Cryptocurrency
Cyber Risk Insurance
Cyber Risk Quantification
Direct Costs
Double Extortion
Factor Analysis of Information Risk (FAIR)
Form 8-K
Form 10-K
Form 10-Q
Identity Theft Protection
Incident
Indirect Costs
Materiality
Moral Hazard
Operational Loss
Public Relations (PR)
Third-Party Coverage
Wire Transfer Scams
With-and-Without Method

Review Questions

1. Why do organizations try to quantify potential and actual losses related to cybersecurity events?
2. What is the purpose of reporting losses related to cybersecurity attacks/incidents?
3. Identify the stakeholders affected by cybersecurity incidents and/or breaches.
4. How do you classify losses associated with cybersecurity incidents?
5. What are the differences between direct and indirect losses?
6. What is the purpose of cyber risk insurance, and how does it work?
7. Explain the direct costs associated with a cybersecurity incident.
8. Explain the indirect costs associated with a cybersecurity incident.
9. What are the differences between operating, compliance, and reputational costs or losses?
10. How do you calculate lost revenue or profits from a cybersecurity incident?
11. What specific operating costs are associated with cybersecurity and cybersecurity incidents?
12. What specific compliance costs are associated with cybersecurity and cybersecurity incidents?
13. What specific reputational costs are associated with cybersecurity and cybersecurity incidents?
14. What is the purpose of materiality? How does the common definition of materiality differ from that relating to cybersecurity?
15. Explain the quantitative models that help organizations estimate potential and actual losses related to cybersecurity.
16. How would you value qualitative losses to an organization?
17. How does a publicly traded organization report cyber losses?

Multiple Choice

Identify the choice that best completes the statement or answers the question.

1. Why might enterprises decide to pay a ransom in a cyberattack scenario?
   a. It is standard operating procedure for all IT issues.
   b. They lack complete backups, or there is a threat of double extortion.
   c. They have excess funds to spend.
   d. They want to support the cryptocurrency market.
   e. To gain publicity and media coverage.
2. Which statement best describes the nature of direct and indirect costs associated with cybersecurity risk factors?
   a. Indirect costs are easily quantifiable and often reimbursed by insurance policies.
   b. Direct costs include losses such as decline in stock price and reputational damage.
   c. Direct costs are always higher than indirect costs for any cybersecurity incident.
   d. Indirect costs are solely related to the physical damages incurred during an attack.
   e. The categorization of costs as direct or indirect can depend on the context and circumstances of the incident.
3. Materiality has become more important in the context of cybersecurity since 2023 because of which of the following?
   a. Public demand for transparency
   b. The increase in cyber insurance claims
   c. New SEC Cybersecurity Rules requiring disclosure of material cyber incidents
   d. International regulations
   e. The decrease in cyberattacks
4. What distinguishes a breach from an incident in cybersecurity terms?
   a. A breach is any unauthorized access of data, whereas an incident causes physical damage.
   b. An incident is always intentional, while a breach can be accidental.
   c. An incident only affects data availability, while a breach affects all aspects of the CIA triad.
   d. An incident involves an unsuccessful attempt to breach data, whereas a breach is successful.
   e. A breach is an incident where data is not just accessed but exfiltrated.
5. Why is valuing the loss from intellectual property theft challenging?
   a. Hackers typically do not target valuable IP.
   b. The theft usually goes unnoticed for years.
   c. Law enforcement often recovers stolen IP quickly.
   d. Intellectual property is often undervalued in financial statements.
   e. Physical evidence of theft is hard to find.
6. What typically happens to audit fees following a cyberattack?
   a. They increase by about 13.5%.
   b. They double.
   c. They decrease by approximately 10%.
   d. They increase by about 5%.
   e. They remain unchanged.
7. What is required by SEC disclosure laws regarding cybersecurity incidents?
   a. A list of all known vulnerabilities and breaches
   b. Detailed financial records of all cybersecurity spending
   c. An annual cybersecurity risk assessment report
   d. Mandatory reporting of all attempted cyberattacks
   e. Reporting of material cybersecurity incidents
8. What core elements of information security are impacted by an incident, according to the CIA triad?
   a. Cryptography, infrastructure, and access
   b. Control, inspection, and authentication
   c. Confidentiality, integrity, and availability
   d. Cooperation, initiative, and agility
   e. Continuity, intelligence, and analysis
9. What role does a PR agency play in the aftermath of a cybersecurity incident?
   a. Restoring lost data
   b. Providing legal counsel
   c. Conducting financial audits
   d. Repairing damaged equipment
   e. Managing communications and mitigating reputational damage
10. Which sectors are particularly vulnerable to cyber threats that can cause temporary or permanent inoperability of infrastructure?
   a. Construction and real estate
   b. Hospitals and utility providers
   c. Retail and fashion
   d. Travel and tourism
   e. Entertainment and media

Case Studies

Sammy Company—Estimating Cyber Insurance Coverage

Sammy Company was so pleased with the work Herron Consulting did in helping it adopt the NIST CSF 2.0 that it has engaged the firm to help it obtain cyber insurance. Natalie Magee has been tasked with working with the company to answer the following questions. Sammy Company does not know if it needs both first- and third-party cyber insurance.

What factors will influence this decision?

Case Requirements

1. Assuming that Sammy Company decides to purchase first-party cyber insurance, what factors should Natalie consider in helping it determine the appropriate amount of coverage?
2. Assuming that Sammy Company decides to purchase third-party cyber insurance, what factors should Natalie consider in helping it determine the appropriate amount of coverage?

Herron Consulting LLP—Calculating Cyber Breach Costs

Chris Mason at Herron Consulting has been contacted by Yellow Subs, as the company has had a ransomware attack. Yellow Subs is unlikely to pay the $500,000 ransom demanded because it believes it has adequate backup data in place to be able to restore operations. However, it also believes that some customer PII has been exfiltrated. The company wants to be able to estimate the potential dollar loss from this incident.

Case Requirements

1. Using an online data breach calculator, what questions will Chris need to ask to provide this estimate?
2. Yellow Subs estimates that it will experience three days of downtime as a result of this incident. What pieces of information does it need to estimate the cost associated with this element of loss?

Martin Pharmaceuticals, Inc.—Calculating and Disclosing Losses (Part 2)

Martin Pharmaceuticals, Inc., is a Fortune 500 drug company with about $60 billion in annual revenue. Its primary source of revenue is patents on specialized vaccines, but the company also has several other drug patents, including for drugs related to diabetes and heart health. The company's CEO, Peggy Lynn, has been encouraging the company to develop new revenue streams, and the research and development team has been working on developing a weight loss drug that appears promising.

In 2024, Martin suffered a ransomware attack that paralyzed operations for 10 days while the Incident Response team worked to restore backups following a decision **not** to pay the ransom. Martin's stock price fell 10% in the days following public disclosure of the attack and estimates that supply chain and operations complications cost the company about $20 million daily and had some lingering effects on operations and reputation Customer and employee data, including Social Security numbers, were exfiltrated, but to date none of this PII has appeared on the dark web. It has also been unclear as to whether IP was stolen, including research on the new weight loss drug. Since the attack, the company has spent $100 million on information system upgrades and cybersecurity governance. Martin has multiple cybersecurity insurance policies, covering extortion, stakeholder notification, and business interruption. The company expects that these policies will cover about 75% of the financial loss from the attack.

As soon as the breach was discovered, Martin contracted with Deloitte, a Big Four public accounting and consulting firm with one of the largest cybersecurity practices globally. The consultants were able to trace the breach to a third-party phishing email that led to the stolen credentials of a high-ranking Martin administrator in the IT department. It is believed that the breach went unnoticed for about two months, until the computer systems were frozen, and the attackers demanded $20 million in ransom.

Martin worked closely with the FBI and other law enforcement following the attack. In August 2025, Anastasiya Bagryanov, a Bulgarian national living in the United States, was arrested. Bagryanov is believed to be a member of the hacker group Robbin Spiders, which was responsible for several of the largest ransomware attacks in 2022 and 2023.

Case Requirements

1. Categorize the possible direct and indirect financial losses stemming from this attack. (Use your best guesses, referencing losses of similar-sized company breaches and documenting your assumptions.)
2. Where will the information about the attack and losses appear in the 10-K report?

Exercises

10-1. The Internet and AI can be useful tools for estimating losses from a cyber breach. Required:
   1. Find a cyber breach calculator on the Internet. What questions does it ask you in determining an estimated loss from a cyber incident?
   2. Using an AI application, develop a prompt to find out what factors should be considered in estimating the loss from a cyber incident. How does this list compare with the questions asked by the cyber breach calculator?

10-2. Table 10-2 in this chapter provides examples of direct and indirect losses from cyber incident damages along three dimensions: operations, compliance, and reputation. Required:
   1. Complete the table with a different example for each cell.
   2. Direct losses are similar to tangible losses and indirect losses are similar to, but not the same as, intangible losses. How are they different? Describe two potential tangible losses and two potential intangible losses from a cyber incident.

10-3. Figure 10-4 in this chapter showed the Commitments and Contingencies Note to the Capital One 2020 Annual Report. Required:
   1. Go to the 2019 Capital One Annual Report. List the sections in the report that relate to the 2019 cyber breach.
   2. Go to the 2020 Capital One Annual Report. List the sections in the report that relate to the 2020 cyber breach.

10-4. Organizations can use ALE to estimate cyber losses from a risk event as a benchmark for allocating resources to control against such an event. Required:
   1. Several companies offer identity theft protection. Estimate your single loss expectancy (SLE) and annualized rate of occurrence (ARO) from identity theft. You may consult online resources for help. Use these values to calculate your annualized loss expectancy (ALE) if your identity is stolen.
   2. Research two companies that provide identity theft protection and the cost of each of the levels of services they offer. Do you think you should purchase identity theft protection based on the cost and ALE? If so, at what level does it make sense in terms of the cost and reduction of risk?

References

Braithwaite, S. (2023, October 24). *ALPHV: Hackers reveal details of MGM cyber attack.* University of Hawai'i—West O'ahu Cyber. https://westoahu.hawaii.edu/cyber/global-weekly-exec-summary/alphv-hackers-reveal-details-of-mgm-cyber-attack/

Chen, H., & Magramo, K. (2024, February 4). *Finance worker pays out $25 million after video call with deepfake "chief financial officer."* CNN. https://www.cnn.com/2024/02/04/asia/deepfake-cfo-scam-hong-kong-intl-hnk

Clorox Company. (2023). *The Clorox Company quarterly report on form 10-Q for the fiscal quarter ended September 30, 2023* (001-07151). US Securities and Exchange Commission. https://www.sec.gov/Archives/edgar/data/21076/000002107623000048/clx-20230930.htm

Collier, K. (2023, November 27). *Emergency rooms in at least 3 states diverting patients after ransomware attack.* NBC News. https://www.nbcnews.com/tech/security/emergency-rooms-least-3-states-diverting-patients-ransomware-attack-rcna126890

Deloitte. (2017, May 17). Seven hidden costs of a cyberattack. *Wall Street Journal.* http://deloitte.wsj.com/riskandcompliance/seven-hidden-costs-of-a-cyberattack-1494993739

Huang, K., Wang, X., Wei, W., & Madnick, S. (2023, May 04). The devastating business impacts of a cyber breach. *Harvard Business Review.* https://hbr.org/2023/05/the-devastating-business-impacts-of-a-cyber-breach

IBM. (2023, July). *Cost of a data breach report 2023.* https://d110erj175o600.cloudfront.net/wp-content/uploads/2023/07/25111651/Cost-of-a-Data-Breach-Report-2023.pdf

IBM. (2024, July). *Cost of a data breach report 2024.* https://cdn.table.media/assets/wp-content/uploads/2024/07/30132828/Cost-of-a-Data-Breach-Report-2024.pdf

PricewaterhouseCoopers. (2023, August). *Making materiality judgments in cybersecurity incident reporting.* https://www.pwc.com/us/en/services/consulting/cybersecurity-risk-regulatory/sec-final-cybersecurity-disclosure-rules/materiality-sec-cybersecurity-compliance.html

Sanicola, L. (2021, May 28). *After Colonial attack, energy companies rush to secure cyber insurance.* Reuters. https://www.reuters.com/technology/after-colonial-attack-energy-companies-rush-secure-cyber-insurance-2021-05-28/

Sophos. (2024, April). *The state of ransomware 2024.* https://www.ccsmedia.com/wp-content/uploads/sophos-state-of-ransomware-2024-wp.pdf

US Securities and Exchange Commission. (1999, August 12). *Staff accounting bulletin No. 99: Materiality.* https://www.sec.gov/interps/account/sab99.htm

Valentiam Group. (2021, April 15). *Valuation of intangible assets: What are the top methods?* https://www.valentiam.com/newsandinsights/intangible-assets-valuation-methods

Velotta, R. N. (2023, September 17). Analyst: MGM losing $4.2M–$8.4M a day because of cyberattack. *Las Vegas Review-Journal.* https://www.reviewjournal.com/business/casinos-gaming/analyst-mgm-losing-4-2m-8-4m-a-day-because-of-cyberattack-2906379/

Chapter 11

Privacy

Chapter Outline

Why Privacy Matters
How Data Are Collected and Used
Privacy Policies
Privacy Frameworks
Privacy Regulation
Privacy and Digital Natives
Auditing Data Privacy

Learning Objectives (LO)

After reading this chapter, you should be able to:

- LO1. Explain why privacy matters and the need for organizations to have privacy policies.
- LO2. Describe the general privacy framework models that organizations use.
- LO3. Compare and contrast the major privacy regulations worldwide.
- LO4. Compare and contrast US privacy regulations with international privacy regulations.
- LO5. Understand the attitude of digital natives toward privacy.
- LO6. Know how auditors evaluate data privacy.

Introduction

Data privacy has received a lot of attention recently. It seems that everywhere you go, whether in person or online, someone wants your data. Some of us give our data freely and talk about anything in front of Alexa and Siri, play quizzes on social media, and are constantly on TikTok or other social apps that may be capturing data. Others are more private, avoiding social media and guarding their PII from others. While individuals may have different attitudes about their privacy, businesses and other organizations do not. They all want personal data, particularly those of their customers.

Information security is concerned with protecting the confidentiality, integrity, and availability (CIA) of data and information. Privacy is also a serious concern, and it is a distinct attribute of data and information that requires protection apart from confidentiality. People often confuse privacy and confidentiality, but the terms do have different meanings. **Confidentiality** is an ethical obligation of trust between two or more parties. **Privacy** is the right, often a legal right, of an individual or entity to have their data and information kept secret. Privacy has become increasingly important in the digital age. For example, your health records used to be recorded in physical files stored in your doctor's office file cabinets. Although they could be stolen if someone broke into the office, or damaged through fire, water, or other elements, they were not exposed in the same way your records are when they are kept digitally. The ability for a patient to access their health data through a digital portal is certainly convenient, but despite security procedures, that data also faces increased threats to CIA and privacy.

Confidentiality The assurance that sensitive data is not disclosed to unauthorized parties, whether intentionally or accidentally.

Privacy The right of individuals to control access to their personal information, communications, and activities. It encompasses the ability to determine what data are collected, how they are used, who can access them, and for how long.

This chapter describes the need for businesses and other organizations to keep data private; how data are collected; privacy policies; privacy frameworks and guidance; and privacy regulations. It provides guidance in creating a privacy policy and describes several privacy management frameworks. We add descriptions of select international and US privacy laws, including the European Union's General Data Protection Regulation, which is viewed as one of the most comprehensive privacy regulations in the world. The chapter concludes with a discussion about privacy and digital natives.

Why Privacy Matters

When we think of privacy, we usually first consider our individual privacy. Protecting your personal data, such as your phone number, health records, financial accounts, earnings, and Social Security number, is important for many reasons. Adversaries can use these data to embarrass you, defraud you, or steal your identity. To help combat that fact, the final chapter in this book provides tips on how you can protect your personal data in cyberspace.

In addition to the need to protect an individual's data, protecting individual data privacy is critical for businesses and other organizations. These entities are charged with keeping their own data and that of others private. This includes the personal data of customers, employees, and other stakeholders. This PII and PHI is a valuable target for threat actors.

As organizations consider data security, it is useful to know the state of those data and their classification based on the level of privacy required. The state of data is that it is either at rest or in motion. **Data at rest** are stored, either on a physical device within your control or in the cloud, where a third party has physical control of the storage device. Physical data storage might be on a USB stick, server, phone or other endpoint device, computer hard drive, or in the cloud. If the device is connected to the Internet, then the threats to privacy increase. **Data in motion** are data in the state of transmission or being processed. As discussed in Chapter 2, a variety of transmission options are available, including optical fibers and wireless connections. Each option has a distinct threat level and potential vulnerabilities.

Data at Rest Digital information that is stored on a physical medium and is not actively moving across networks or being processed.

Data in Motion Digital information that is actively moving from one location to another, whether across the Internet or a private network or between devices and applications.

You can classify data in many ways. The US government categorizes sensitive information by the degree to which national security could be compromised by theft or leaking. The three levels of classification it uses are confidential, secret, and top secret, with the last requiring the most security to ensure privacy. A classification system that is often used by businesses and other organizations that collect, process, store, and transmit data uses four groupings: public, internal, confidential, and restricted. **Public data** require no security because they are available to anyone and may be shared without restriction. **Internal data** are available and shared within an organizational unit. These data may be memos, budget data, strategies, or other data of value to insider stakeholders. **Confidential data** constitute a more restricted category of internal data. They may be shared only with authorized users. For example, a human resources department maintains confidential data about employees. The classification of data requiring the most security is **Restricted Data**. These data are highly sensitive, and their security is paramount. An example would be user passwords.

Public Data Information that is freely available to the public and does not require any specific access controls.

Internal Data Data generated and used within an organization, not intended for public distribution.

Confidential Data Information that is meant to be kept secret and is protected from unauthorized access.

Restricted Data Information that is highly sensitive and subject to strict access controls to protect its confidentiality.

> **CASE IN POINT**
>
> **Encryption is an important element in securing data privacy. Signal is a free, open-source app that encrypts messages so that they cannot be read if intercepted. In addition to end-to-end encryption, the app has a disappearing message feature that will cause messages to automatically delete after a period. Although the app came under some criticism for use in discussing US military plans in 2025, its features make it generally a good choice for messaging sensitive information.**

Business organizations are concerned about the privacy of stakeholder data for both ethical and legal reasons. Breaches of stakeholder data, such as customer personal and financial information, can result in distrust and sometimes legal damages. In addition, investors and creditors may look to the **environmental, social, and governance (ESG)** score of a company when making decisions about whether to invest in debt or equity if sustainability and ethical practices are of concern. Privacy may be thought of as a part of the social pillar of ESG, and a data breach can negatively impact this score. Using ESG as a framework, MCSI is a public company offering tools to support investment decisions. Its ESG ratings assign an industry-relative letter based on how a company manages sustainability risks and opportunities versus its peers. Figure 11-1 shows the rating index.

Environmental, Social, and Governance (ESG) Criteria that measure an organization's impact on environmental sustainability, social responsibility, and governance practices.

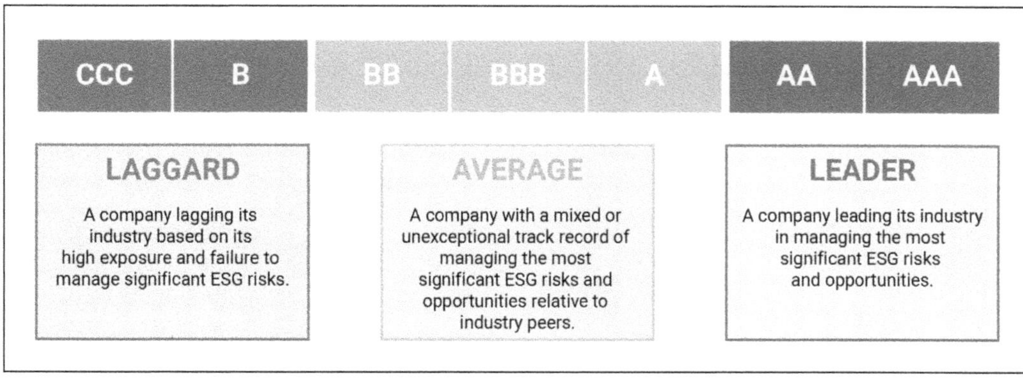

Figure 11-1. MCSI ESG Scoring

How Data Are Collected and Used

Businesses love data. Collecting data from customers allows organizations to make better decisions about product design, pricing, market trends, and advertising. The more data a company can collect from a customer transaction or even a potential customer's visit to a website, the better it can make decisions that affect every part of the organization, from production to marketing to budgeting. Customer databases are valuable intangible assets that can appear on a company's financial statement if the company buys them directly or gains them through a merger, partnership, or acquisition.

Organizations may buy customer lists and other data from **list brokers** that specialize in collecting email and mailing list data and maintaining databases that can be used for targeted marketing. **Data brokers** are similar, but they offer more services, including data analytics. Data are so valuable that a company may purchase another organization just to acquire its customer database, although that can sometimes backfire, as in the following case.

List Broker A marketing intermediary who rents, sells, or manages access to contact lists on behalf of clients seeking to reach specific target audiences.

Data Broker A company or individual that collects, aggregates, processes, and sells/licenses personal and behavioral data regarding individuals or organizations. Most do not have a direct relationship with the data subject.

> **CASE IN POINT**
>
> **In 2015, General Wireless, a subsidiary of hedge fund Standard General LP, purchased RadioShack's customer data along with other assets. The sale faced some opposition from state officials and other regulators over privacy concerns. RadioShack's own privacy policy included the promise that the company would not sell or rent customer PII. While the sale did go through eventually, the data sold were more limited due to the intervention of the US Federal Trade Commission (Meyer et al., 2015).**

Valuable customer data are not just names, email addresses, phone numbers, credit card data, and Social Security numbers. The data that may matter most are data about likes and dislikes and purchase histories. **Data mining** is the process of taking the raw data an organization collects and turning it into valuable information managers can use in making decisions. The tension is that customers may wish to keep their data private, while organizations want as much of it as they can get.

Data Mining The process of discovering patterns, correlations, trends, or useful information in large data sets using statistical, algorithmic, and/or machine learning techniques.

Cookies Small text files stored on a user's computer by websites they visit, used to remember login details, preferences, and other information.

Online websites collect data from you every time you enter their URL. These data collections are called **cookies.** A cookie is a small text file containing identification about your computer. Your web browser saves the cookie so that your device will be recognized in the future. The websites you visit will then recognize you on future visits. Businesses might use cookies to collect data about your viewing habits. They can use this information to deliver customized advertisements to you and perhaps to make your browsing experience more efficient. There are many types of Internet cookies. Some last only as long as your session of interacting with a website, whereas others are stored and stay on your computer unless you clear them. You can do this on your computer by accessing your browser settings and considering options in your browser history. Figure 11-2 shows some types of Internet cookies.

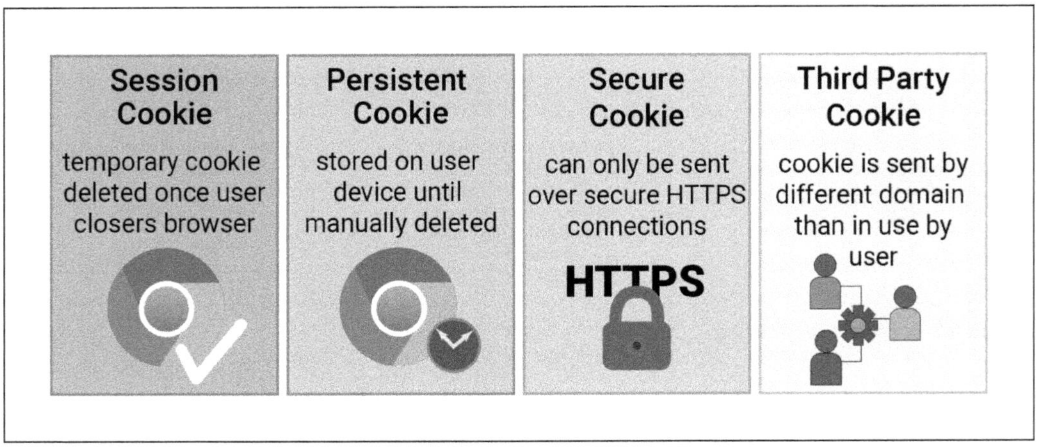

Figure 11-2. Types of Internet Cookies

You likely have noticed that on some websites a pop-up box appears each time you access it, asking if you will Allow or Block cookies. You have that privacy choice thanks to data privacy laws. Cookies are relatively harmless, and they do not contain malware. However, they do track user browsing history, which may be exposed in a cyberattack.

CASE IN POINT

Brian Walshe was arrested and is being prosecuted for allegedly killing his wife, Ana. Included in the evidence against Mr. Walshe are details of his online searches. His Google searches included prompts such as "Can you be charged with murder without a body?" and "Dismemberment and the best ways to dispose of a body." In court, prosecutors presented a timeline of his search history before and after his wife's disappearance (Levenson & Sgueglia, 2023).

Privacy Policies

Privacy Policy A statement by an organization that outlines how it collects, uses, and protects personal data.

Almost every commercial entity has a privacy policy. When you access a business website, you are generally linked first to a long company **privacy policy** that seems to be in the way of access to the data or services you are seeking. Many regulations control these policies, and many companies, understanding stakeholder privacy concerns, are promising not to share data or only to do so in a limited way. These policies, then, ensure compliance with laws and regulations, protect an entity against litigation for stakeholder privacy violations, and work to provide general data security. Unfortunately, these goals for privacy protection result in policies that are long and difficult to digest, even if you do read them. Even the most honest among us may have, at least on occasion, noted that we have read and understood a lengthy policy, when in fact we have not.

> **CASE IN POINT**
>
> **A 2019 Pew Research Report found that 9% of Americans surveyed said they always read a company's privacy policy. The same survey reported that 36% never read them. Of those who do read them, only 22% follow them all the way through. It is much more likely that stakeholders read them partially or just glance at parts of the policy. The good news is that most of those surveyed reported that they understood the policies they read. Unfortunately, while the policies are not widely read, most Americans surveyed also expressed low confidence that their data privacy is protected by businesses (Auxier et al., 2019).**

A business's privacy policy should explain *why* data are collected and the general security measures taken to protect them. Privacy policies usually have multiple parts. At a minimum, each section will describe what personal information is collected, how it is collected, how the data or information is used, whether stakeholder data is shared with third parties, and user rights. An entity's legal counsel might draft a policy, or a small business may elect to draft its own using an online template or by consulting with a vendor that specializes in creating website policies. Policies will vary with general governmental and industry requirements. The best policies and practices describe the privacy trade-offs for consumers and other stakeholders with whom they interact. For example, a retailer might request your email address and offer to send you coupons or discount codes. Table 11-1 summarizes the typical components found in a comprehensive privacy policy.

Table 11-1. Sections in a Comprehensive Privacy Policy

Privacy Policy Section	Brief Description
Introduction and Scope	Brief introduction with purpose and intended users
Data Collected	Types of PII collected and how collected (e.g., automatic vs. input)
How Data Are Used	E.g., to enhance experience, personalized advertising, etc.
How Data Are Shared	E.g., with third parties like service providers and advertisers, etc.
Rights of Users	To access, change, or delete data
Data Retention	How long data are stored and criteria for retaining periods
Data Security	How data are protected from unauthorized access, loss, or misuse
Opt-Out Procedures	Ability to opt out of services or data collection and how to do it
Contact Information	How users can contact user support for privacy-related issues
Children's Privacy	If services directed to children, compliance with protective regulations (Children's Online Privacy Protection Act)
Legal Compliance	How policy complies with regulations such as the GDPR

Although privacy policies may be laborious to read, they can be quite important to consumers, particularly when a company collects highly personal data.

Privacy Frameworks

Privacy frameworks are voluntary guidelines that can help an organization to create a privacy strategy, manage privacy management risk, and comply with industry or government regulations. To implement a privacy framework, an organization will first identify and assess its privacy risks. This process involves understanding current privacy practices and policies, privacy objectives, and regulatory requirements. The organization will then identify gaps between the desired state and the current profile and select a framework to implement. Once the framework is implemented, the organization should monitor its effectiveness and make improvements as needed. Here we describe four popular privacy frameworks.

AICPA Privacy Management Framework (PMF)

The AICPA, together with the Canadian Institute of Accountants, first published a privacy framework in 2003. The voluntary framework was revised in 2009 and renamed Generally Accepted Privacy Principles. Then in 2020, the AICPA released an updated version called the Privacy Management Framework (PMF). The latest version incorporates technology changes, changes in data privacy standards and legislation, and updates to the AICPA Trust Services Criteria. Organizations can use the PMF to create and maintain a comprehensive information privacy program that allows them to manage privacy compliance obligations and risks.

The PMF is intended as guidelines to assist organizations that collect, create, use, store, process, and transmit PII. They may be businesses, not-for-profit organizations, or government agencies. The PMF recognizes the global nature of digital transactions and considers the need for organizations to comply with international as well as home country regulations.

AICPA's PMF consists of nine components, each of which is supported by an objective and criteria that allow for management of privacy risk and regulatory compliance. Table 11-2 describes the components, each of which includes detailed guidance in the form of "points of focus." The framework also includes a mapping to the AICPA Trust Services Criteria and the EU's General Data Protection Regulation.

Table 11-2. AICPA Privacy Management Framework Components

AICPA PMF Component	Description
Management	Responsibility and accountability for PII privacy policies and procedures
Agreement, Notice, and Communication	Formal arrangements and appropriate notifications and consent for subject PII
Collection and Creation	Limited to identified purposes and ongoing communications
Use, Retention, and Disposal	Limited to identified purposes, retention only as necessary for stated purposes, and secure disposal
Access	Data subjects can access, update, and correct data
Disclosure to Third Parties	Only for purposes identified and with explicit consent
Security for Privacy	PII is protected against unauthorized access, alteration, deletion, and disclosure
Data Integrity and Quality	Maintain accurate, complete, and relevant PII for intended purposes
Monitoring and Enforcement	Monitor compliance and have process for handling disagreement

Fair Information Practice Principles (FIPPs)

Fair Information Practice Principles (FIPPs) date back to a 1973 report from a US Department of Health, Education, and Welfare Advisory Committee. This set of principles guides how the US government handles information privacy. The principles have been used in developing US federal and state legislation, including the US Privacy Act of 1974 and the EU's Global Data Protection Regulation (GDPR), and they have been incorporated in privacy policies of organizations globally. As with other privacy frameworks, FIPPs are guidelines and are not mandated. Table 11-3 describes the principles.

Table 11-3. Fair Information Practice Principles (Federal Privacy Council, 2025)	
FIPP Principle	**Description**
Access and Amendment	Individuals should be granted access to their PII and the chance to amend or correct it.
Accountability	Agencies are accountable for compliance with these principles and privacy requirements.
Authority	Agencies should only create, collect, use, process, store, maintain, disseminate, or disclose PII when authorized and should identify the authorization.
Individual Participation	Agencies should involve individuals in use of PII and seek consent for any use or storage. There should be established procedures for complaints or inquiries.
Minimization	Only necessary data will be collected, stored, and processed.
Purpose Specification and Use Limitation	Data should be used only for their original intended purpose.
Quality and Integrity	Data should be accurate, complete, and current.
Security	Personal data should be protected from unauthorized access, disclosure, or use.
Transparency	Individuals should be informed about the collection, disclosure, and use of their data.

NIST Privacy Framework

The NIST Privacy Framework, first introduced in 2020, is a voluntary tool created to assist users in managing privacy risk in alignment with their strategic mission. The framework is flexible and may be customized. Organizations in all industries and of any size may use it to understand, assess, prioritize, and communicate their privacy management policies and actions. The framework notes the importance of assessing and managing privacy risk by mitigating, accepting, avoiding, or transferring it.

NIST's Privacy Framework aligns with the NIST Cybersecurity Framework and is similarly structured, as shown in Figure 11-3.

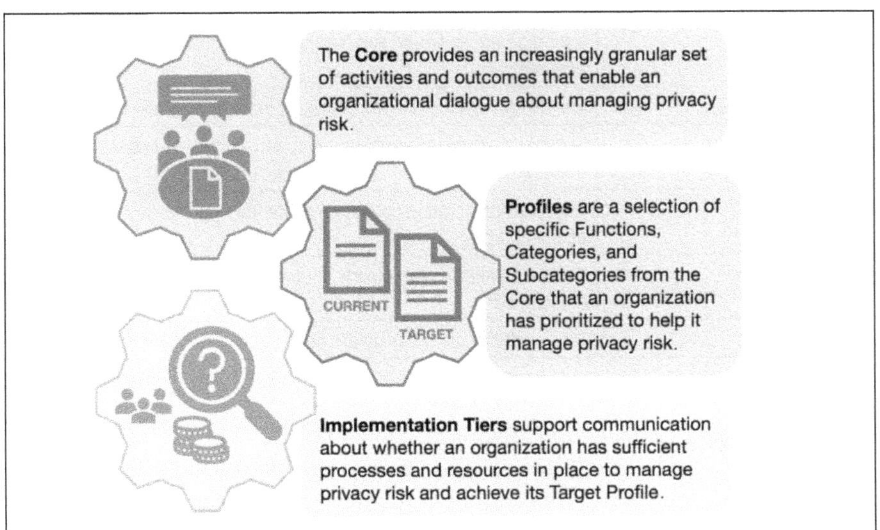

Figure 11-3. NIST Privacy Framework 1.1 Components

Implementation Tiers represent levels of risk management maturity, and, as with the CSF, the four tiers are partial, risk-informed, repeatable, and adaptive. Current and Target Profiles allow organizations to prioritize their actions and outcomes to align with their organizational mission, values, and risks. The core includes five functions, which are slightly different from those in the CSF. The core functions in the Privacy Framework are identify, govern, control, communicate, and protect. Within each function are categories (groups of privacy outcomes) and subcategories (specific outcomes or activities).

> **EXAMPLE**
> **One core function within the NIST Privacy Framework is CONTROL-P (CTP): Develop and implement appropriate activities to enable organizations or individuals to manage data with sufficient granularity to manage privacy risks. A category within this function is Data Processing Management (CT.DM-P): Data are managed consistent with the organization's risk strategy to protect individuals' privacy, increase manageability, and enable the implementation of privacy principles (e.g., individual participation, data quality, data minimization). One of the subcategories is CT.DM-P9: Log records are determined, documented, implemented, and reviewed in accordance with policy and incorporating the principle of data minimization (NIST, 2025, p. 28).**

Privacy by Design

The intent of Privacy by Design is to incorporate data protection into the design of IT infrastructure and business practice as a given. Similar to the Security by Design principles promoted by CISA, the approach moves from a reactive to a proactive posture. Dr. Ann Cavoukian, currently the Executive Director at Global Privacy & Security Design Centre in Toronto, Canada, first introduced the principles of Privacy by Design in the 1990s. Since that time, the approach has gained global recognition as a best practice, and many privacy and security regulations have incorporated the principles. Table 11-4 describes the seven foundational principles of the framework.

Table 11-4. Privacy by Design Principles

Privacy by Design Principle	Description
Proactive not reactive	Prevent privacy breaches before they occur.
Privacy as the default setting	User data are collected and retained only as necessary; data protection is automatic.
Privacy embedded into design	The design of systems and practices incorporates privacy considerations.
Full functionality	Security and privacy together are positive objectives and do not require a trade-off.
End-to-end security	Privacy consideration spans the entire life cycle of data from collection to disposal.
Visibility and transparency	Systems and practices clearly communicate how data are collected, used, and shared.
Respect for user privacy	Privacy decisions focus on the needs of individual users.

Privacy Regulation

Protection of privacy has long been a concern, but the digital world has created threats to privacy that have led to more protections and legal and industrial regulations. This section of the chapter examines national and regional laws and provides examples of select specific industry sector regulations.

For individuals and businesses trying to navigate legal privacy protection requirements, the number and complexity of the regulations presents a challenge. For example, a company operating in the United States but with global customers could be subject to many different national, state, and industry laws. As the world tries to figure out how best to protect the rights of individuals, everyone is likely to face a bumpy road.

Governments have long been concerned with the protection of individual privacy. The US Constitution, which became effective in 1789, did not specifically call out privacy protections, but subsequent Supreme Court adjudication has found that privacy protection is embedded in the first, fourth, and fifth amendments. In 1914, the United States established the **Federal Trade Commission (FTC)**, the primary governmental body with privacy protection oversight. The **US Privacy Act of 1974** established a Code of Fair Information Practice, which regulated personally identifiable information collection, maintenance, use, and distribution by federal agencies. Beginning in the 1990s, the United States and other countries began enacting legislation protecting personal data stored and transmitted digitally.

Federal Trade Commission (FTC) A US government agency responsible for protecting consumers and ensuring a competitive market by preventing unfair, deceptive, or fraudulent practices.

US Privacy Act of 1974 A federal law that governs how federal government agencies collect, maintain, use, and share personally identifiable information (PII) about individuals.

In 1986, the United States enacted the Electronic Communications and Privacy Act (ECPA), which protects the privacy of voice calls, emails, and other electronic communications. Not only does it cover communications as they occur, but the Stored Communications Act section concerns computer communications that are not in transit. The act has been used to prosecute those who intercept or access electronic communications without authorization, as in the following case.

> **CASE IN POINT**
>
> **In 2005, Elaine Cioni was having an extramarital relationship with another married individual. After two years, she suspected that her married boyfriend was having affairs with other women, and she began harassing him with phone calls, using a distorted voice. An FBI investigation found that she had used spoofcard.com, an app that lets you disguise your voice and the source of your communication. She also used yourhackerz.com (now disabled) to access her boyfriend's emails. Elaine Cioni was convicted of a felony, but the sentence was later reduced on appeal to a misdemeanor.**

International Privacy Laws

Many countries have enacted national laws governing data privacy, and while that does not always have an immediate effect on the typical business executive, it is important to understand these laws if the organization wants to do business in that country or area. Similar to financial laws in the United States (FCPA, banking laws, etc.) organizations are barred from doing business without evidence of compliance with established privacy laws.

European Union

In 1995, the European Union adopted the Data Protection Directive, which regulated processing of personal data. This legislation was superseded by the **General Data Protection Regulation (GDPR)** in 2018. Although the GDPR is a European Union regulation, its impact is global, and it is the most far-reaching privacy regulation in the world. The rule requires any organization that offers goods or services to or processes personal data of EU citizens or residents to comply with its privacy and security standards. The regulation specifically enhances personal privacy rights, increases the duty to protect data, requires reporting of data breaches, and includes significant penalties for noncompliance. The types of data protected are basic identity information; web data, such as IP address and location; health and genetic data; biometric data; racial/ethnic data; political opinions; and sexual orientation. Table 11-5 describes the protection and accountability principles with which entities processing personal data must comply under the GDPR (GDPR, 2016).

General Data Protection Regulation (GDPR) A regulation in EU law on data protection and privacy for all individuals within the European Union and the European Economic Area. It also addresses the transfer of personal data outside the EU and EEA.

Table 11-5. GDPR Protection and Accountability Principles

GDPR Protection and Accountability Principles	Description
Lawfulness, fairness, and transparency	Processing must be lawful, fair, and transparent to the data subject.
Purpose limitation	You must process data for legitimate purposes.
Data minimization	You should collect and process only as much data as absolutely necessary for the purposes specified.
Accuracy	You must keep personal data accurate and up to date.
Storage limitation	You may only store personally identifying data for as long as necessary for the specified purpose.
Integrity and confidentiality	Processing must be done in such a way as to ensure appropriate security, integrity, and confidentiality (e.g., by using encryption).
Accountability	The data controller is responsible for being able to demonstrate GDPR compliance with all of these principles.

EXAMPLE

GDPR Article 5(1)—Principles relating to processing of personal data states: "Personal data shall be . . . adequate, relevant, and limited to what is necessary in relation to the purposes for which they are processed." This means that a doctor's office is not allowed to sell a patient's information to a medical supply company, because the patient has given it to the doctor only for recordkeeping. Further distribution is forbidden unless permission is obtained directly from the patient.

The GDPR not only imposes data protection requirements on businesses, but it also provides individuals with a variety of rights with respect to their privacy. Figure 11-4 identifies these eight rights.

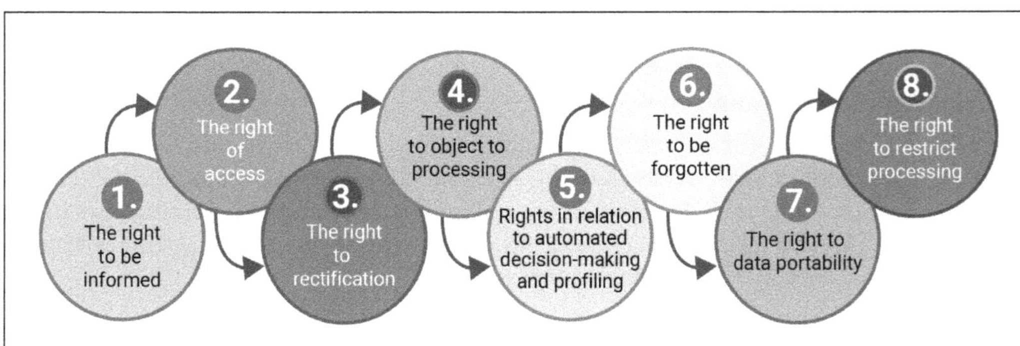

Figure 11-4. Individual Privacy Rights Under the GDPR

The GDPR is somewhat vague and leaves a lot to interpretation, making it difficult to gauge penalties for noncompliance. For example, the law requires companies to provide a "reasonable" level of protection over PII, but knowing what is reasonable is subjective. Penalties and fines are steep and can run to millions of dollars, plus compensation for any damages, so companies are incentivized to do all they can to comply. This includes adhering to the mandatory breach reporting requirement. Organizations that the regulation covers must report a breach of personal data to a Data Protection Authority within 72 hours of their awareness of the security breach.

CASE IN POINT

In 2020, Luxembourg's privacy regulator fined Amazon for noncompliance with the GDPR. The fine, which was upheld on appeal, was for €746 million. Amazon is not alone. WhatsApp, Facebook, and Google have all been fined for noncompliance with the GDPR. These fines can stem from a variety of compliance failures, including an absence of due care in protecting data. For example, British Airways was fined $26 million for failure to maintain sufficient security measures over systems, networks, and data when it suffered a data breach.

One of the features of the GDPR that impacts just about every online user is its regulation of cookies. The regulation requires all websites covered by the law to allow users from Europe to decide if they wish to activate cookies that would collect and track personal data. This includes all websites that European users might see, many of which are the same websites users in other parts of the world see. As a result, users around the world will almost always see a popup box asking if they wish to allow or deny cookies when accessing a site. Depending on the answer, some of the website features may be inaccessible.

China

China has three laws intended to provide data security and privacy. The **Personal Information Protection Law (PIPL)** became effective in 2021, and it is considered one of the world's strictest privacy regulations. Also, in 2021, China enacted the Data Security Law. These two regulations built on the 2017 Chinese Cybersecurity Law.

Personal Information Protection Law (PIPL) China's first comprehensive law specifically focused on protecting personal information.

The PIPL is aligned with the EU's GDPR. The law applies to personal information that is "related to identified or identifiable natural persons," and these data do not need to be enough to identify a specific person. The law applies to all data activities of entities that include the personal information of Chinese individuals, and that includes activities of enterprises outside China that seek to provide products or services to those within China. Violations of the law involve steep penalties.

United States Privacy Legislation

While the United States has some general federal legislation governing privacy, including the **Electronic Communications Privacy Act (ECPA)**, much has been left to industry sectors and individual states to regulate privacy and data security. This may change if the **American Data Privacy and Protection Act (ADPPA)** becomes law. This federal data privacy legislation was drafted in 2022, and, if passed, it would provide a comprehensive data privacy framework that provides individual data privacy protection.

Electronic Communications Privacy Act (ECPA) A US federal law that protects wire, oral, and electronic communications while they are being made, in transit, and when stored on computers.

American Data Privacy and Protection Act (ADPPA) A proposed US federal privacy law that aims to establish comprehensive national standards for how organizations collect, process, and share personal data.

CASE IN POINT

In the absence of federal privacy legislation, several states have enacted their own privacy regulations. Sometimes, the states file an action together. In 2022, a 40 US state coalition reached a privacy violation settlement with Google for $391.5 million. The settlement was a result of a four-year investigation into how Google tracked user locations. The app stored user locations even when an individual opted out of location tracking. The settlement also requires Google to make more disclosures about location tracking.

With respect to US state laws, California was the first state to regulate privacy, and it was followed by privacy legislation in Colorado and Virginia. Connecticut and Utah enacted privacy legislation in 2022, and by 2025, 20 US states had adopted some form of privacy regulation. The state adoption of comprehensive consumer privacy protection clearly has momentum.

California

California Consumer Privacy Act (CCPA) A state-level data privacy law that gives California residents significant rights over their personal information and imposes transparency and compliance obligations on certain businesses.

The California privacy law may be thought of as the gold standard. It was passed in 2018 and became effective in 2020. The **California Consumer Privacy Act (CCPA)** was influenced in part by the GDPR and has become a model for other states, such as Virginia's Consumer Data Protection Act.

The CCPA defines personal information as information that identifies, relates to, or could be linked reasonably to a person or household. This includes name, Social Security number, email address, browsing history, purchase records, geolocation data, biometrics, and other characteristics that could be used to create a personal profile.

The CCPA is designed to protect consumers who are California residents, and it provides them with the rights shown in Figure 11-5.

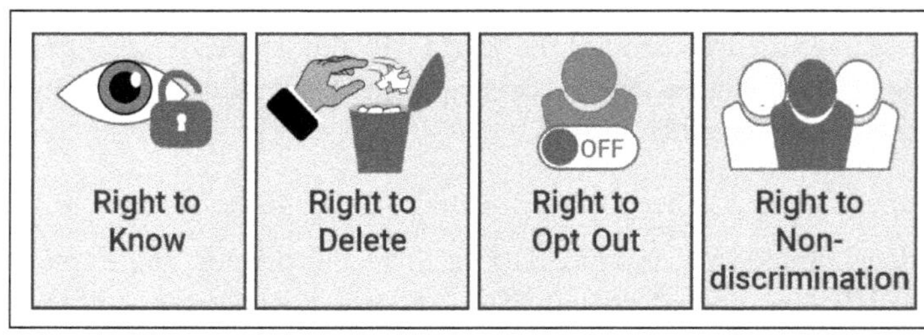

Figure 11-5. Individual Privacy Rights Under CCPA

Businesses have several obligations under CCPA, primarily protecting the individual rights of consumers. They must ensure that the personal data they gather are collected, stored, processed, and, if applicable, sold in compliance with the law. Consumers may individually or as a class sue businesses for a violation of their rights, but to be awarded monetary damages, the consumer must prove that the business did not maintain reasonable security measures. Willful violations of the law can result in significant monetary damages, as the total amount of fine is calculated per individual violation.

EXAMPLE

Individuals can only explicitly opt in to an agreement to have their data made available to third-party organizations for marketing purposes; the data cannot be automatically entered into a marketing database.

The CCPA requires notification of a data breach for any company doing business in California that owns, licenses, or maintains the personal information of California consumers. The breach of consumer personal information must be reported as soon as possible after a business realizes it has occurred.

CASE IN POINT

The first year of the CCPA saw dozens of cases of cited violations, but there were no publicly announced fines levied. Two examples of unnamed companies in violation are a car dealership and a grocery store chain. The car company did not notify consumers who test drove vehicles that it had collected their information. In addition, its privacy policy omitted information about rights under the CCPA. The grocery chain failed to notify consumers that it was collecting information about them in its loyalty program and did not provide them a notice that explained that their participation in the program was in exchange for their personal information (Irei, 2021).

In 2023, California amended the CCPA with the California Privacy Rights Act (CPRA). This legislation expanded the rights of California residents, including giving them the right to correct their personal information, limit the use of sensitive personal information, and opt out of automated decision-making. The CPRA also established the California Privacy Protection Agency as an enforcement arm.

Virginia

In 2021, Virginia became the second US state to enact privacy legislation. The **Virginia Consumer Data Protection Act (VCDPA)** shares some commonalities with the GDPR and the CCPA but also has a few differences. Unlike the CCPA, the Virginia law does not have an automatic revenue threshold, so a large company not meeting the defined personal data management criteria is not subject to the law. Another difference is that the Virginia law has a few more carve-outs, including one for employee and business-to-business personal data. The rights granted to individuals under the law are like those in other regulations. Under the VCDPA, an individual has the right to access, delete, correct, and move personal data. Consumers may also opt out of providing data and may appeal a decision to disallow one of the stipulated rights. There are limitations on the data a business (controller) may collect, and security measures to protect data are required. Only Virginia's Attorney General may enforce the law, and penalties are $7,500 per violation.

Virginia Consumer Data Protection Act (VCDPA) A state-level data privacy law enacted in March 2021 that establishes privacy rights for Virginia residents and compliance obligations for businesses that process their personal data.

Colorado

The third state to enact privacy legislation was Colorado. The **Colorado Privacy Act** became effective in 2024 and gives Colorado residents the right to opt out of targeted advertising and prevent sales of personal data and some kinds of profiling. The law has similarities to those of California and Virginia but is somewhat narrower than California in its applicability.

Colorado Privacy Act A state-level law that grants Colorado residents specific rights over their personal data and imposes obligations on businesses that collect or process this data.

Industry-Specific Legal Privacy Requirements

Many industries have specific data security and privacy protection requirements. Four industries that typically collect an abundance of data about their consumers and other stakeholders are educational institutions, medical care providers, financial service institutions, and credit card processors. Chapter 7 discussed the PCI DSS framework for credit card processors. This section examines the laws specific to educational, medical, and financial services organizations.

The Family Educational Rights and Privacy Act (FERPA)

The **Family Educational Rights and Privacy Act (FERPA)** was signed into law in 1974, making it one of the oldest comprehensive privacy protection statutes in the United States. The law applies to all public and private schools as well as any educational agencies receiving funds from a US Department of Education program.

Family Educational Rights and Privacy Act (FERPA) A US federal law that protects the privacy of student education records. It applies to all schools and educational institutions that receive funding from the US Department of Education.

FERPA protections vary, depending on the classification of information. "Directory information" includes a student's name, address, phone number, dates of attendance, and other identifying information that would not cause harm if shared. A school may disclose directory information without the consent of students or parents, unless specifically asked in writing not to do so.

Parents, or students who are 18 years or older ("eligible students"), have privacy protection rights regarding educational records and PII. These rights include the right to inspect and review education records the school maintains and the right to request data corrections. The educational institution may not release educational information without consent from parents or eligible students, with some exceptions. Figure 11-6 shows a flyer from the US Department of Education created by its Student Privacy Policy Office's Privacy Technical Assistance Center.

> **EXAMPLE**
>
> **Sometimes, a student is unhappy with a final grade in a course. Occasionally, that student's parents will call the professor directly to discuss the grade or, unfortunately, have an attorney call the professor. Under FERPA, a professor may not discuss the grades of an adult student (over the age of 18) with others without authorization. The student needs to contact the professor directly.**

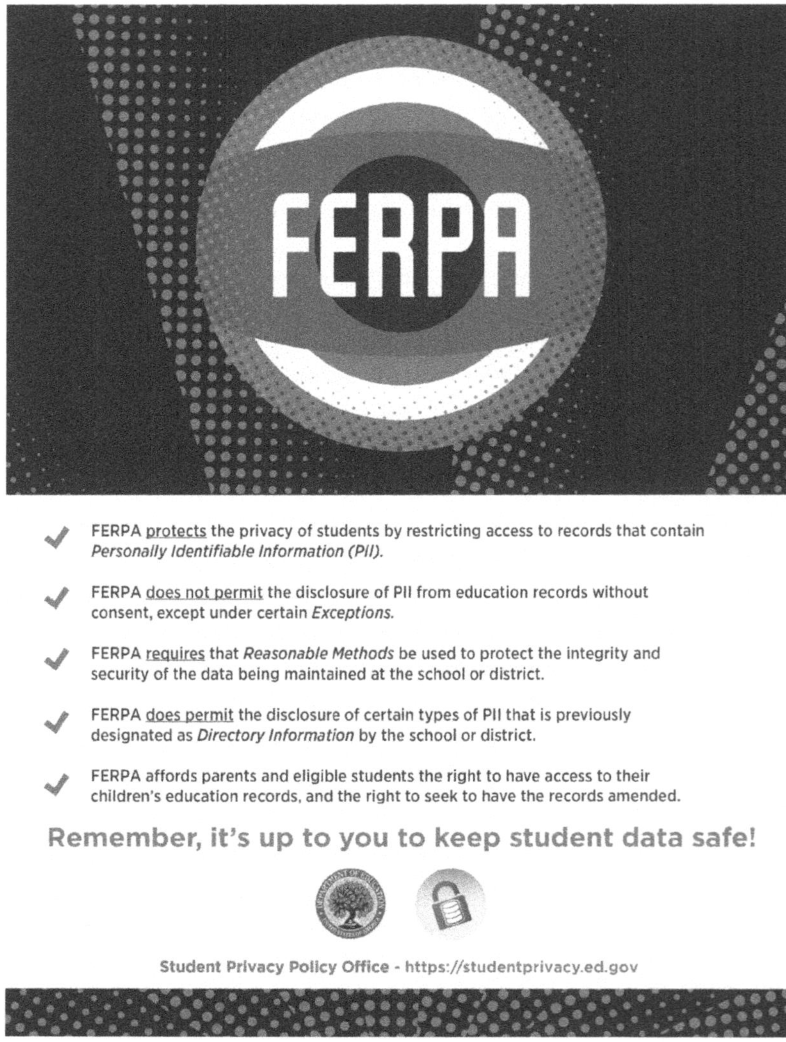

Figure 11-6. A Sample FERPA Flyer (US Department of Education)

In institutions of higher education, eligible students may choose to waive some of their privacy rights. For example, a student may waive the right to have their educational performance kept private from their parents. The student may also waive the right to view an instructor recommendation, understanding that the recommendation may carry more weight if it has been kept private from the subject of it. FERPA rules can be difficult for educators to navigate and typically require some training to ensure unwitting violations.

Health Insurance Portability and Accountability Act (HIPAA)

Health Insurance Portability and Accountability Act (HIPAA) A US law designed to provide privacy standards to protect patients' medical records and other health information.

The **Health Insurance Portability and Accountability Act** became US law in 1996. This law provides medical patients with the right to protect their data from disclosure without their knowledge or consent. Most of us know the law as being the one that requires us to sign that special form in our doctor's office, making it perhaps the best known privacy law. HIPAA is to medical records what FERPA is to student educational records.

HIPAA covers protected health information (PHI), which is data relating to an individual's health, healthcare provision to an individual, and individual healthcare payments. These data include any medical record data that can identify an individual, as well as data created, used, or disclosed during the course of medical diagnosis or treatment. HIPAA compliance is required of **covered entities**. These entities include mostly healthcare providers, health plans, and hospitals, but they do not include all types of organizations that may have your PHI, as in the following case.

Covered Entities (HIPAA) The organizations and individuals directly subject to HIPAA's Privacy, Security, and Breach Notification Rules.

> **CASE IN POINT**
>
> **23andMe is a genetic testing company that collects an individual's most personal data—their DNA—yet the company is not covered by HIPAA. While customers need to opt in to give consent for their data to be shared, a 2023 data breach highlighted the potential for PII to be taken. This breach contributed to financial distress for the company and such a significant decline in value that the company needed to sell its assets. Its most valuable asset was its database of customer DNA. While there were some assurances that the privacy policy would protect an individual's data, many people who had submitted their genetic material for testing sought to delete their data prior to a sale.**

HIPAA is known to most of us as privacy legislation, but it is actually much broader in scope. The law includes "portability" in its name because it is intended to provide healthcare insurance coverage to employees who move from one job to another. The law was drafted to promote healthcare industry efficiency through standard coding that would reduce documentation burdens.

HIPAA has three rules to protect patient data. The **HIPAA Privacy Rule** describes when PHI may be disclosed. This rule gives individuals the right to request to see the information their healthcare providers maintain about them. The **HIPAA Security Rule** describes the protocols required for safeguarding electronic PHI. These standards protect electronic PHI created, received, used, or maintained by a covered entity, which is any entity handling PHI that is covered under HIPAA. It was not until 2009 that the third rule, the **HIPAA Breach Notification Rule**, was introduced. This rule requires covered entities to notify individuals and the US Department of Health and Human Services of a data breach that has compromised PHI.

Figure 11-7 shows the steps a covered entity needs to take to be HIPAA compliant.

HIPAA Privacy Rule A core part of the Health Insurance Portability and Accountability Act of 1996 (HIPAA) that sets national standards for how protected health information (PHI) is used and disclosed.

HIPAA Security Rule A part of the Health Insurance Portability and Accountability Act of 1996 (HIPAA) that sets national standards for protecting electronic protected health information (ePHI), complementing the HIPAA Privacy Rule.

HIPAA Breach Notification Rule A federal regulation under the Health Insurance Portability and Accountability Act (HIPAA) that requires covered entities and their business associates to notify affected individuals, the US Department of Health and Human Services (HHS), and, in some cases, the media when there is a breach of unsecured protected health information (PHI).

Figure 11-7. HIPAA Compliance (Source: Healthcare Compliance Pros)

Graham-Leach-Bliley Act (GLBA)

Apart from health records, consumers are generally most interested in protecting the privacy of their financial information. Enacted in 1999, the **Gramm-Leach-Bliley Act (GLBA)** included

Gramm-Leach-Bliley Act (GLBA) A US law requiring financial institutions to explain their information-sharing practices and protect sensitive data.

many financial services industry reforms. Among them was the protection of individual financial privacy. The US federal law requires consumers to be informed about how their financial data are shared and requires companies to provide data security. GLBA applies to any business that is "significantly engaged" in providing financial products or services. Figure 11-8 summarizes the legislation.

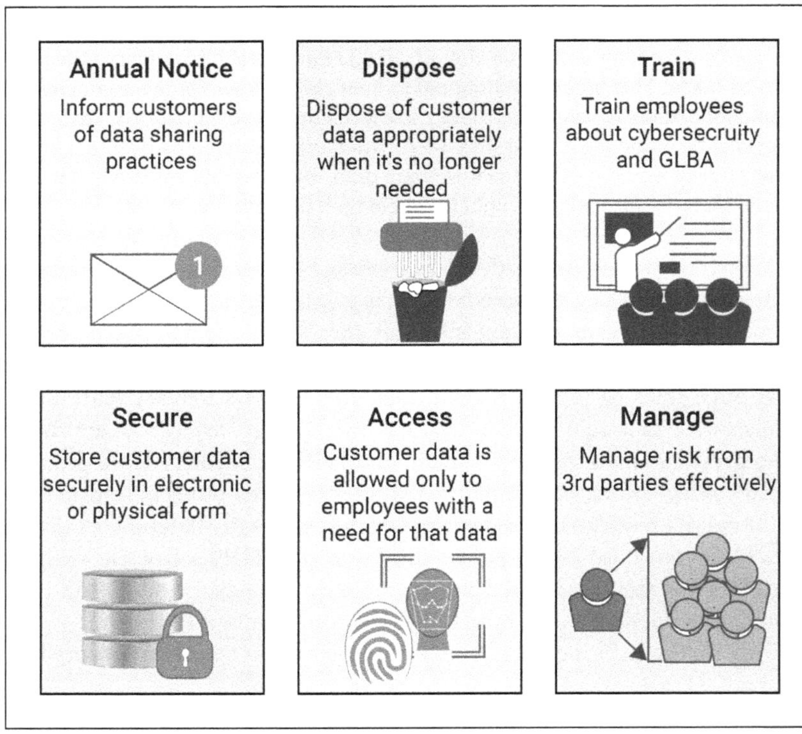

Figure 11-8. Gramm-Leach-Bliley Act Requirements

GLBA Privacy Rule A federal regulation under the GLBA of 1999 that requires financial institutions to protect the privacy of consumers' nonpublic personal information (NPI) and provide them with clear notices of their privacy practices.

One of GLBA's two rules is the **Privacy Rule**. (Chapter 9 emphasized the Safeguards Rule.) The Privacy Rule includes most PII and transaction data, such as bank account numbers. It also covers data that an entity might obtain during a transaction with consumers, such as in a lending agreement or credit report. The rule requires visible notice about an institution's policies and practices for sharing consumer information. Customers should, with some exceptions, have the right to opt out of having nonpublic personal information shared with nonaffiliated third parties.

Financial penalties for noncompliance with GLBA are significant, with a potential civil fine of $100,000 per violation. Officers and directors of covered institutions also might be criminally prosecuted and subject to personal liability for a $10,000 civil fine per violation.

Privacy and Digital Natives

Digital Native A person who has been exposed to digital technology from an early age and is generally comfortable using digital tools such as computers, smartphones, the Internet, and social media.

The digital age brought with it new ways to access individuals' private information. **Digital natives** have grown up with the Internet and are used to their information being "out there." That does not mean that privacy does not concern them, but it does mean that they are likely to be online more than older generations. But as users increasingly spend time online engaged with websites, such as social media and gaming platforms, hackers and scammers have found increasing opportunities for obtaining and monetizing users' personal information.

CASE IN POINT

One of the first cases that demonstrated how personal information could be exploited in a digital environment is that of Cambridge Analytica. Cambridge Analytica Ltd. was a British political data consulting firm. The company used data modeling and profiling to send targeted political messages to specific audiences. In 2014, the company harvested data from Facebook users to create psychological profiles of them and subsequently influence their behavior. Interestingly, the Cambridge Analytica scandal was not considered a data breach because the personal data was not stolen. Instead, the data could be taken without permission because of the state of privacy protections in the Facebook platform as it existed at the time, which allowed third parties to access data from users' Facebook friends. The scandal prompted enhanced privacy protections on social media platforms. One article noted that it led to a "great privacy awakening" (Lapowsky, 2019).

Individuals do have ways to protect themselves, though they do not always take advantage of them. Consumers today seem to be willing to give out their information without reading privacy policies or asking why they are being asked to give a vendor their birth date or mobile number. Brick-and-mortar retailers ask for phone numbers and email addresses at checkout. Physical and online vendors trade loyalty cards with perks for personal data.

Users *can* make changes to default settings to help guard their privacy. Apple, Google, Microsoft, Facebook, and others all have settings for personal information such as location and cookies, and they can be reset to protect privacy. (Chapter 13 covers personal identity protection in depth.)

Auditing Data Privacy

As noted, data privacy is important enough worldwide for there to be laws enacted specifically addressing this issue. While breaking these laws does not (normally) result in jail time for the perpetrators, the laws do have serious financial fines attached to them, with the possibility of punitive damages being attached to settlements if the breach of privacy was particularly egregious. It is therefore a good practice for organizations to either audit their own data privacy or have an independent third party perform a data privacy audit.

The areas relating to a data privacy audit are similar to a normal audit and encompass the following steps:

1. Define audit scope and objectives.
2. Identify data assets.
3. Review privacy and policy notices (including data consent, rights, and retention).
4. Evaluate legal and regulatory compliance.
5. Review existing controls and safeguarding measures.
6. Evaluate compliance and identify needed changes.
7. Document findings, provide recommendations, and implement needed improvements.

A privacy audit is similar to a controls audit that would take place as part of a financial audit. However, the applicable privacy laws (the GDPR, individual state requirements, etc.) in jurisdictions where the company does business or collects data are the primary focus, as well as how those law requirements are being met. Although most of the laws do not explicitly require privacy audits (with China's PIPL being a noted exception), it is generally considered a functional requirement for organizations to conduct regular reviews to confirm compliance with the applicable laws.

Summary

Privacy and cybersecurity are closely related. Privacy involves keeping personal and proprietary data safe from unauthorized access. This chapter described different classifications of data in terms of levels of desired security. The chapter went on to discuss data security and privacy policies. Countries around the world are increasingly aware of the importance of privacy and are enacting laws and creating regulations and governance to protect individual and organizational data privacy. Some of these laws are governmental, and others are prescribed by industry sectors. The chapter describes examples of both. Many individuals and businesses today do not take the steps to protect their data, or at least they don't enact protections until it is too late. Hopefully, a better understanding of the need for privacy will alert individuals and organizations to do a better job securing data. Chapter 13 provides guidance on personal data protection.

Chapter Review

Key Terms

American Data Privacy and Protection Act (ADPPA)
California Consumer Privacy Act (CCPA)
Colorado Privacy Act
Confidential Data
Confidentiality
Cookies
Covered Entities (HIPAA)
Data at Rest
Data Broker
Data in Motion
Data Mining
Digital Native
Electronic Communications Privacy Act (ECPA)
Environmental, Social, and Governance (ESG)
Family Educational Rights and Privacy Act (FERPA)
Federal Trade Commission (FTC)
General Data Protection Regulation (GDPR)
GLBA Privacy Rule
Gramm-Leach-Bliley Act (GLBA)
Health Insurance Portability and Accountability Act (HIPAA)
HIPAA Privacy Rule
HIPAA Security Rule
HIPAA Breach Notification Rule
Internal Data
List Broker
Personal Information Protection Law (PIPL)
Privacy
Privacy Policy
Public Data
Restricted Data
US Privacy Act of 1974
Virginia Consumer Data Protection Act (VCDPA)

Review Questions

1. Why does privacy matter?
2. What types of data are important with respect to privacy?
3. Who collects data on individuals and organizations? Are there any inherent issues with these practices?
4. How are data collected on individuals and organizations?
5. Explain the use of an organizational privacy policy.
6. What are the major sections of a privacy policy?
7. What are the major privacy frameworks?
8. What are the similarities (and differences) between the major privacy frameworks?
9. Explain the significant privacy regulations and compare the international laws with laws in the United States.
10. What are the main industry legal privacy requirements?
11. What are the similarities (and differences) between the main industry privacy requirements?
12. What is a digital native? Why is this classification important?
13. How do you audit data privacy?

Multiple Choice

Identify the choice that best completes the statement or answers the question.

1. How do confidentiality and privacy differ in the context of information security?
   a. Confidentiality refers to the ethical obligation of trust, while privacy is a legal right to keep data secret.
   b. Confidentiality and privacy are synonymous terms with no difference.
   c. Privacy involves protecting data from unauthorized access, whereas confidentiality does not.
   d. Privacy is only concerned with digital data, while confidentiality applies to physical data.
   e. Confidentiality is a legal requirement, whereas privacy is an ethical choice.
2. What are the main reasons for protecting individual privacy?
   a. To increase social media engagement
   b. To ensure freedom of speech
   c. To prevent embarrassment, fraud, or identity theft
   d. To enhance advertising strategies
   e. To comply with public relations policies
3. What is the significance of the 2022 US Supreme Court Ruling *Dobbs v. Jackson* in the context of Internet privacy?

a. It highlighted the risk that search histories could lead to criminal prosecutions.
b. It declared that all Internet search histories are private under all circumstances.
c. It resulted in the outlawing of cookies on websites.
d. It mandated that all search engines erase their logs every 24 hours.
e. It increased the legal protections for digital data collection.

4. What information do privacy policies typically contain?
a. Only the company's financial data
b. Details about the company's stock options
c. What personal information is collected, how it is used, and if it is shared with third parties
d. Employee personal contact information
e. Competitive marketing strategies

5. Why has the digital world intensified the need for privacy regulations?
a. It has simplified data collection and reduced privacy concerns.
b. Digital advancements have eliminated all previous privacy threats.
c. The digital world has introduced new privacy threats, necessitating more protections.
d. Digital technologies only affect privacy in certain geographic regions.
e. The digital world has decreased the overall data available for collection.

6. Which state was the first to enact a privacy law in the United States?
a. Virginia
b. Colorado
c. California
d. Connecticut
e. Utah

Case Studies

Yellow Subs—Privacy Policy

Herron Consulting has been engaged to help Yellow Subs draft a privacy policy. Many customers purchase their food from the company online, either for takeout or delivery. Natalie has been assigned to draft the policy. She knows that online templates and tools are available to help her in completing this assignment, but she has decided that the best place to start is to consult AI.

Case Requirements

1. Create an AI prompt to draft a privacy policy for Yellow Subs. What elements did you include in your prompt? Can you think of additional data you could include in the prompt to create a better draft?
2. Once you have obtained a draft of a privacy policy for Yellow Subs from AI, compare it with the sections described in Table 11-1. Does it include all of the sections listed in the table? If not, which sections were omitted, and why do you think this occurred?
3. Are there any changes or enhancements Natalie should make to the AI-generated policy prior to submitting it to the client? If so, describe them. If not, do you believe the policy draft is ready to be submitted to the client?

Uber Technologies, Inc.—Privacy Complaint

In 2014, Uber experienced a cyber breach in which hackers obtained the PII of the company's drivers. In 2016, Uber suffered a second, more extensive cyber breach in which hackers accessed the PII of millions of both drivers and riders. Rather than disclose the second breach, the company's Chief Security Officer paid the hackers $100,000 in bitcoin and asserted that no PII had been accessed. The US Federal Trade Commission reached a settlement with Uber in 2018 concerning allegations that the company had miscommunicated with consumers about its privacy and data security practices.

Case Requirements

1. Research the Uber privacy case. Identify three specific complaints made about Uber's privacy policies and practices.
2. Review the 2018 Uber settlement with the FTC and the FTC decision (Complaint #1523054). What was Uber required to do in the decision with respect to a privacy program, third-party privacy assessments, and covered incidents?

Ohio Southern University—Internal Audit and Privacy

Nancy Tran is an internal auditor at Central New Jersey University. The state has just passed a new data privacy law, the New Jersey Data Protection Act. The new law shares many of the features of other state laws, including Colorado's. The act grants consumers the right to access, correct, delete, and opt out of certain processing. Along with other requirements, controllers are obligated to practice data minimization and provide security. Nancy will need to conduct an audit to assess whether the university is complying with this new legislation.

Case Requirements

1. The New Jersey Data Protection Act is like the state regulation in Colorado but differs in some respects from the Iowa Consumer Data Protection Act. How is New Jersey's law different from Iowa's? Why do you think states adopt different requirements in their data privacy legislation?
2. As part of her audit, Nancy needs to understand how the organization is complying with data minimization. What steps might she take to accomplish this?
3. As part of her audit, Nancy needs to ensure that the university can respond to the rights of consumers specified in the law. What steps might she take to accomplish this?

Exercises

11-1. Colleges and Universities have extensive PII and PHI to protect. Find your school's Privacy Statement or Privacy Policy and answer the following questions.Required:
1. What PII data does your school collect?
2. What does the Privacy Statement or Policy say about how PII data are used?
3. What does the Privacy Statement or Policy say about how PII data are shared?
4. Do-Not-Track is a public-private initiative that allows a user to activate a flag in their browser software to notify websites that they do not want to have their browsing behavior tracked. Does your school's Privacy Statement or Policy note that they would agree with this request?

11-2. The EU's General Data Protection Regulation (GDPR) became effective in 2018. It includes strict penalties for violation of its data privacy and security standards. While it is an EU law, it has had global impact, not only because many businesses in every country have customers in the EU but also because it has influenced other country and territorial laws and regulations. Access the regulation and answer the following questions. Required:
1. What specifically does the GDPR say about consent?
2. What are the special consent considerations for a child under 16 years of age?
3. Does the regulation specifically mention cookies? If not, how has it influenced the collection of PII online?

11-3. Both the AICPA's Privacy Management Framework (PMF) and the Fair Information Practice Principles (FIPP) provide guidance on creating a privacy risk management program. Required:
1. Compare and contrast the components in the PMF with the principles described in the FIPP.

11-4. We are frequently asked to accept or deny a cookie policy when our Internet browser accesses a website. Required:
1. Log on to your computer and check your browsing history. How many items are there in your history? How many cookies and other site data have been collected from you in the past 24 hours?
2. What do you think will be the impact to you if you delete your browsing history and cookies?
3. Access a new retail company online. Read the cookie policy. What kinds of cookies does the site use? What features will you lose if you deny permission to install cookies?

References

Auxier, B., Rainie, L., Anderson, M., Perrin, A., Kumar, M., & Turner, E. (2019, November 15). *Americans' attitudes and experiences with privacy policies and laws.* Pew Research Center. https://www.pewresearch.org/internet/2019/11/15/americans-attitudes-and-experiences-with-privacy-policies-and-laws/

Federal Privacy Council. (2025). *Fair information practice principles (FIPPs).* Retrieved November 13, 2025, from https://www.fpc.gov/resources/fipps/

General Data Protection Regulation (GDPR). (2016). https://eur-lex.europa.eu/eli/reg/2016/679/oj

Irei, A. (2021, September 30). *10 CCPA enforcement cases from the law's first year.* TechTarget Network. https://www.techtarget.com/searchsecurity/feature/10-CCPA-enforcement-cases-from-the-laws-first-year

Lapowsky, I. (2019, March 17). How Cambridge Analytica sparked the great privacy awakening. *Wired.* https://www.wired.com/story/cambridge-analytica-facebook-privacy-awakening/

Levenson, E., & Sgueglia, K. (2023, January 18). *These are the Google searches Brian Walshe made before and after killing his wife Ana Walshe, according to prosecutors.* CNN. https://www.cnn.com/2023/01/18/us/brian-walshe-ana-walshe-google-searches

Meyer, C. D., Heuga, M. P., Jennings, E., & Manglik, G. (2015, August 12). *Transferring customer data in an asset sale.* Pillsbury Winthrop Shaw Pittman LLP. https://www.pillsburylaw.com/en/news-and-insights/transferring-customer-data-in-an-asset-sale.html

National Institute of Standards and Technology (NIST). (2025, April 14). *NIST privacy framework 1.1.* https://doi.org/10.6028/NIST.CSWP.40.ipd

Chapter 12

Geopolitics and Cybersecurity

Chapter Outline

Geopolitical Cybersecurity Risks
Nation-State Threat Actors
Threats to Infrastructure
Healthcare Systems at Risk
Manufacturing and Supply Chain Disruption
Spreading Misinformation and Hacking Elections

Learning Objectives (LO)

After reading this chapter, you should be able to:

- LO1. Understand how geopolitical cyber risks impact organizations.
- LO2. Evaluate nation-state actors posing cyber risk to the United States and their motivations.
- LO3. Understand how hacktivism creates cyber risk for organizations.
- LO4. Be familiar with cyber threats to infrastructure, including communication systems, education, governments, monetary systems, the power grid, and water supplies.
- LO5. Be familiar with cyber threats to healthcare systems, manufacturing and supply chains, and elections.

Introduction

Although this book focuses primarily on cybersecurity as it relates to business organizations, accountants and business leaders must also consider geopolitical cybersecurity risks as part of an overall risk assessment. These risks may have significant revenue and expense implications that may even result in issues about whether the organization is a going concern. This is particularly the case in industry sectors that are considered critical infrastructure, such as electric utilities and manufacturing. Beyond the concern of accountants and business leaders with respect to financial performance, our nation's economy and political system face threats that can lead to political unrest with economic implications.

This chapter describes digital risks at the global economy level rather than at the individual, firm, or industry level. It provides background on adversarial nation-states posing the most significant threat to the United States and its allies, recognizing that almost all nations pose cyber threats to others. One focus of nation-states' cyberwarfare or cyber espionage efforts is attacking the infrastructure of their opponents. Nation-state cyber threats to infrastructure include attacks on educational institutions, government entities, the power grid, transportation modes, and the water supply. Cyberattacks on healthcare systems are particularly insidious, as they can even result in loss of life, in addition to causing chaos and instilling doubt in a country's institutions. Furthermore, the assets of healthcare entities, including PHI, allow nation-states to gather data and create profiles of individuals, as well as making them a popular target for cybercriminals, as discussed in Chapter 11. Two other types of geopolitical cyberattacks are disruption of the supply chain and attacks on election systems, both of which can destabilize the attackers' opponents.

Geopolitical Cybersecurity Risks

In 2020, Robert J. Samuelson, a former business and economics writer for the *Washington Post*, warned:

> The consequences of a massive cyberattack could make the disruptions caused by the pandemic seem like child's play. There might be simultaneous assaults on the nation's power, communication, financial, and transportation networks. People would stumble about in a cyber fog with public and private communications channels, from email to cable TV, disabled or overwhelmed. (Samuelson, 2020)

If it sounds scary, that is because it is! In 2018, David Sanger published *The Perfect Weapon*, a book warning about how cyberweapons have changed geopolitics and threaten the survival of nations by potentially taking down infrastructure, disrupting manufacturing and supply chains, crippling financial systems, and creating turmoil through the spread of disinformation (Sanger, 2018). In this book, he describes Stuxnet as the first **cyberweapon** and how nation-states deploy cyberweapons in much the same ways they might deploy other weaponry, except that sometimes the source, reach, and nature of the threat are more difficult to assign. Building an aircraft carrier and developing nuclear weapons may be beyond the reach of many nations, whereas cyberwarfare poses a much lower barrier to entry. And while apocalyptic scenarios where a nation dismantles a portion of the infrastructure of an enemy nation, such as its power grid, are daunting, disinformation is an even lower cost way to engage in cyberwarfare.

Cyberweapon A software tool or system designed specifically for the purpose of causing harm in cyberspace, often used in cyberwarfare.

Global Threat Landscape The evolving array of risks and malicious activities that impact cybersecurity, geopolitical stability, economies, and societies worldwide. It encompasses cyber threats, terrorism, disinformation campaigns, climate-related risks, and emerging technologies that may be exploited for harm.

It is important for those concerned with cybersecurity to keep up to date with global geopolitical threats. Many security organizations offer **global threat landscape** reports. These reports describe the total environment of cybersecurity threats, including the types of attacks and the prevalence of each type, the current active threat actors, and nation-state activity.

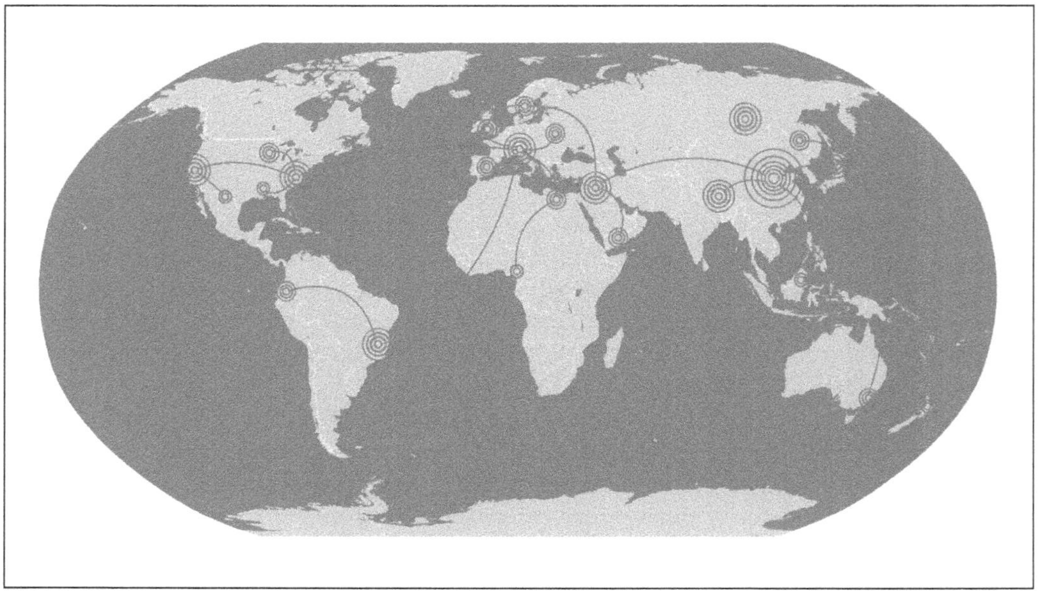

Figure 12-1. A Static Cyber Threat Map

CASE IN POINT

CrowdStrike is a US cybersecurity company with global reach. Its *2025 Global Threat Report* includes a table showing individual adversary names, such as Crane, Hawk, and Bear, and their nation-state affiliations. The report also shows the global regions by intrusion percentage, with the United States targeted most by interactive intrusions (53%). The report concludes by noting that Russia's war against Ukraine would focus on intelligence collection and that China-nexus threat actors will leverage cyber investments for prolific global activity (CrowdStrike, 2025 p. 43).

Cyber Threat Map Also known as a cyberattack map. A real-time map of the computer security attacks that are going on at any given time.

Another source of data about global cyber activity are **cyber threat maps**, which track global cyberattacks. Graphic visualizations of cyberattacks around the globe demonstrate the frequency and worldwide incidence of cyber threats. Many cybersecurity companies produce these maps. Some of them are in real time, such as the one shown in Figure 12-1. This one, from Check Point, summarizes the top nations targeted, the most targeted industries, and frequently observed malware types.

Nation-State Threat Actors

It is often difficult to tell the difference between attacks from nation-states and general cybercriminals. Russia often uses cybercriminals as proxies for attacks on its adversaries. Many nations hack into other nations' systems and do nothing but maintain access to those systems for use at a later date. A ransomware attack on an infrastructure organization could be a criminal taking advantage of a security weakness, or it could be a nation-state test run to gauge the reaction of an opponent. Script-kiddies or social hackers are unlikely to be the bad actors in these cases, as they generally require a high skill level. A nation-state might corrupt an insider of an organization in a target country to gain access. Most nation-state cyberattacks are either by criminals who may be protected from prosecution within their home country or by government-sanctioned adversaries.

Chapter 3 identified the governments most likely to wage a cyberattack against the United States as China, Iran, North Korea, and Russia. Each of these countries has differing motivations, and, beyond these four, almost every country with an active military considers cyberweapons as tools in their military arsenal. The United States also engages in cyberwarfare in both defense and offense. Both the US Army Cyber Command and the National Security Agency (NSA) use cyberspace operations to defend networks in the United States and influence global cyber activities. Their offensive actions are meant to deter foreign cyber aggression against US interests. Cyber incidents may be a weapon that precedes a military invasion or attack, as was the case with Russia and Ukraine. Escalating tensions between China and Taiwan are playing out in cyberspace as well.

Table 12-1 summarizes the motives that China, Iran, North Korea, and Russia may have in launching a cyberattack against the United States. Other countries may have similar or different motives, ranging from financial gain to political disruption and instability or chaos.

Table 12-1. Nation-State Threats to the United States and Their Motivations

Country	Possible Motivation for Cyberattack
China	Stealing IP and trade secrets; political and economic disruption
Iran	Retaliation for perceived aggression and economic sanctions; disruption of critical infrastructure
North Korea	Financial gain; causing reputational damage and promoting its political agenda
Russia	Retaliation for perceived aggression and economic sanctions; disrupting critical infrastructure

The following sections of this chapter delve into more depth around the cyber agendas of China, Iran, North Korea, and Russia, though they are not always clear cut.

CASE IN POINT

While they are allies and have pledged, both in 2009 and 2015, not to promote cyberattacks on each other, there is significant evidence that China views Russia as a target for cyber espionage. Since the beginning of the war between Ukraine and Russia, groups affiliated with China and China's cyberwarfare directorate have targeted Russia. Many believe that China is trying to augment their information regarding actual battlefield expertise, which they do not have compared to Russia. As Russia has never acknowledged any of the attacks that have occurred, it is difficult to determine if the attacks have been successful; however, recently obtained F.S.B. (Russia's internal security/intelligence agency) documents indicate that Russia is concerned and even go so far as to classify China as an "enemy," at least in cyberspace (Bird, 2025).

China

In terms of purchasing power, China is the world's largest economy, and it is likely to overtake the US economy in size sometime in the next decade or so. The US Cybersecurity and Infrastructure Security Agency (CISA) posits that the People's Republic of China "engages in malicious cyber activities to pursue its national interests" (CISA, 2025). CISA warns that China has targeted and continues to threaten a variety of industries and government units to obtain intellectual property (IP) and data from critical infrastructure entities. CISA is not alone in expressing concerns posed by China as a cyber threat. According to the *2025 Annual Threat Assessment of the US Intelligence Community*, "China has demonstrated the ability to compromise U.S. infrastructure through formidable cyber capabilities that it could employ during a conflict with the United States" (Office of the Director of National Intelligence, 2025, p. 9).

The president of the People's Republic of China, Xi Jinping, has emphasized the importance of cybersecurity in national defense and encouraged the development of talent and technology to further this goal. These capabilities also provide the country with strategic advantages in offensive cyber intrusions as well, particularly against perceived enemies and rivals, including Taiwan and the United States. In recent years, Chinese hacker groups have made significant cyber intrusions into US companies and government entities. Two of the groups active in 2024 were Salt Typhoon and Volt Typhoon. The next two cases describe attacks by each group.

CASE IN POINT

In 2024, the United States discovered that China had infiltrated its telecommunications infrastructure. Salt Typhoon, an allegedly state-backed hacker group, breached multiple US telecom systems, including AT&T and Verizon. It appeared that the breach targeted specific high-profile customers, including political leaders. While Chinese officials denied culpability, Senator Mark Warner of Virginia, who chaired the Senate Intelligence Committee, warned of the seriousness of this breach, as the intruders had the capability to listen to audio calls and obtain records from thousands of customers.

CASE IN POINT

Volt Typhoon is a Chinese nation-state-backed hacker group that uses advanced persistent threat techniques. According to the FBI and the CISA, Volt Typhoon compromised the IT environments of several critical infrastructure organizations in communications, energy, transportation, and water system sectors in the United States (CISA, 2024). Volt Typhoon is particularly adept at stealth and can "live off the land," which means that rather than introduce its own malware, the attacker is able to compromise existing tools and processes to perform malicious actions.

Iran

Iran itself was the target of a joint cyberattack effort, the Stuxnet worm launched by the United States and Israel, which was intended to slow its nuclear program by causing centrifuges to spin out of control without the action being detected. Stuxnet is often referred to as the first incidence of global cyberwarfare. Figure 12-2 depicts the phases of Stuxnet.

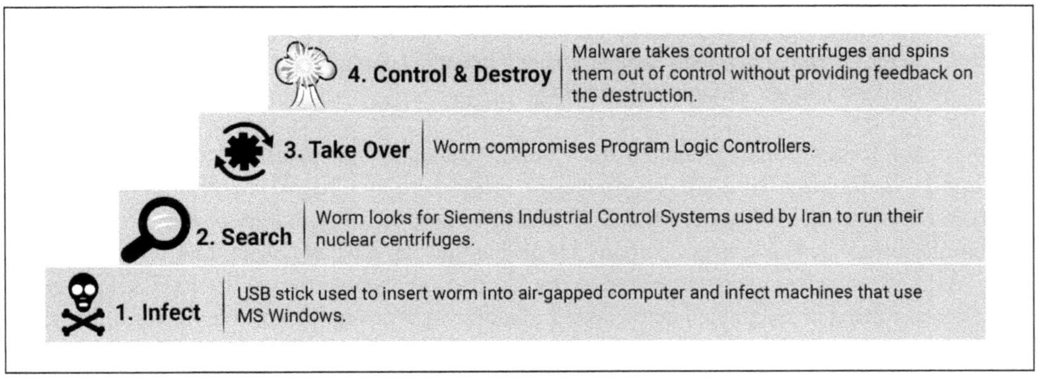

Figure 12-2. The Stuxnet Worm

While Iran was the target of the first cyberwarfare attack, it has also been home to many threat actors that have staged attacks against US industries, including those in critical infrastructure sectors. Iran's Islamic Revolutionary Guard Corps (IRGC) is believed to be behind these efforts. In addition to attacks in the United States, Iran has used cyberattacks within its own borders and in other international arenas where it seeks to influence thinking or engage in espionage.

> **CASE IN POINT**
> **In 2014, relations between Iran and Albania deteriorated following Albania's acceptance of members of an exiled Iranian militant opposition group known as the People's Mujahideen Organization of Iran (MEK). In 2022, Iran was seemingly behind multiple cyberattacks on the MEK and Albanian government networks, which included leaving anti-MEK messages on computers. Although Iran has denied the attacks, these cyber intrusions disrupted diplomatic ties between the two countries.**

While politically motivated state-sponsored attacks are one prong of cyberwarfare against nonallied nation-states, countries including Iran also may turn a blind eye to cybercriminals operating within their countries. These criminals attack entities within perceived enemy nations for monetary reasons. In 2022, for example, Assistant US Attorney General Matthew Olsen from the Justice Department's National Security Division noted that the Government of Iran had become a safe haven for cybercriminals (Office of Public Affairs, 2022). Criminals in safe harbor countries can use ransomware and other malware to extort payment for compromised data.

North Korea

The Democratic People's Republic of Korea, commonly referenced as North Korea, stages cyberattacks globally to generate revenue, gather information, and sometimes disrupt other governments for political ends. Targets of their nation-state cyber program include financial institutions, critical infrastructure, cryptocurrency exchanges, and healthcare entities. The funding from these attacks is thought to be used to pay for nuclear capability development and weapons programs. Given the economic sanctions on the country imposed by many parts of the global community, North Korea may view nation-state sponsored cyberattacks as a revenue source.

CASE IN POINT

One of the most widespread cyber incidents to date was the WannaCry ransom malware deployed in 2017, which both the United States and the United Kingdom assert was created by a cybercriminal group with North Korean connections. This ransom incident impacted several hundred thousand computers in about 150 countries, resulting in hundreds of millions of dollars in damages. The WannaCry worm targeted Microsoft computers by encrypting data and demanding Bitcoin payments if the target wanted the data decrypted. While the attack infected many companies, industries, and nations, it was particularly damaging to the UK National Health Service, where it infected hospitals and medical offices, often rendering patient records inaccessible.

As economic sanctions against North Korea increase, so do hacker intrusions from state-sponsored cyberattacks. In 2022, cybersecurity researchers at Chainalysis asserted that North Korea-backed attacks took more than $1.7 billion in cryptocurrency. Part of that stems from a $100 million theft at blockchain company Harmony that the FBI has blamed on Lazarus Group (a.k.a. APT38), a hacker gang connected to the North Korean government. In 2023, CISA, along with several other US federal agencies, issued an advisory warning about attacks stemming from North Korea on healthcare organizations and other infrastructure entities. In addition to outright thefts, the country also uses cyberwarfare to achieve political ends, particularly targeting South Korean government offices and infrastructure.

CASE IN POINT

2024 appears to have been another good year for state-sponsored threat actors in North Korea, as they stole about $1.34 billion in cryptocurrency, most likely to fund military programs, including weapons development. One of the attacks was on the Japanese crypto exchange DMM Bitcoin. The $308 million theft ultimately led to the shutdown of the exchange. The FBI and several other groups attributed the attack to North Korean threat actors in the Trader Traitor group. Figure 12-3 describes the stages of this attack (Lakshmanan, 2024).

Figure 12-3. Stages in 2024 DMM Bitcoin Cyberattack

Russia

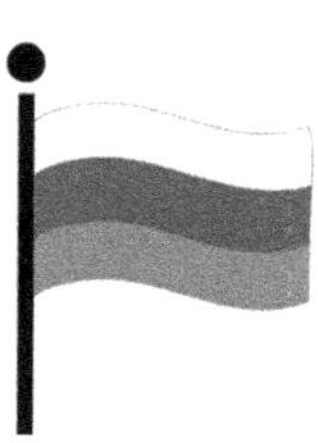

Russian cyberattacks on other nations have multiple motives. Hackers in Russia engage in cyberwarfare aimed at espionage, financial gain, harm to infrastructure, and destabilization of foreign governments. Bad actors have operated from Russia to attack supply chains, reap cryptocurrency rewards from ransomware attacks, shut down power grids in other countries, and steal documents and IP. The US Office of the Director of National Intelligence in its *2024 Annual Threat Assessment* noted that "Moscow views cyber disruptions as a foreign policy lever to shape other countries' decisions and continuously refines and employs its espionage, influence, and attack capabilities against a variety of targets" (Office of the Director of National Intelligence, 2025, p. 16).

> **CASE IN POINT**
>
> In 2016, the Petya malware, named for a Soviet satellite in a James Bond movie, was used in a series of ransomware attacks. The malware, which spread through compromised email attachments, encrypted a file system such that Windows systems were stopped from booting up. Victims were asked for payment in Bitcoin to obtain a decryption key. In 2017, NotPetya, a nonransomware variant, began to attack important Ukranian websites, including government ministries, banks, and utilities. From there, the malware went on to attack enterprises in more than 60 other countries, causing billions of dollars in damages. The United States has blamed the Russian hacker group Sandworm, which is affiliated with the Russian Main Intelligence Directorate (GRU), for the NotPetya cyberattacks.

The Russia-Ukraine war began in 2014, when Russia annexed Crimea, a militarily strategic area located within Ukrainian territory. The war escalated in 2022, when Russia invaded and occupied additional parts of Ukraine on the pretense of defending ethnic Russians from Ukrainian aggression. In 2015, Russia, through cybercriminal proxies, launched a cyberattack on Ukraine's power grid that knocked out electric power to some parts of the country for up to six hours. As a result, there was an expectation that cyberwarfare would be a significant part of the conflict. Leading up to and following the 2022 invasion, Russia-backed hackers attacked websites, stole and erased data, spread disinformation, and interfered with communications, but the cyber incidents to date were not devastating. Ukraine has defended itself through security and backup policies and procedures, including multifactor authentication.

> **CASE IN POINT**
>
> Couples seeking to marry in Ukraine at the end of 2024 may have had to wait because of a Russian attack on state services. The attack, one of the largest yet in the Russia-Ukrainian conflict, targeted multiple state registers that stored business records, voter information, tax records, and other citizen data. Following the intrusion, Ukraine temporarily suspended the registers.

Hacktivism

Chapter 3 briefly described hacktivists as individuals who use computer technology to advance a social cause or political agenda. While hacktivists may embrace domestic causes, their activities frequently cross international borders. Over time, **hacktivism** has evolved to become sophisticated in recruiting and targeting opposition. The term was first used in 1996, when the words *hack* and *activism* were combined by a member of the hacker group Cult of the Dead Cow (CDC). The group launched several digital protests, often targeted at companies, including a 1990 stunt where they aimed to force Microsoft to fix security flaws.

Hacktivist An individual or group that uses hacking techniques to promote political or social agendas.

The CDC (not to be confused with the US Centers for Disease Control) likely influenced the hacking group Anonymous. Anonymous, a decentralized hacktivist group, formed in the early 2000s and became famous in early 2008 when they staged a digital attack against the Church of Scientology. The protest was partly digital, with the inclusion of distributed denial of service (DDoS) attacks, but the group also used annoying phone calls and other disruptions intended to punish the church for its perceived denial of free speech. Anonymous continued to operate in many countries, including Iran. In 2009, the group protested over both the operation and the results of the Iranian presidential election. Throughout the 2010s and into the 2020s, Anonymous has continued to be active in launching attacks against individuals, organizations, and nations they believe to be corrupt. The following are some of their hacks:

- A teen in California who ran an anti-swearing website (2009)
- Websites run by Tunisia and Zimbabwe over censorship issues (2010)
- The Oregon "Tea Party" Facebook page for using a slogan ("We are Legion") that Anonymous uses (2010)

- Pro-copyright and antipiracy organizations, law firms, individuals (2010)
- Malaysian government websites to retaliate for web censorship (2011)
- Rural law enforcement data (PII, training files, informant information, etc.) as part of their "Shooting Sheriffs Saturday" campaign (2011)
- Syrian government email, with 2.4 million emails given to Wikileaks (2012)
- The Federal Reserve (2013)
- The government of Thailand to influence plans to strengthen cybersecurity laws (2016)
- Freedom Hosting II (a dark web hosting company), revealing that 50% of their clients' websites were for child pornography and other illegal activity (2019)
- DDoS attacks on Minneapolis government websites to protest the death of George Floyd (2020)

In 2022, following the death of Mahsa Amini in police custody, the group launched a series of attacks that disrupted Internet access throughout Iran. They also claimed to have accessed data held by the Iranian Parliament, and in one of their more notorious actions, they interrupted state TV news with an image of Ayatollah Khamenei surrounded by flames.

Hacktivists do not typically practice cyberterrorism. Their attacks are political or social, but rather than trying to raise fear, they typically work to raise awareness. For example, hacktivists might target government websites to protest election outcomes or policies. They also hack targets that will garner them widespread publicity. Common hacktivist tactics are summarized in Table 12-2.

Table 12-2. Hacktivist Tactics and Motivations

Tactic	Motive
Data breach	Steal and release sensitive information
Denial of service	Disrupt website of organization or government they oppose
Media campaign	Use social media platforms to spread messages and raise awareness
Website access	Access government-censored websites or access private sites
Website defacement	Replace website content with political messages

While hacktivists operate in every country around the globe, hacktivists are particularly active in uprisings, such as the Arab Spring in the early 2010s, and in nation conflicts, such as the Ukraine-Russia war and the war in Gaza. Hacktivists have been particularly active during the Russia-Ukraine conflict, taking cyber militancy to a new level. Anonymous and other hacktivist groups hacked Russian state TV stations to spread messages about the war and to communicate pro-Ukrainian messages. Various hacktivist groups have stolen information, interfered with news broadcasts, and leaked stolen data. The following case provides another example that illustrates the creativity hacktivists sometimes employ to get their messages across.

CASE IN POINT

In 2022, the group Anonymous took responsibility for hacking into the Russian cab company Yandex Taxi. The hackers caused a large traffic jam by sending the cabs to the same location at the same time. The traffic gridlock, which was publicized on social media platforms, snarled traffic and generally caused havoc lasting for several hours.

Infrastructure The fundamental physical and organizational systems and facilities needed for the operation of a society, business, or service. It provides the backbone that supports economic activity, public services, and daily life.

Threats to Infrastructure

One of the most frightening objectives of cyberwarfare is disruption of an enemy country's infrastructure. A nation's **infrastructure** is made up of all the systems and services that are vital to a country's economy and way of life (Figure 12-4). For example, transportation infrastructure, which includes roads, railways, airports, and the like, allows for the movement of goods and people in support of economic activity. Infrastructure is expensive and capital intensive and generally includes communication systems, monetary systems, the power grid, transportation, and the water supply. Education and government functions are also part of a nation's infrastructure.

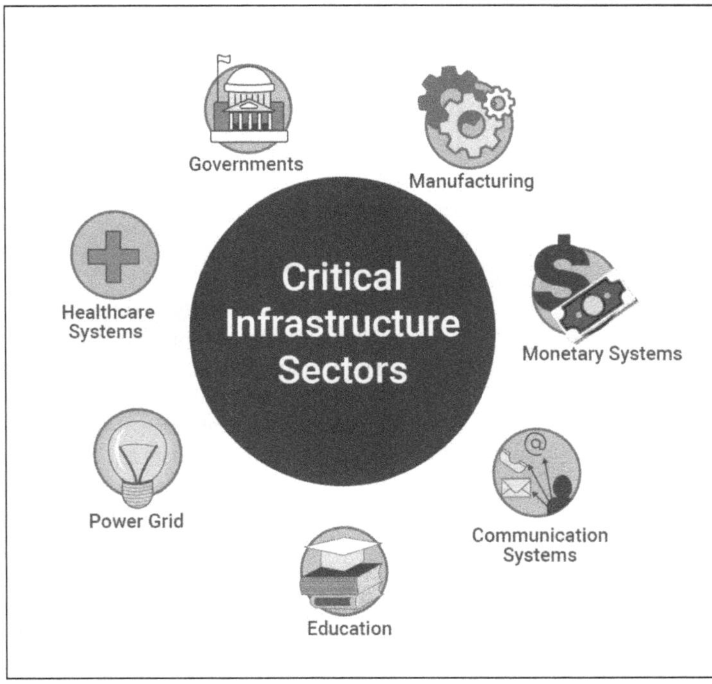

Figure 12-4. National Infrastructure Sectors

Communications Systems

It is difficult to imagine going through a day without a mobile phone, a digital tablet, a computer, or access to the Internet. Communications infrastructure makes the world and economies go round. It consists of both wireless and wired components, including satellites, fiber optic or other wired connections, and the radio frequency spectrum. A hack by a foreign nation-state that disrupts any of these communication channels in a sovereign country would bring the economy to a halt, as individuals and businesses could not transact.

While a large scale cyberattack on a nation's phone and Internet service would be difficult, it can happen, possibly through sabotage of satellites or ground stations. In 2023, more than 5,000 satellites were in orbit around the earth, with the majority of them being small satellites and many enabling communications. Satellites are increasingly commercial ventures, and they may be owned by governments, businesses, or individuals. The number of satellites is expected to grow exponentially, which means that the threat of a cyberattack will increase apace.

CASE IN POINT

In 2022, both the United States and the European Union blamed Russia for an attack on a satellite network in Ukraine and some neighboring European countries. The wiper malware attack destroyed Viasat customers' satellite modems and routers. It appeared that the intent of the attack was to disrupt communications in Ukraine prior to Russia's invasion of that country. Wiper malware, as its name suggests, is meant to "wipe out" or erase data. This particular attack cut off Internet access for some civilians, and, in damaging Viasat customer routers, it disconnected communications to more than 5,000 German wind turbines.

Wiper Malware Malicious software designed to delete or "wipe" data from a system, rendering it irrecoverable.

Nation-states often view an attack on an enemy's communication systems as providing a significant advantage in any ramp-up to military actions. They might infiltrate communication systems to access messages, as was the case in the 2024 telecommunications hack by Salt Typhoon described earlier in this chapter.

Education

Cyberattacks on education have tended to be either for-profit, with ransomware dominating, or DDoS attacks that disrupt services. They are aimed at both K–12 education and colleges and universities. Nation-state attacks may occur in higher education for many reasons, including the motive of stealing **intellectual property (IP)**.

Intellectual Property (IP) Creations of the mind, such as inventions, literary works, designs, or symbols, that are legally protected from unauthorized use.

Cyber Espionage The act of using cyber techniques to gain unauthorized access to confidential information, typically for political or economic gain.

Spying and stealing data is particularly a threat for research universities where medical, scientific, and engineering scholars are engaged in discovery. **Cyber espionage** involves both theft and surveillance. Foreign nation-states may steal data for economic or military advantage. The hallmark of these attacks is stealth, whereby the bad actor gains a foothold in the target system and maintains a low profile in the network while gathering data and information. The global pandemic, when nations were competitively researching and developing vaccines and treatment drugs, saw an increase in this type of activity.

CASE IN POINT

In 2019, the University of Hawaii's Applied Research Laboratory (ARL) was among departments and labs at more than 27 universities targeted in a spear-phishing attack that may have originated in 2017. China was blamed for the malware insertion at the lab, which is a Navy-sponsored University Affiliated Research Center. Research at the laboratory includes study of ocean environments, astronomy, environmental sensors, and renewable energy. Researchers at the university frequently share data with other academics, and a number of messages that included malware appeared to have originated from the ARL. The ultimate target may have been the Woods Hole Oceanographic Institution, which specializes in marine science and engineering (Volz, 2019).

Another motive might be targeting universities for political views or DEI initiatives, which might be viewed by groups as being divisive. Even in these instances, however, the tactics are likely to involve exfiltration of data, including research and credentials of faculty and staff.

Governments

In cyberwarfare, the goal is likely to be to disrupt the functioning of a foreign nation-state. Disabling government functions and discrediting government officials are efficient ways to accomplish that goal.

Center for Strategic and International Studies (CSIS) A nonpartisan think tank in Washington, DC, that provides research and policy recommendations on international security, foreign policy, economics, and technology.

The **Center for Strategic and International Studies (CSIS)**, a bipartisan, nonprofit organization that conducts policy research with an aim to ensure national security and prosperity, has tracked major cybersecurity incidents affecting nation-states since 2006. Table 12-3 summarizes some listed incidents from Spring 2025 (CSIS, 2025).

Table 12-3. Selected Incidents from the Center for Strategic and International Studies' Significant Cyber Incident Site

Date	Summary of Incident
March 2025	Hackers in Iran conducted cyber espionage campaigns against government entities in Iraq and telecommunications in Yemen.
March 2025	Companies linked to a Chinese tech firm targeted newly laid off US federal workers via job site recruitment ads.
April 2025	Hackers linked to Algeria launched a cyberattack against the National Social Security Fund in Morocco.
April 2025	Unidentified hackers spied on US bank regulator emails at the Office of the Currency for more than a year.
April 2025	US Cyber Command found Chinese malware on partner networks in several Latin American nations.
April 2025	North Korean cyber spies posed as remote workers and targeted European defense and government organizations for multiple reasons, including data theft, espionage, and extortion.

While malware and DDoS attacks are common approaches to hacking government entities, spreading information or disinformation is another way of destabilizing regimes. Disinformation campaigns can be a strategic approach to influence politics or public opinion in foreign nations.

EXAMPLE

There are many examples of this type of activity, going back as far as the Cold War, when Russia launched INFEKTION, a disinformation campaign that attributed creation of the AIDS virus to the United States.

Deepfakes are another way that bad actors can spread misinformation in attempts to topple a government regime. These computer-generated videos, images, or recordings showing well-known government officials saying or doing things that are either controversial or incendiary are a growing threat, as AI technology makes them appear more convincing. For example, former US presidential candidate and Secretary of State Hilary Clinton has been featured in multiple deepfakes, saying things that make no sense and perhaps even engaging in pornographic acts. In 2022, a deepfake video purporting to show Ukraine President Volodymyr Zelenskyy telling his military troops to surrender in their fight against Russia made the rounds on social media.

Deepfake A piece of synthetic media in which a person's likeness is replaced with another's or is manipulated in other ways, often used maliciously to create fake videos or audio clips.

The good news is that technology is being used to help viewers spot deepfake videos, such as lip sync deepfakes or face swap deepfakes. In a lip sync video, a person's mouth is manipulated to fit different audio so that they can appear to be saying something other than what they have actually said. The face swap is easy to understand—one person's face (head) is put in place of someone else's. Researchers use technology to track facial movements to identify the fake versus the real.

Monetary Systems

When Willie Sutton was asked why he robbed banks, he famously replied, "'Cause that's where the money is." Central banks are found in almost every country, and globally there are more than 200 of them. The Federal Reserve System is the central bank of the United States. The system has several functions, including monetary policy, stability, ensuring safe and sound financial institutions, consumer protection, and services that facilitate financial transactions. The system has trillions of dollars in financial assets, primarily securities. The Division of Information Technology within the Federal Reserve System has several information security functions, including a **zero-trust program**, meaning no user, device, or application is trusted by default.

Zero Trust A security model that assumes no implicit trust and requires verification for every access request.

The Federal Reserve's oversight of the US financial system recognizes the threats accompanying increased digitization in the system and recommends that an organization protect against the inherent risks. In 2021, the Federal Reserve Board, along with the Federal Deposit Insurance Corporation and the Officer of the Comptroller of the Currency, issued a ruling requiring cybersecurity incident notification for banks and their service providers. The rule also required notification be made to affected customers.

Many threat actors, from criminal gangs to nation-states to hacktivists, are motivated to launch cyberattacks against financial institutions and central banks. Central banks and large private banks are frequently a hacking target, and for that reason, and despite the fact that most have sophisticated and comprehensive cybersecurity programs, there have been lapses.

Computers and money have an enhanced relationship with the advent of cryptocurrency. Chapter 2 of this book discussed blockchain technology and cryptocurrency in some detail. As we noted there, the mere existence of cryptocurrency has enabled many cyber threats because of its pseudonymity. Cryptocurrency is a global currency, so there is no need for exchange when it crosses national borders. A problem for cybersecurity is that this allows hackers to move stolen funds from one country to another, making stolen funds denominated in cryptocurrency difficult to recover.

CASE IN POINT

In 2022, hackers stole almost $4 billion from cryptocurrency platforms. Many of the hackers were thought to be working to help the North Korean government obtain funds

to support its military initiatives, including its nuclear program. Chainalysis researchers reported that one of the ways that hackers were attacking decentralized finance platforms was by using tools that allowed users to move their cryptocurrency across blockchains. As an example, one such attack, on the blockchain company Harmony, netted hackers about $100 million (Greig, 2023).

The Power Grid

We take it for granted that when we flip a light switch, the lights will come on. A nation's power grid is a complex network of connected transmission and distribution lines that make up the system that transmits electricity to consumers. In the United States, the power grid is divided into three regional sections: Eastern, Western, and Texas. Other countries may have just one central grid.

Both cyber and physical attacks can disable or disrupt a nation's power grid. In 2024, for example, the power went out on the Iberian Peninsula for several hours, though the exact cause was unclear. Power failures have a cascade effect on other infrastructure, as communications, water supply, and transportation rely on electricity to function.

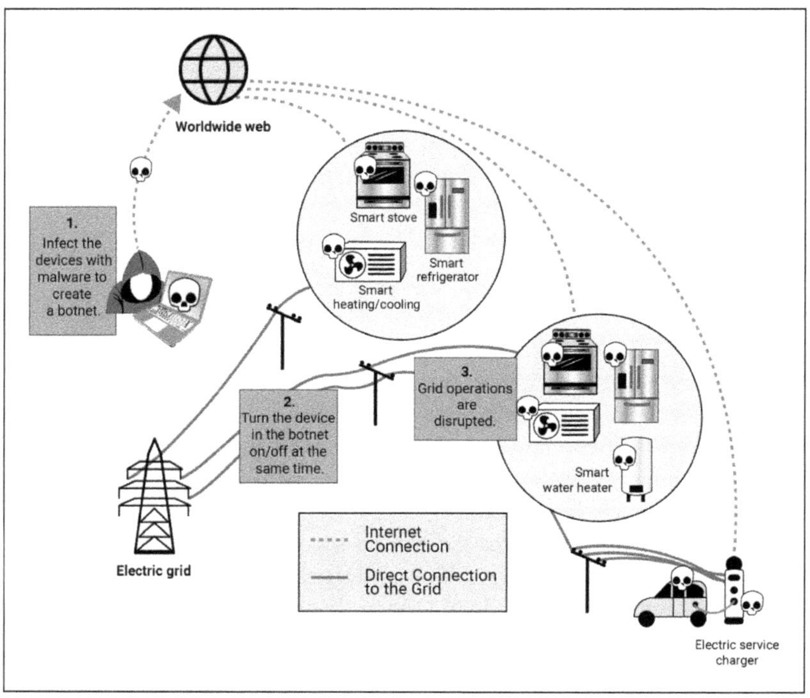

Figure 12-5. A Cyberattack on an Electricity Grid

Cybersecurity is a major concern for the electric utility industry. This has long been the case, but the global community became more attuned to the risks following the 2015 attack on Ukraine's power grid. Although cybersecurity poses significant risk to our electric grid, surges in power supply and physical attacks on substations are also important risks.

From a geopolitical perspective, a cyberattack on an enemy nation's power grid would be a good way to create chaos. The United States Government Accountability Office (GAO) acknowledges the cybersecurity risks to the US power grid and notes that it is "an attractive target for cyberattacks from US adversaries—such as nations like China and Russia" (US Government Accountability Office, 2022). The GAO developed the illustration in Figure 12-5 to show how a threat actor might attack a power grid.

A power grid presents many possible points of attack, particularly as the Internet of Things (IoT) introduces more connections.

EXAMPLE

A hacker could exploit a vulnerability in a homeowner's smart thermostat that has an Internet connection. Obtaining access via a weak password would allow the hacker to install malware to intercept network traffic. They then might be able to obtain banking credentials from connected devices and transfer funds from the homeowner's bank account.

Distributed Energy Resources (DERs) Small-scale electricity generation or storage technologies that are located close to where electricity is used, rather than at a large, centralized power plant.

The proliferation of **distributed energy resources (DERs)**, decentralized power supplies such as microgrids or energy storage batteries, has significantly increased cyberattack risk. Another source of concern are possible attacks on the many hardware and software components in the grid itself, including **supervisory control and data acquisition (SCADA)** networks. SCADA computer systems are used to collect and analyze data from sensors that monitor equipment.

Supervisory Control and Data Acquisition (SCADA) A system used for monitoring and controlling industrial processes and infrastructure.

> **CASE IN POINT**
> **What would happen if the power grid went down? How fast could power be restored? These are questions best answered through simulating a cyber incident. For several years, the US Defense Department's Defense Advanced Research Projects Agency (DARPA) oversaw simulated attacks on the power grid on Plum Island in New York. GridEx, a biennial simulation conducted by the North American Electric Reliability Corporation (NERC), brings together a variety of stakeholders to conduct drills and exchange information.**

NERC is the nonprofit regulatory organization that develops reliability standards for North America's power grid. In 2008, the US Federal Energy Regulatory Commission approved the **NERC Critical Infrastructure Protection (CIP) reliability standards**. Chapter 7 described these standards in some detail.

NERC Critical Infrastructure Protection (CIP) Reliability Standards A set of cybersecurity and physical security requirements developed by the North American Electric Reliability Corporation (NERC). They are designed to protect the bulk electric system (BES) in North America from cyber threats, physical attacks, and other disruptions.

Transportation

People depend on a nation's public transportation infrastructure to get where they need to go and to move goods, including necessary food and medicine, from one point to another. Nation-state bad actors can do harm to another country's supply chain or even cost lives by interfering with rail, air, shipping, or trucking movements.

> **EXAMPLE**
> **As an act of cyberterrorism, a nation-state hacker could compromise the logistics software of a large national trucking company to disrupt routing and communications. They could manipulate GPS software to create accidents or divert traffic. The ensuing chaos would cause delivery delays and destabilize supply chains.**

The mission of the US Department of Homeland Security's Transportation Services Administration (TSA) is to "Protect the nation's transportation systems to ensure freedom of movement for people and commerce." In a 2018 report, the TSA issued a cybersecurity road map designed to incorporate cybersecurity into its strategy and "improve the TSA's cybersecurity posture" (US Department of Homeland Security, 2018). The report goes on to note that public transportation systems are threatened by nation-states and nonstate actors motivated by espionage, money, and politics or ideology. The US is not alone in facing cyber threats to transportation systems. In 2023, the **European Union Agency for Cybersecurity (ENISA)** published its first threat landscape report for the transportation sector. The report noted that nation-state attacks mostly impacted maritime or governmental transportation authorities.

European Union Agency for Cybersecurity (ENISA) The EU's official agency responsible for cybersecurity.

> **CASE IN POINT**
> **Cyberattacks against the aviation industry are perhaps the most frightening of threats to transportation, recalling the horror of the physical attacks in the United States on 9/11 in 2001. Nation-state adversaries have many possible avenues for hacking into aviation systems, including reservation systems, air traffic management and control, passport systems, navigation systems, and flight control systems. There have been several attacks on airlines and airports in recent years. For example, in 2022, Taiwan Taoyuan International airport suffered a DDoS attack, and a DDoS attack briefly disrupted air traffic at several large US airports as well.**

Water Supply

Just as we expect the lights to come on with the flick of a switch, so do we anticipate that clean, potable water will flow from a faucet in our homes when we open a spigot. Water is a natural resource, but it needs to be treated to become suitable for drinking and bathing. Water treatment involves adding chemicals and removing contaminants. Public water treatment is generally done at a municipal level, and water treatment plants may be managed by public or private water utilities, or some combination of both.

The American Water Works Association calls cybersecurity mission-critical for water utilities. While the energy industry has universal cybersecurity standards, to date they do not exist for US water utilities.

The most frightening risks to water supply are contamination and denial of service, which could render the water supply inaccessible. While we mostly think of contamination as someone physically adding a chemical to water, a cyberattack on a water supply treatment plant in Florida in 2021 occurred as a hacker obtained remote access to the system and changed the levels of certain chemicals that the plant regularly used in a way that made the water poisonous. While no one was harmed in this instance, thanks to a watchful operator, the attack highlighted the importance of cybersecurity resilience for our water supply. One such response was CISA's development of an

Table 12-4. Cyber Risks to Water Supply Function

Risk to Water Supply Function	Description
Network complexity	Hundreds of diverse components in operational technology are difficult to map.
System maintenance	Operational technology components may not all be patched and maintained as needed.
Network segmentation	IT networks can be used to target operational technology networks and systems that have not been properly segmented.
Data	Threat actors could access IT systems to steal sensitive data and access sensitive systems.
Ransomware	Ransomware attacks on IT systems could disrupt operations in connected operational technology systems.

infographic* illustrating cyber risk resources for the water and wastewater systems sector. Table 12-4 summarizes the risks listed in the CISA publication.

Healthcare Systems at Risk

Cybersecurity may not always protect against a risk to "life and limb." In 2020, a woman in Germany was in an ambulance on her way to a hospital that was undergoing a ransomware attack. Her subsequent death following the diversion delay is widely viewed as the first death stemming from a cyberattack.

Hospitals, medical offices, healthcare insurance companies, and other components of the healthcare industry are ripe for cyberattacks for many reasons. In addition to PII, they also maintain **protected health information (PHI)**. Gone are the days when medical offices used a large chunk of their space storing paper records. Today, patients have "portals," and medical records are (mostly) digitized. PHI includes data about health status, health history, vaccine records, and medications, and all this information is identifiable with a specific individual. Because of the especially important sensitivity of these data, in 1996, the United States enacted the **Health Insurance Portability and Accountability Act (HIPAA)**, which is discussed in Chapter 11.

Protected Health Information (PHI) Any information about health status, provision of healthcare, or payment for healthcare that can be linked to an individual.

Health Insurance Portability and Accountability Act (HIPAA) A US law designed to provide privacy standards to protect patients' medical records and other health information.

In addition to hosting highly sensitive data, the healthcare industry is vulnerable to cyber risk for other reasons. They include a wide and complex attack surface, in many cases outdated technology, and the need for remote access to data.

Nation-state attacks on healthcare systems are an increasing threat. In addition to PHI, healthcare organizations maintain medical research and other desirable data that could be used either

* www.cisa.gov/sites/default/files/2023-02/infographic-supply-water-national-critical-function-102021-508.pdf

to make money (cybercriminals) or to be weaponized (nation-state). While these data may not be sought by a nation-state specifically to cause political chaos or disruption, foreign nation-states may harbor criminal groups that seek the monetary value from the rich data that this industry sector maintains.

> **CASE IN POINT**
> **In 2022, the US Treasury Department, CISA, and the FBI, issued an alert to healthcare sector organizations about North Korean state-sponsored ransomware attacks. In these attacks, the adversaries used Maui ransomware to encrypt servers that maintained healthcare data and provided services. In 2023, an updated advisory noted that North Korea-based hackers were staging ransomware attacks to obtain funds to support the nation's priorities and political agenda.**

Manufacturing and Supply Chain Disruption

To put it simply, the supply chain is everything from seed to table, or raw material to goods in the hands of customers. Manufacturing transforms raw materials into products. and a disruption in manufacturing can impact the entire supply chain. It might also be impacted in another sense, as was the case with the SolarWinds hack. In that case, a cog in the supply chain (SolarWinds software) was embedded with malware, which was then distributed to customers. Supply chain attacks of this kind are devastating because they can impact hundreds if not thousands of organizations. Think about how you allow automated updates to your cell phone. Now imagine that an update contains malware. This is how the SolarWinds scenario evolved—the update software, Orion, was infected. Figure 12-6 shows the stages of the SolarWinds attack, which has been attributed to a Russian nation-state adversary.

Industrial Control Systems (ICS) Systems used to monitor and control industrial processes, such as manufacturing, power generation, or water treatment.

Programmable Logic Controllers (PLCs) Industrial digital computers used to control machinery and processes, such as in manufacturing environments.

Manufacturers can be easy to hack into because they make use of so much technology. **Industrial control systems (ICSs)** come in many shapes and sizes and may be monitored and managed by **programmable logic controllers (PLCs)** or SCADA systems. ICSs consist of hardware and software that manage the machinery used in manufacturing, and human interaction with those machines. Nation-state attacks on manufacturers may have the intention of disrupting the supply chain for political purposes, obtaining financial gain, or stealing secret formulas or processes. Bad

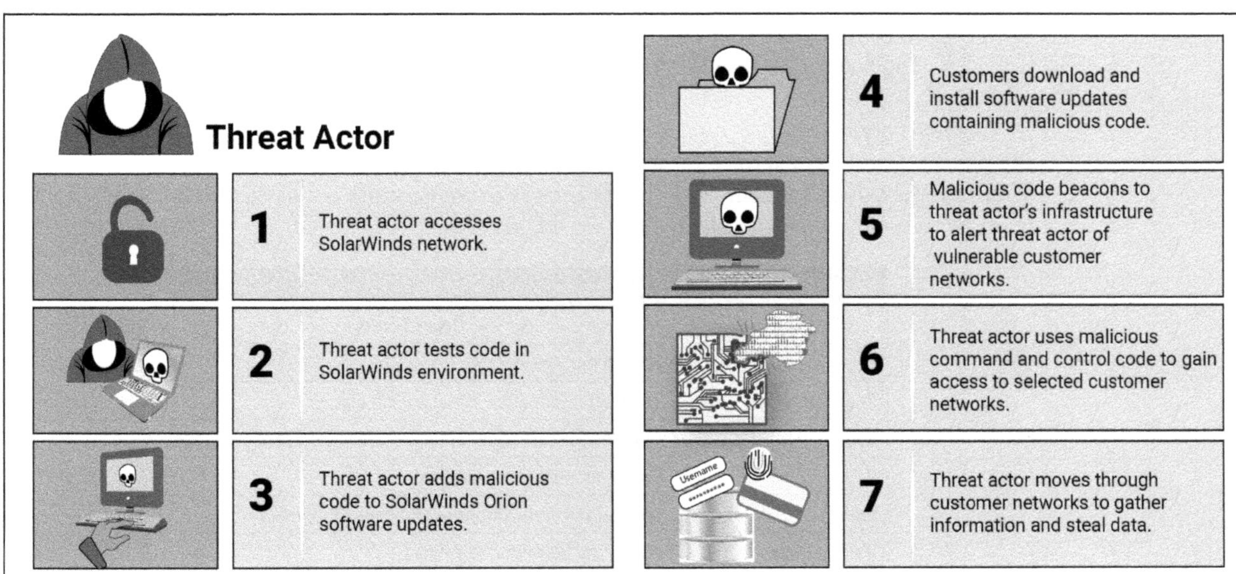

Figure 12-6. SolarWinds Cyberattack Path

Cyber Mercenaries Individuals or groups hired to conduct cyberattacks on behalf of others for financial or political gain.

actors can be hacktivists, criminals or criminal groups, sponsored government hackers, or **cyber mercenaries**.

> **CASE IN POINT**
> **The Atlantis Cyber-Army (a.k.a. AIG, for "Atlas Intelligence Group") is an example of a group offering hackers for hire or cyber mercenaries. This criminal gang recruits experienced hackers to carry out a variety of attacks, including DDoS and leaked databases offered for sale. Researchers assert that the group focuses on governments globally, including the United States, Pakistan, and Israel (Montalbano, 2022).**

Some manufacturing sectors are more likely than others to be the focus of geopolitical cyberattacks. Disruption to industries that manufacture weaponry or vital global exports can cause harm from adversarial countries, and bad actors staging those attacks are likely to be government sponsored. Industries that are subject to hacktivist attacks are those that go against a particular social agenda, such as those in the fossil fuel industry due to climate change concerns.

Spreading Misinformation and Hacking Elections

Spreading propaganda has been an essential tool in warfare throughout the ages. Convincing your population that the "other" population is (insert pejorative adjective here) and working against "our side" has always been seen as a viable strategy for winning popular support during a conflict. Likewise, spreading propaganda to the "other" group to undermine their leadership has also been used extensively in conflict. When the information spread is incorrect, perhaps without intent, it is called misinformation, and it has been used to cause damage in many different ways.

> **CASE IN POINT**
> **In November 1943, Allied leaders met and determined to open a second front in France, dubbed "D-Day (Operation Overlord)," which would have the Allies attack German positions in Normandy. For this to work, Allied commanders needed the attack to be a surprise and to prevent Germany from quickly reinforcing the Normandy area. To facilitate this, the Bodyguard operation was enacted with the goals of luring military assets away from Normandy and tricking Germany into thinking that the Normandy attack was diversionary. Through a series of leaks, physical deception, and quite a bit of luck, the Allies were able to convince Germany that their attack would be at Calais (the closest physical point to England) rather than Normandy (Adams, 2024).**

Elections are a necessary function in nations governed as democracies or republics. Open and fair elections are characterized by access to voting and tallying and reporting of votes submitted. Unfortunately, there are many ways for bad actors to compromise public elections. Table 12-5 lists some of the ways a cyberattack on elections could occur.

Table 12-5. Cyber Threats to Elections

Election Stage	Cyber Threat
Registration	Altered or deleted data
Registration	Incorrect information about polling places
Registration	DDoS attacks on registration or voter information websites
Voting	Compromised voting machines
Voting	Distribution of misinformation about candidates
Vote tallying	Compromised voting machines
Vote tallying	Theft of voting data
Reporting	Disinformation about voting results

Given the multiple potential threats to election systems, many countries have taken steps to defend against both physical and cyberattacks. In the United States, a CISA website states:

> The American people's confidence in the value of their vote is principally reliant on the security and resilience of the infrastructure that makes the Nation's elections possible. Accordingly, an electoral process that is both secure and resilient is a vital national interest and one of CISA's highest priorities. (CISA, n.d.)

The US Department of Homeland Security, which oversees CISA, has made election infrastructure protection a part of the nation's critical infrastructure, recognizing that harm to it would be devastating for the nation.

Many of the cyber threats toward voting systems and behavior concern hardware and software systems, but technologies that make deepfakes, synthetic identities, and other means of spreading **disinformation**, which is intententional misinformation, about candidates are also of concern. Social media exacerbates the problem, which may be thought of as "hacking the mind" rather than "hacking the machine." Advances in AI are likely to make the spread of this purposeful inaccurate information worse, as it will be increasingly difficult to differentiate fact from fiction.

Disinformation False or misleading information spread deliberately to deceive.

CASE IN POINT

In 2018, when Brett Kavanaugh was undergoing Senate hearings prior to his appointment to the US Supreme Court, a *New York Times* article that asked readers to submit examples of misinformation found multiple instances linked to the confirmation process. One example was an altered photo showing George Soros, a noted liberal, with a woman who had accused the court nominee of sexual assault. In reality, the picture was of Mr. Soros with a different woman named Lyudmyla Kozlovska (Roose, 2018).

Summary

This chapter looked beyond business to examine cyber threats to national security. The chapter began with a brief examination of the outsize risk that these threats pose. We next described groups of adversaries that are particularly threatening to the interests of the United States, though most every country weaponizes cyber to some extent. A nation's infrastructure is particularly vulnerable to global hacking, with some industries particularly targeted. The chapter briefly described threats to communication systems, education, governments, monetary systems, the power grid, transportation, and the water supply. The chapter concluded with global threats to healthcare systems, supply chains, and elections.

Chapter Review

Key Terms

Center for Strategic and International Studies (CSIS)
Cyber Espionage
Cyber Mercenaries
Cyber Threat Map
Cyberweapon
Deepfake
Disinformation
Distributed Energy Resources (DERs)
European Union Agency for Cybersecurity (ENISA)
Global Threat Landscape
Hacktivist
Health Insurance Portability and Accountability Act (HIPAA)
Industrial Control Systems (ICS)
Infrastructure
Intellectual Property (IP)
NERC Critical Infrastructure Protection (CIP) Reliability Standards
Programmable Logic Controllers (PLCs)
Protected Health Information (PHI)
Supervisory Control and Data Acquisition (SCADA)
Wiper Malware
Zero Trust

Review Questions

1. What are geopolitical cybersecurity risks?
2. How do geopolitical risks differ from normal cybersecurity risks?
3. Describe each of the bad actors discussed in this chapter. Compare and contrast their motivations.
4. What is critical infrastructure?
5. Describe each component of critical infrastructure and how it has been or could be attacked.
6. How do healthcare systems figure in to geopolitics as it relates to cybersecurity?
7. How can supply chain disruptions further a nation-state attack on another country?
8. Is election hacking an attack on a foreign entity? Why or why not?
9. How does spreading disinformation become a cybersecurity issue?
10. How does an attack on elections or election integrity impact geopolitical risk?

Multiple Choice

Identify the choice that best completes the statement or answers the question.

1. What is a potential consequence of a large-scale cyberattack on a nation's communication systems?
   a. Expansion of satellite networks
   b. Temporary improvement in satellite connectivity
   c. Economic halt due to disruption of individual and business transactions
   d. Enhanced security in digital communications
   e. Increased efficiency in communications
2. Which of the following is **not** listed as a reason for the vulnerability of the healthcare industry to cyber risks?
   a. Outdated technology in many facilities
   b. Extensive databases of PII and PHI
   c. The necessity for remote access to data
   d. The decentralized nature of healthcare systems
   e. Limited financial resources to upgrade systems
3. What was the primary intention of the Stuxnet worm, which targeted Iran?
   a. To steal financial data from Iran
   b. To enhance Iran's nuclear capabilities
   c. To slow Iran's nuclear program
   d. To improve Iran–US relations
   e. To support Iran's IT infrastructure
4. What type of technology is identified as increasing the cyberattack risk because of its connectivity?
   a. Traditional coal-powered plants
   b. Basic electrical transformers
   c. Old-fashioned meters
   d. Nondigital refrigeration units
   e. Distributed energy resources
5. What is a primary reason for cyberattacks on higher education institutions?
   a. To steal intellectual property for economic or military advantage
   b. To encourage more open-source research sharing
   c. To disrupt online courses for entertainment
   d. To improve educational standards
   e. To test cybersecurity systems within educational institutions
6. What is one of the ways bad actors can compromise public elections?
   a. Encouraging voter turnout
   b. Conducting public awareness campaigns about voting
   c. Enhancing the security of voting machines
   d. Increasing voter registration
   e. Altering information on polling places
7. What are the potential motives for nation-states to engage in cyberattacks?
   a. To promote global peace
   b. To enhance educational programs
   c. To disrupt an enemy nation's economy
   d. To support environmental initiatives
   e. To improve international trade relations
8. What dangerous potential do deepfakes represent in the context of cyberwarfare?
   a. They provide clear facts.
   b. They can spread misinformation.
   c. They are exclusively used for educational purposes.
   d. They help verify the authenticity of information.
   e. They are used to enhance the reputation of political figures.

Case Studies

Connect Us—Implementing Zero Trust

Sean Silver is a cybersecurity consultant working in the assurance and risk management division of a large global public accounting firm. One of the company's clients, Connect Us, is a software company that offers video conferencing to clients, which range from individuals to large business users. The company's software is available on both macOS and Windows operating systems, as well as on mobile phones. Connect Us is concerned that the company may be vulnerable to a nation-state cyberattack or a ransomware attack by a foreign cybercriminal. It implemented the NIST Cybersecurity Framework four years ago and believes it has a mature cybersecurity stance. However, the company is concerned about escalating cyber threats in its industry and has hired Sean to implement a zero trust cybersecurity program.

Case Requirements

1. How would a zero trust program differ from a mature cybersecurity program?
2. What action steps will Sean take in implementing zero trust at Connect Us?
3. An important element of a zero trust program is identity and access management (IAM). How would IAM be different for a company that has adopted zero trust versus one that has not?

Kaseya Supply Chain Cyberattack

In early July 2021, the IT management company Kaseya was hit with a ransomware attack. The threat actors, the Revil ransomware group, exploited a vulnerability in Kaseya's Virtual System Administrator (VSA) software, a remote monitoring and management application that Kaseya sells to its customers. The attackers installed a fake update in the VSA software so that customers updating their software would introduce malware into their systems. They were able to successfully encrypt the data of dozens of Kaseya customers and hundreds of downstream businesses. Using Internet search and other tools such as AI, answer the following questions. Be sure to corroborate your answer by consulting multiple sources.

Case Requirements

1. When and how was the ransomware attack discovered by Kaseya?
2. What were Kaseya's initial actions to control the attack?
3. Why is the Kaseya attack so important?
4. What do you think Kaseya did right and wrong prior to and immediately following the attack?

Blue Water—A Readiness Assessment Case

Blue Water is a privately held wastewater treatment company that operates in three midwestern US states. The company's primary service is removing contaminants from community wastewater with biological, chemical, and physical processes, and then returning the water to the environment.

Blue Water, as an infrastructure company, is concerned about cybersecurity threats from nation-states. Although its size may keep it under the radar, the company's management knows that affecting the water supply is one possible way that cyberterrorism can occur. The US Environmental Protection Agency (EPA) has a Water Sector Cybersecurity Evaluation Program under which companies in this sector may obtain a free cybersecurity evaluation assessment. The assessment requires input from management and staff, as well as a significant amount of documentation about systems, policies, and procedures.

Jack Martin, the CEO of Blue Water, has contracted with its accounting and consulting firm to provide a readiness assessment and gather the materials needed to obtain the EPA evaluation assessment. The Cybersecurity and Risk Management practice within the accounting and consulting firm has assigned the work to Ava Nickels, a senior associate.

Case Requirements

1. Is cybersecurity for a wastewater treatment client different from that necessary to protect other infrastructure industry sectors from a nation-state threat actor? Explain your answer.
2. The first thing Ava should do is develop an understanding of specific cyber threats facing the water treatment sector. Describe two of them.
3. Both CISA and the EPA have many resources available to assist in cybersecurity programs for organizations in the water supply and treatment sector. Visit the appropriate websites and, in a few sentences, describe two of them.
4. As part of the readiness assessment, Ava will need to talk with individuals and gather documentation. Describe three specific action items in some depth. As an example, one might be to interview the CEO about risk appetite and risk tolerance. Another example would be to gather documentation describing identity access management policies and procedures.

Exercises

12-1. Stuxnet is widely viewed as the first cyberweapon. The 2016 documentary *Zero Days* provides detailed insights into this cyber worm, including its development and purpose. The movie is available on multiple streaming platforms. Watch the documentary and answer the following questions. Required:
1. What was the motivation behind the development of Stuxnet?
2. Was Stuxnet successful in achieving its purpose? Explain your answer.
3. What are the dangers associated with launching a cyberweapon? How has Stuxnet evolved since its initial deployment?

12-2. A nation's power grid may be its most strategic target in cyberwarfare. Ukraine is an example of a country that has suffered multiple physical attacks and cyberattacks on its power grid, beginning in 2015. Imagine that your country's power grid is successfully attacked by a nation-state, and power is lost for several days. Required:
1. Describe all the ways this event would impact your daily life.
2. What precautions should you take so that such an attack would cause the least disruption to you personally?
3. How would such an attack further the cause of an enemy nation-state?

12-3. Many large cybersecurity consulting companies produce annual reports on cyber trends and threats. An example is Check Point (https://www.checkpoint.com/). Its report *The State of Cybersecurity 2025* includes a Global Analysis. Download this or a similar report and answer the following questions. Required:
1. What does the report say about the nature of geopolitical cyber threats?
2. Does the report identify specific nation-states that pose a threat to the United States? If so, what are they, and what is or are the motive(s) of the threat actor(s)?
3. The report should list the incidence of global attacks on specific industry sectors. What are the top five industry sectors identified as frequent targets of cyberattacks?

References

Adams, M. (2024, October 2). *How the Allies duped the Nazis for D-Day with Operation Fortitude*. Obscure Histories. Retrieved September 29, 2025, from https://www.obscurehistories.org/post/how-the-allies-duped-the-nazis-for-d-day-with-operation-fortitude

Bird, C. (2025, June 19). China hackers escalate efforts as Russia war rages in Ukraine. *The New York Times*. https://www.nytimes.com/2025/06/19/world/europe/china-hackers-russia-war-ukraine.html

Center for Strategic and International Studies (CSIS). (2025). *Significant cyber incidents*. Retrieved September 29, 2025, from https://www.csis.org/programs/strategic-technologies-pro gram/significant-cyber-incidents

CrowdStrike. (2025). *CrowdStrike 2025 global threat report*. Retrieved September 29, 2025, from https://www.crowdstrike.com/en-us/global-threat-report/

Cybersecurity and Infrastructure Security Agency (CISA). (n.d.). *Election security*. Retrieved September 29, 2025, from https://www.cisa.gov/topics/election-security

Cybersecurity and Infrastructure Security Agency (CISA). (2024, February 27). *Cybersecurity advisory alert AA24-038A: PRC state-sponsored actors compromise and maintain persistent access to U.S. critical infrastructure*. https://www.cisa.gov/news-events/cybersecurity-advisories/aa24-038a

Cybersecurity and Infrastructure Security Agency (CISA). (2025). *People's Republic of China threat overview and advisories*. Retrieved September 29, 2025, from https://www.cisa.gov/topics/cyber-threats-and-advisories/nation-state-cyber-actors/china

Greig, J. (2023, February 2). *Last year was the worst on record for crypto hacks, as North Korean groups cash in*. The Record. https://therecord.media/last-year-was-the-worst-on-record-for-crypto-hacks-as-north-korean-groups-cash-in

Lakshmanan, R. (2024, December 24). *North Korean hackers pull off $308M Bitcoin heist from crypto firm DMM Bitcoin*. The Hacker News. https://thehackernews.com/2024/12/north-korean-hackers-pull-off-308m.html

Montalbano, E. (2022, July 21). *Hackers for hire: Adversaries employ "cyber mercenaries."* ThreatPost. https://threatpost.com/hackers-cyber-mercenaries/180263/

Office of Public Affairs. (2022, September 14). *Three Iranian nationals charged with engaging in computer intrusions and ransomware-style extortion against U.S. critical infrastructure providers*. https://www.justice.gov/opa/pr/three-iranian-nationals-charged-engaging-computer-intrusions-and-ransomware-style-extortion

Office of the Director of National Intelligence. (2025, March). *Annual threat assessment of the U.S. intelligence community*. https://www.odni.gov/files/ODNI/documents/assessments/ATA-2025-Unclassified-Report.pdf

Roose, K. (2018, November 4). We asked for examples of election misinformation. You delivered. *The New York Times*. https://www.nytimes.com/2018/11/04/us/politics/election-misinformation-facebook.html

Samuelson, R. J. (2020, August 2). The Internet's threat is more about national security than the economy. *The Washington Post*. https://www.washingtonpost.com/opinions/the-internets-threat-is-more-about-national-security-than-the-economy/2020/07/31/352f96ca-d348-11ea-9038-af089b63ac21_story.html

Sanger, D. E. (2018). *The perfect weapon: War, sabotage, and fear in the cyber age* (1st ed.). Crown Publishers, an imprint of the Crown Publishing Group.

US Department of Homeland Security. (2018). *TSA cybersecurity roadmap 2018*. https://www.tsa.gov/sites/default/files/tsa_cybersecurity_roadmap.pdf

US Government Accountability Office. (2022, October 12). Securing the U.S. electricity grid from cyberattacks. *WatchBlog: Following the Federal Dollar*. https://www.gao.gov/blog/securing-u.s.-electricity-grid-cyberattacks

Volz, D. (2019, March 5). Chinese hackers target universities in pursuit of maritime military secrets. *Wall Street Journal*. https://www.wsj.com/articles/chinese-hackers-target-universities-in-pursuit-of-maritime-military-secrets-11551781800

Chapter 13

Protecting Individuals in Cyberspace

Chapter Outline

Cybersecurity Risks for Individuals
Types of Digital Scams
Personal Cyber Hygiene
Staying Informed
What to Do When You Have Been Hacked

Learning Objectives (LO)

After reading this chapter, you should be able to:

- LO1. Evaluate the risks individuals face in cyberspace.
- LO2. Be familiar with common types of digital scams, including bank account, credit card, and identity theft; job, online investment, online shopping, and romance scams; and wire fraud.
- LO3. Adopt cybersecurity hygiene practices that protect your data, your electronic devices, and your privacy.
- LO4. Know the types of services with which you can contract that can help you protect yourself in cyberspace.
- LO5. Know what to do in the event of a data breach, identity theft, or other cyber incident affecting you personally.
- LO6. Stay informed of trends in cyber threats and incidents that may impact you or your data.

Introduction

If you mention cyber scams in a group of people, each person will have a story to share. Anyone with a digital presence has been phished or pretexted, had a credit card compromised, or been scammed in some way. The US **Internet Crime Complaint Center (ic3.gov)** received complaints of $12.5 billion in losses in 2023, with almost 300,000 complaints of phishing attacks (Federal Bureau of Investigation, 2024).

Internet Crime Complaint Center (ic3.gov) An initiative run by the FBI that serves as the central hub for the public to report cyber-enabled crimes.

Most of this book concerns cybersecurity as it relates to accounting and business entities. The previous chapter took a global perspective, and this chapter looks at cybersecurity as it impacts individuals. This chapter begins with a discussion of who is at risk, then it describes the various types of digital scams and offers some specific advice on protection. Good cyber hygiene is a practice that can help protect against cybercrimes, and the chapter provides insights on how to practice it, how to stay informed about scams, and what to do if you are a victim of a cyberattack or scam.

Cybersecurity Risks for Individuals

A common misconception is that the elderly are the group most at risk for cyber scams and identity theft. This age group may be targeted, but the overriding characteristic of victims is vulnerability. Just as hackers seek vulnerabilities in the digital world, they are also looking for weaknesses in the physical realm. An AARP study in 2021 found that stressful life events made people most susceptible to fraud or scams of all types (Shadel, 2021). When you are tired at the end of a work day or suffering the loss of a loved one, you may be at your most vulnerable for hackers to take advantage.

Different population segments are more likely than others to be victims of specific types of cyber scams. The elderly may be at risk due to a lack of tech savvy, but they are also less likely to have a digital presence. Seniors who know enough to be online, but not so much that they practice good cyber hygiene, can be particularly vulnerable. Younger adults may be overconfident and less worried about their privacy, making them prey to online scammers. Young people are more used to sharing information, as they have grown up in a digital environment. We commonly refer to those born after 2000 as **digital natives**. The following case illustrates that younger people are the most likely to report cybercrimes.

Digital Native A person who has been exposed to digital technology from an early age and is generally comfortable using digital tools such as computers, smartphones, the Internet, and social media.

CASE IN POINT

The FBI maintains a website where individuals can report cybercrimes (www.ic3.gov). **According to its 2023 report** (Federal Bureau of Investigation, 2024), **it received complaints from 150,548 individuals under the age of 40 and 101,068 from those over 60. Although the younger group reported the crimes more frequently, the older group suffered the most in financial losses ($3.4 billion for over 60 versus $1.56 billion for those under 40).**

Types of Digital Scams

Hackers have many ways to target victims in the digital world, and the dollar amount of losses varies as well. The following list is not exhaustive, but it does cover the common, current types of cyber scams.

Skimmer A malicious device or software that secretly captures payment card data, either through physical attachments on ATMs/POS machines or code injected into online checkout pages.

Keylogger Software or hardware devices that track and record a user's keystrokes.

Man-in-the-Middle (MitM) Attack A cyberattack in which an attacker secretly intercepts and relays communication between two parties who believe they are directly communicating with each other.

SIM Swap A type of fraud in which an attacker transfers a victim's phone number to a new SIM card to gain access to their accounts.

Bank Account Theft

Online banking is convenient, but when you access your bank account over the Internet, it exposes your financial assets to digital scams. It is scary to think about a bad actor accessing your accounts and wiping out your life savings, but there are many stories of individuals who have had this happen to them.

Hackers can get into your bank accounts in a few ways. A bad actor might modify an ATM with a **skimmer** that captures your input data. The skimming device is a counterfeit card reader that scans your data and stores it, along with your PIN. Figure 13-1 illustrates how skimmers work.

Hackers can steal money from your financial institution by accessing your online communications with your bank using a **key logger** or launching a **man-in-the-middle (MITM) attack**. A hardware key logger is attached to your computing device, and it records your keystrokes. Software keyloggers may present as legitimate software applications or run in the background of a software program, again capturing keystrokes. An MITM is a form of eavesdropping, where a hacker intercepts communications between a user and a web application (e.g., your bank). Yet another way for a cybercriminal to access your bank account is through a **SIM swap**. In this case, the hacker uses your identity details to convince your network provider to transfer your current phone number to their SIM card. This process allows the hacker to intercept your SMS code when the bank provides it to you so you can access your bank account.

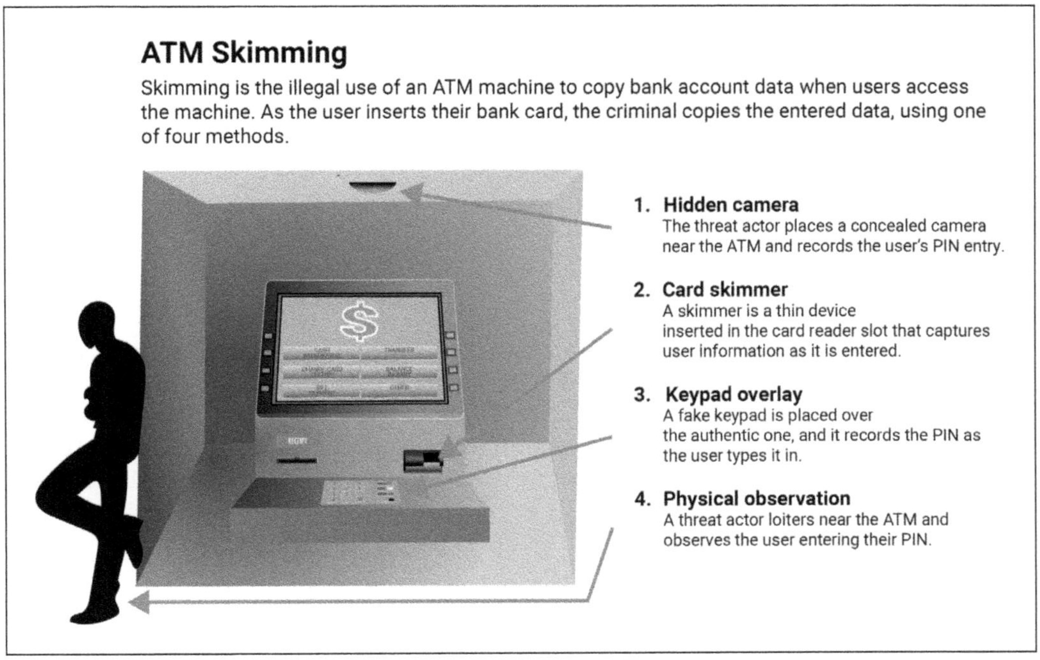

Figure 13-1. ATM Skimming Process

When we think of hackers stealing our data, we often think of sophisticated technological techniques. However, social engineering, or trickery, is perhaps one of the easiest ways for threat actors to steal from personal bank accounts, as shown in the following case.

> **CASE IN POINT**
>
> **While driving home from dinner one evening, Gail was surprised to get a call from her bank. The caller said that someone was trying to access her account, and they were in the process of transferring out several thousand dollars. The person on the phone said they just needed some account information from her to be sure that they were speaking to the account owner, and they would then stop the transfer. Gail had had a small glass of wine at the restaurant and was tired at the end of a long day, so she gave the caller the information they sought. After she got home, she realized her error and fortunately called her bank and put a block on her accounts.**

Credit Card Theft

How many times have you had to get a new credit or debit card because yours has been compromised? According to Security.org, about two thirds of credit card holders have experienced credit card theft, so the answer to the question is that some people have avoided this, but a majority have not (Frew & Petrino, 2025). The good news is that even if your card is stolen, it is unlikely to have much of a financial consequence to you, the cardholder, because of fraud detection measures at the bank that issues your credit card. It is inconvenient to have to change to a new credit card, but even when the charges are processed, the bank will reverse them and replace your card. The US 1974 Fair Credit Billing Act generally limits your loss from credit card fraud to $50, but you do need to report the fraudulent transactions.

Banks and other credit card issuers are increasingly sophisticated in their ability to spot anomalous cardholder behavior. On occasion, you may be unable to process a payment because security software has tagged unusual purchase behavior. Usually, you are sent a text message verifying your identity, and this is all part of security efforts to protect your accounts.

EXAMPLE

A credit card holder rarely travels outside their home state, and most of their credit card purchases are about $50. A purchase on the card appears for a luxury item costing $500 at a retail store in a distant state. A message is sent to the purchaser to check the legitimacy of this transaction. Credit card companies identify suspicious transactions by combinations of unusual amounts, geolocation, time of day, type of merchant, transaction frequency, and type of purchase.

Identity Theft

Identity Theft Identity theft and identity fraud are terms used to refer to all types of crime in which someone wrongfully obtains and uses another person's personal data.

Proving you are you may be an inconvenience, but a hacker who steals your information might be able to identify as you in an online environment and do significant damage. **Identity theft** is the use of your personal details without permission. In the past, when transactions took place in physical locations, you could show your driver's license or other photo ID, and your friendly banker or an official would be able to verify your identity. It was always possible that the ID was fraudulent, but the incidence of identity theft was much less common in the physical world than it is in today's digital environment.

CASE IN POINT

After May 7, 2025, you could not board an airplane for domestic travel in the United States, or access certain federal facilities, without a REAL ID or passport. The US REAL ID Act created security standards for state-issued driver's licenses and identification cards. The act was a result of the 9/11 commission, which pushed for the increased reliability and security of these identification documents. The newly issued REAL ID cards are more difficult to forge or counterfeit than predecessor credentials.

The growth of fraud in the digital age is stunning. In 2024, the US Federal Trade Commission (FTC) received 2.6 million fraud reports, with total estimated losses of about $12.5 billion (US Federal Trade Commission, 2025). The FTC noted that this included 1.2 million reports of identity theft.

Why does a bad actor assume someone else's identity? It is most often to obtain financial resources. With your personal details in hand, a fraudster can open credit cards or take out loans in your name, obtain government benefits, buy things with your credit cards, compromise your health insurance, or claim a tax refund in your name.

CASE IN POINT

One of this book's authors opened a letter from the IRS one day to find out that they were the victim of income tax identity theft. Fortunately, the IRS has sophisticated software in place that flags fraudulent tax returns claiming refunds. As a result, they try to verify the taxpayer's identity prior to issuing the refund and issue a personal identification number (PIN) that must be submitted with future tax returns to prevent issuing refunds to identity thieves.

Signs of Identity Theft

Rather than be caught unaware that your identity has been compromised, you can look for the clues included in the following list to recognize this has occurred.

1. Financial
   - Unexplained bank withdrawals
   - Unexplained credit or debit card purchases
   - A denied credit or debit card purchase
   - Receiving unfamiliar banking or credit card statements
   - Calls from debt collectors for payment of debts that are not yours

2. Health
   - You are denied health insurance for reasons unrelated to your personal health condition.
   - A medical claim is denied because you have supposedly reached your limit.
   - You receive bills for medical services you did not use.
3. Data
   - You are notified of a data breach at a company that maintains your PII.
   - You stop receiving mailed bills.
4. Government
   - The IRS notifies you that your tax return has been filed (and it has not been!).
   - You fail to receive a Social Security payment or other expected income or investment returns.

Job Scams

Online scammers use any ploy they can to entice users to click on a link containing malicious code. The promise of employment is an offer that can lure members of a vulnerable group, those seeking a first job or a better one, to take the bait. Threat actors may send a job offer via email or text message to thousands of devices in hopes that one or more naive or desperate users will bite. The message shown in Figure 13-2 is an example of one sent to one of the authors via text.

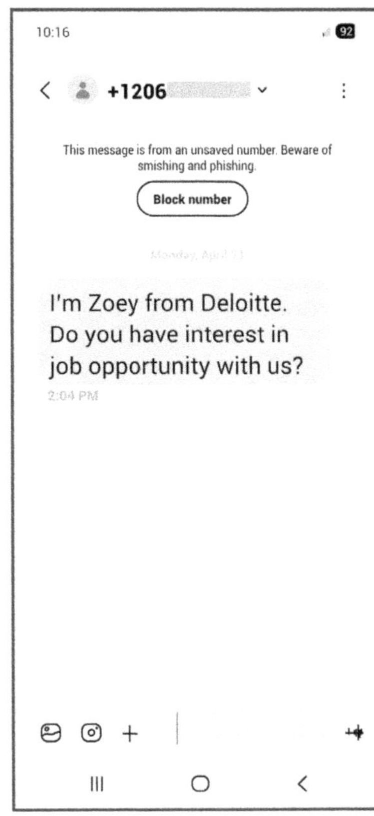

Figure 13-2. SMS Scam Job Offer

Professional social media or employment digital platforms such as LinkedIn can also be used in elaborate scams.

> **EXAMPLE**
>
> **An individual seeking a job posts their resume on LinkedIn or a similar site and is then contacted by an online scam artist. The scammer offers potential employment and sets up an online interview or engages in further communication via email or text. Once the victim is "hired," they are asked for personal information, which may include bank account information, ostensibly so that they can have their wages deposited directly into their accounts. There are many variations on the scam, and the increasing sophistication of these scams may make them difficult to detect.**

During the COVID pandemic, many of us needed to work from our home. Bad actors found that the promise of remote work was particularly enticing. There are many signals that your dream remote job offered online might be fake. The old adage, "If it sounds too good to be true, it probably is," applies in this case. Some telltale signs can indicate an employment scam, such as lack of experience requirements or the promise of high pay for not much work. You should verify any job offer where you are asked for personal details or to send any money in advance. You can do this by checking with the company directly via their publicly available contact information, or through other means if the company is not known. You should never pay in advance for the promise of a job.

> **CASE IN POINT**
>
> **A well-known case of a job scam involved Sandi Pounder. She applied for a job through LinkedIn and was pleased when she was told she had been hired as a remote data analyst. She then received an email that provided her with a list of equipment she would need to have in her home to be able to do the work. The list included a computer, printer, and so forth. All she had to do was purchase these items from the employer. Fortunately, when**

she opened the official employment letter, she found that it had been addressed to someone else, so she avoided the scam.

Online Investment Scams

Investment Scams Fraud schemes where attackers trick people into investing money in fake or nonexistent opportunities.

Fraudulent "Get rich quick!" schemes have been around forever, but they have exploded in the digital environment. **Investment scams** promise high returns with low risk, and they may include testimonials and faked data to lure you in. Testimonials are easy to falsify by using paid actors or even friends and relatives of the bad actor. Credentials are easy to fake as well, so you should not trust that an online personality has a professional certification unless you have verified them with the credentialing authority.

Social media platforms are good places for scammers to introduce and refine a variety of schemes. Anyone can falsely advertise themselves as an investment expert and manufacture data to support their promises. Media hype around huge returns to investors in cryptocurrency has made scams in Bitcoin and other crypto assets a lucrative fraud scheme. Another way for hackers to gain financial assets from investment fraud is to offer investment training or coaching. Ads for finance and real estate seminars that promise to help you gain expertise quickly and with little effort may be fraudulent, which you only learn after you have paid up front for the training. Social media can also be used to spread misinformation and rumors designed to entice investments.

Ponzi Scheme A fraudulent investment scheme in which returns are paid to earlier investors using the capital of new investors, rather than from legitimate profit earned by the operation of a business.

CASE IN POINT

Ponzi schemes actually precede their namesake, Charles Ponzi, but Charles was the fraudster who perfected the scheme. Some of the first known such frauds date back to the 19th century. In a Ponzi scheme, investors are promised high returns. The fraudster then pays out these high returns to early investors, relying on new money for these payouts. It is basically a pyramid scheme in which the thief needs to constantly lure in new money to pay the returns. The digital environment allows would-be Ponzi schemers to use social media, email, and text messaging to advertise these investments. Possibly the largest Ponzi scheme ever was perpetrated by Bernie Madoff from the late 1980s to the end of 2008.

Figure 13-3. Charles Ponzi

Investment scams are just a way to part people from their money. If you are asked for funds at any time when dealing with a potential investment, do research and hang on to your wallet. Cryptocurrency has helped threat actors devise a type of investment scam labeled as "pig butchering." In this type of investment fraud, the scammer convinces victims to deposit increasing amounts of cryptocurrency into fake investment accounts. The term pig butchering references the fattening up of the victim before the slaughter. The following example describes the steps in this type of cryptocurrency investment fraud.

EXAMPLE

The scammer identifies a potential victim, perhaps via social media, and builds trust through communications. The threat actor then pitches a lucrative investment opportunity, most likely asserting their own financial expertise. The victim is told to open a cryptocurrency account and transfer money from their bank to this account. The threat actor has the victim open an account in the investment and move the crypto there. The scammer may encourage increasing amounts of investment. Finally, when the victim tries to withdraw funds, they will find they are unable to do so because the threat actor has taken over the account or moved the funds.

Online Shopping Scams

According to Forbes, online retail sales will be about 26% of all purchases in 2026. In 2023, global e-commerce was expected to top $5 trillion. More than half of all people worldwide shop online, and most do it on their smartphone (Baluch, 2023). Hackers know this, and online shopping scams have grown along with e-commerce. Fake websites and misuse of social media are two ways bad actors conduct online shopping scams.

CASE IN POINT

In the early 2000s, one of the authors of this book was looking to buy a new laptop. They found a top-of-the-line IBM laptop for sale on eBay for $1,500. Retail for this laptop at IBM was closer to $3,000, but the seller explained that the models were purchased from IBM as part of a bulk deal, and they were passing on the savings to the buyer. The author suggested paying using an escrow company, where the money would be held by a third party until the purchase was verified. The seller responded with a number of increasingly angry emails demanding electronic payments for the sale. The sale ultimately fell apart because the seller finally realized that they would not be paid until the author received the laptop.

A common form of online shopping scam is where a hacker offers products for sale, takes your money, and does not deliver. They can do this on a social media platform or through false advertising. To avoid these scams, consumers can limit purchases to known retailers and buy directly from a seller's website rather than by clicking a link on a social media platform. Buyers also need to check spelling to be sure they have the "real" website of a retailer and not that of an imposter.

Scammers do not always look for a victim to buy something in an online retail scam. Instead, they may use a fake ad from a popular brand to entice potential victims to click on a link to obtain a coupon, which may lead to a malware infection rather than a great deal. There are often clues that the deal is not legitimate. For example, in the Father's Day scam offer in Figure 13-4, you might note that Chick-fil-A is not open on Sundays!

Figure 13-4. Fake Online Coupon Offering

Romance Scams

You can be whoever you want to be on the Internet, and vulnerable widows, widowers, and other individuals looking for love can be fooled into believing they have found their soulmate online. According to the US Federal Trade Commission, almost 70,000 individuals experienced a **romance scam** in 2022, with a median loss of about $4,400 (Fletcher, 2023). Online romance bad actors can never seem to meet in person, with popular excuses that they are on a foreign military base or working on an oil rig. A romance scammer may be juggling multiple relationships at the same time. As an example, Kofi Osei was incarcerated for stealing from hundreds of victims after creating fake online dating profiles (US Attorney's Office District of Massachusetts, 2023).

Romance Scam A type of fraud where an attacker feigns romantic interest to gain the victim's trust and steal money or personal information.

Perpetrators of digital romance scams often do their research. They can learn a lot about their target victim by combing for information posted online, particularly on dating websites. They may pose as your perfect match on a dating site, or a potential victim could receive an unsolicited contact via email or on social media. Typically, the romancing fraudster communicates for a while, sharing information and sounding perfectly suited for the victim to the point where a relationship forms. Once the relationship reaches a certain comfort level, the scammer comes up with a sudden need for funds or offers an investment opportunity or any other type of excuse to get the victim to part with their money, usually in cryptocurrency or via electronic transfer. It is sad and somewhat remarkable that otherwise smart and even skeptical individuals can fall for these frauds, but, again, vulnerability can turn the normally savvy into cyber victims.

> **CASE IN POINT**
>
> **Kate Kleinert was a widow for several years before she met Tony, a handsome surgeon who wanted to be her Facebook friend. He quickly became romantic and professed that he was in love while they chatted on Google Hangouts. He posed as a surgeon working in Iraq for the United Nations. He had two children and loved dogs. As the relationship progressed, Tony's children asked Kate for money, and Kate sent gift cards. Tony was finally going to meet Kate in person, but on the way, he said he ran into trouble and was in jail. This of course meant that Kate needed to send money. By the time Kate realized that Tony was a fraud, she had lost almost $40,000 (Rohde & Ianzito, 2021).**

Wire Fraud

Wire Transfer Fraud/Scheme Fraudulent schemes that trick individuals or organizations into transferring money to the scammer's account.

Wire fraud, or **wire transfer fraud**, does not require an Internet connection, but it does require telecommunication of some sort, and the digital world offers many opportunities for hackers to get unsuspecting victims to misdirect wired funds. A scammer can use different methods to perpetrate wire fraud electronically. A common scenario is where someone poses as a legitimate third party and redirects a payment to the bad actor's account rather than to that of the intended recipient. This can happen to businesses, but also to individuals. In a business scenario, an accounts payable clerk may believe they are paying an invoice but instead are sending the money to an imposter. Phishing scams are another way to commit this type of crime. Individuals transferring money online can also be fooled by phishing emails or other means to send money to scammers.

> **EXAMPLE**
>
> **Dave found a vacation beach house rental online at a great price. He messaged the host through the rental platform and received an email that looked official, including the logo for a well-known vacation rental company. The host said that he needed to secure the rental by making a bank transfer for the full deposit. He wired the funds but received no confirmation. When he contacted the company, he was told they had no record of the transaction and he had likely been dealing with an imposter.**

Another common type of wire fraud occurs when payments are intercepted.

> **CASE IN POINT**
>
> **The most expensive purchase an individual is likely to make in their lifetime is for a home. The Butschkys sold their Naples home and expected to get the proceeds. Unfortunately, an online scammer changed the wiring instructions without their knowledge, and the money went directly to the criminal. In this instance, the scammer posed as the Butschkys' son and asked the closing agent to change the transfer instructions (Layden, 2018).**

While checks and cash also carry some risk, digital payments up the stakes. Consumers increasingly use payment systems such as Venmo, Zelle, and PayPal to pay for purchases and transfer money to family members or friends. Although these payment systems offer some security, individuals can still be manipulated into transferring money through them to hackers. As an example, a criminal could ask for payment of a purchase prior to delivery and then disappear and fail to deliver. Identity theft can also figure here, where a scammer poses as someone you know and seems to need money urgently. You should carefully research any transfer of money (e.g., contact the person directly and verify the request) and never do it in a hurry.

Personal Cyber Hygiene

You can protect yourself against most of the scams we have described by practicing good cyber hygiene. You can take many specific precautions, such as not sending money to someone you do not know, or shopping at known retailers online, but some of the most effective precautions are just general practices that nullify the majority of cyber threats you will encounter. It is just as important to practice good hygiene online as it is to shower and brush your teeth as part of your daily routine. **Cyber hygiene** is the actions you take to ensure that you keep your personal information and resources safe from cyber threats. Just as you might lock the doors to your residence, you need to put guards around your data and online resources. You can do many things to stay safe, and there are also things you should *not* do. The following are some best practices with respect to cyber hygiene.

Cyber Hygiene The practices that users of computers and other devices follow, and the steps they take, to maintain system health and improve online security. These practices are often part of a routine to ensure the safety of identity and other details that could be stolen or corrupted.

Authentication

To prove you are really you requires some form of authentication. This might be what you have, what you know, or who you are. Currently, **passwords**, a form of what you know, remain the most common method of authentication for online accounts. Most of us log in to our financial, email, retail, entertainment, and other personal accounts using a password. Since knowledge of your password can allow hackers to obtain your PII and even your money, it is important to choose and manage passwords carefully. The best passwords are unique to specific accounts and are lengthy. Many security experts recommend that longer passwords afford more safety than shorter passwords with special characters. You might think in terms of a **passphrase** rather than a password. The phrase could be a line in a song you like or any character string that it is unlikely anyone would guess but is easy for you to remember. Cybercriminals can use brute force and other techniques to guess passwords, but you can make it more difficult for them. Unfortunately, the advent of quantum computing and AI sophistication weakens passwords as a secure form of authentication (Foisy, 2025).

Password A secret word or phrase used to authenticate a user and provide access to a system or resource.

Passphrase A longer, more secure version of a password, typically composed of multiple words or a sentence.

Many of us maintain dozens of online accounts, but having a unique password for each one makes them difficult to manage. A solution is to use a **password manager**. Figure 13-5 shows the logos of some of the popular password manager vendors.

Password Management Software Software that helps users generate, store, and manage complex passwords securely.

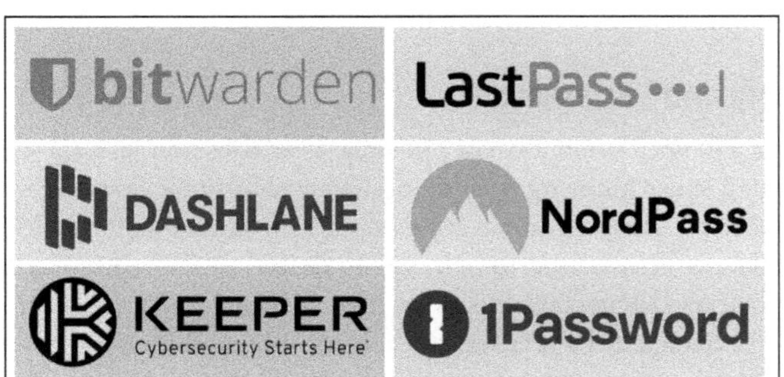

Figure 13-5. Popular Password Managers

A password manager is a software application that generates and stores passwords for all your online accounts. You pay a fee and create a master password that allows you to access your password account. There are advantages to using this tool for managing passwords, and cybersecurity professionals recommend it, although it does have a single point of failure. You must remember your master password and ensure that the provider you choose is going to keep your information safe. Without knowledge of your master password, you cannot access your individual account passwords. It is important to keep the master password safe and to be sure it would be available to your heirs in the event you are incapacitated. Unfortunately, while password managers are a good option for individuals, they are not immune from hackers.

CASE IN POINT

LastPass, a popular password manager application, suffered two data breaches in 2022. The first breach allowed hackers to obtain information that allowed them to install a keylogger on a company engineer's computer. Hackers were able to obtain user account information, including billing and email addresses, phone numbers, and IP addresses. Bad actors also obtained data in customer password vaults, some of which was encrypted.

Multifactor Authentication (MFA) A security process that requires users to provide two or more authentication factors to verify their identity.

Biometric Authentication A method of verifying identity using unique biological characteristics, such as fingerprint scanning, facial recognition, or retina scans.

Inherent Authentication Authentication based on characteristics that are intrinsic to the user, such as biometrics (e.g., fingerprint scanning or facial recognition).

Given how easy it is for adversaries to crack passwords, most online sites that hold sensitive data allow users to opt for **multifactor authentication (MFA).** MFA requires two or more forms of authentication to access the site. When a user logs into a website, they are asked to provide a password and another identifying characteristic, such as a code sent to email, a text message code, or an answer to one or more security questions. MFA is one of the best ways for individuals and businesses to protect data, but some types of MFA provide more security than others. E-mail and text message authentication can be bypassed by hackers who have access to phone numbers and email addresses.

Biometric or **inherent authentication** is the process of verifying an individual by their unique characteristics. These may be physical characteristics, including a person's fingerprints, voice, iris, or face. They can also be behavioral, using machine learning analysis to detect patterns of behavior and actions (e.g., typing speed). Unfortunately, even biometrics are not perfect in authenticating an individual. Hackers can use fake computer-generated fingerprints, facial photographs, and other workarounds to bypass biometric authentication, though it is more challenging than guessing a password.

CASE IN POINT

Which is more secure—facial recognition or fingerprints? The answer is that "it depends." A photograph may bypass facial recognition, but when Apple moved from TouchID to FaceID to unlock an iOS device, the company asserted that it was more secure and had other advantages because of Apple's advanced technology. According to Apple, "The probability that a random person in the population could look at your iPhone or iPad Pro and unlock it using Face ID is less than 1 in 1,000,000 with a single enrolled appearance whether or not you're wearing a mask" (Apple Inc., 2024).

Avoiding Phishing and Variations

Most cyberattacks begin with some form of phishing. Whether it's phishing messages for a broad population or targeted spear phishing or whaling communications aimed at specific high-profile targets, social engineering that gets an unsuspecting user to click on an email link is how hackers gain a foothold. Phishing emails used to be easier to detect, as they contained misspellings, poor grammar, or other markers suggesting they were fake. Hackers today are more sophisticated, and they use a variety of enticements to get users to click, including offering job opportunities or prizes or using scare tactics. Thanks to AI, an email in a foreign language can easily be written with perfect grammar and flow, or a QR code can be read and lead you to a fraudulent website. Figure 13-6 is an example of a phishing or quishing effort. (It's actually a training email that was sent to one of the authors.)

Rather than clicking on links in email or text messages, the better action is to go to a website directly and verify the authenticity of the message. For example, if you get an email seemingly from Amazon or UPS telling you there is a problem with your delivery, rather than clicking on the embedded link, visit the appropriate website and check to see if there is a legitimate problem.

CASE IN POINT

In the United States, the Social Security Administration (SSA) advises account holders to guard against phishing scams by making sure that emails from them contain .gov in the

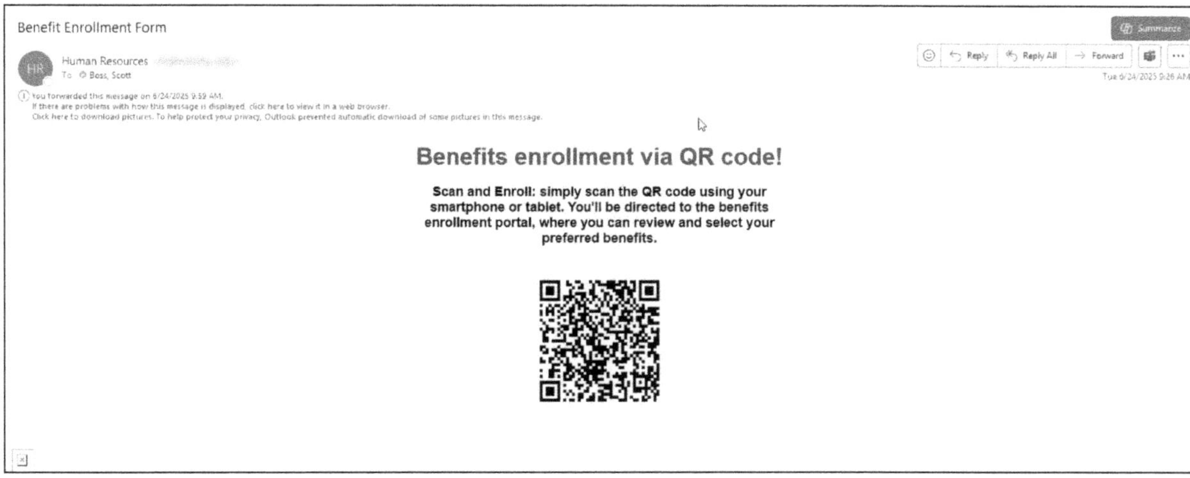

Figure 13-6. Phishing (Quishing) Email with QR Code

> **online address and are sent from** no-reply@ssa.gov. **To verify a web address within an email, they advise that users should hover over it to be sure that the link is authentic and contains .gov. If there is any doubt that an email from Social Security is legitimate, the SSA says that you should not click on the link and should report the incident to the US Computer Emergency Readiness Team (US-CERT) (which has now been folded into the Cybersecurity and Infrastructure Security Agency [CISA], mentioned previously in this book).**

Pretexting The use of a fabricated story, or pretext, to gain a victim's trust and trick or manipulate them into sharing sensitive information, downloading malware, sending money, or otherwise harming themselves or the organization for which they work.

Phishing emails can be difficult to spot. A hacker may use **pretexting** to gain trust and trick a user into responding or engaging in some way. Pretexting involves research to find out information about target victims that can help to make communications sound more legitimate. The bad actor can find personal information online or on social media and use that information to create a false scenario. For example, a hacker can claim to be someone who graduated from your same high school or used to work at the same company.

AI provides cybersecurity experts with advanced tools to thwart attacks, but hackers can also deploy AI to increase the success of online fraud. AI, such as ChatGPT, can craft phishing messages that sound legitimate. Although many AI programs will refuse to generate these messages, clever individuals can work around them. As an example, a student in a cybersecurity class was able to have an AI program generate a phishing email by asserting that it was a class project. AI can be trained to copy the writing style of an individual, such as a company executive, to increase the likelihood of enticing someone to take an action benefiting the hacker.

Smishing is like phishing in every way except that it involves text messages rather than email messages. Once hackers have your phone number, they can scam you with fake messages like the one in Figure 13-7.

As with phishing messages, smishing is about getting the victim to click on a link or reply to the text. Frequently, workplaces or organizations will send their stakeholders warnings about current scams.

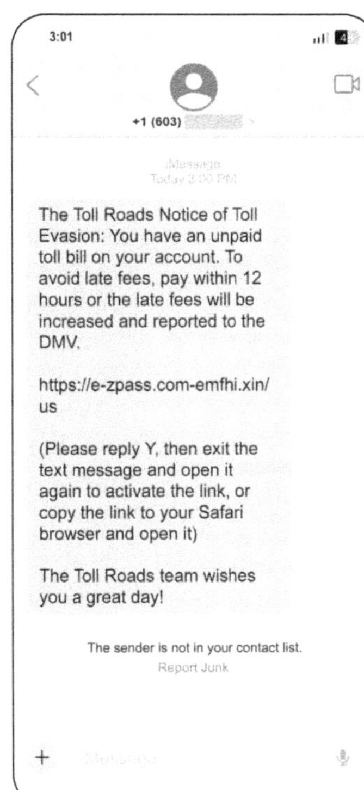

Figure 13-7. Smishing Toll Scam Message

Protect Your Accounts and Devices

In addition to good password management and using MFA wherever possible, you can use a few other practices to enhance your digital security. **Antivirus software** detects software viruses or malware and prevents it from infecting your system.

Antivirus Software A computer program used to prevent, detect, and remove malware. Antivirus software was originally developed to detect and remove computer viruses, hence the name.

Firewall A security system that monitors and controls incoming and outgoing network traffic based on predetermined security rules.

Antispyware Applications Software that detects and removes spyware, a type of malicious software that gathers information about a person without their knowledge.

Virtual Private Network (VPN) A technology that creates a secure, encrypted connection over a less-secure network, such as the Internet.

Security Settings Configurations that control access to resources and determine how security features are applied in a system.

Privacy Settings User-configurable settings that determine how personal data is shared and managed in an application or service.

Antivirus programs can also remove viruses, worms, trojans, and other malware from your system. Antivirus programs have become more complex over time, as has the software hackers use to try to infect your devices, including mobile devices, such as phones, tablets, and laptops, as well as desktop computers. Modern antivirus software can analyze your typical online behavior to detect new viruses, in addition to matching message signatures with known malware. Since bad actors are continuously inventing new ways to access your devices, good cyber hygiene dictates continuously updating security software.

In addition to installing antivirus packages on devices for protection, **firewalls** and **antispyware applications** can provide additional online security. Firewalls are configured to block unwanted communications. Some operating systems include built-in firewalls. To be effective, you need to be sure that the firewall in these systems is turned on. While antivirus software scans for viruses and other malware, antispyware software is looking at your hard drive to see if hackers have installed spyware or adware. Spyware tracks and monitors user activity, with the goal of stealing information without detection. Many information security packages today bundle together antivirus, firewall, and antispyware software, along with other safety features.

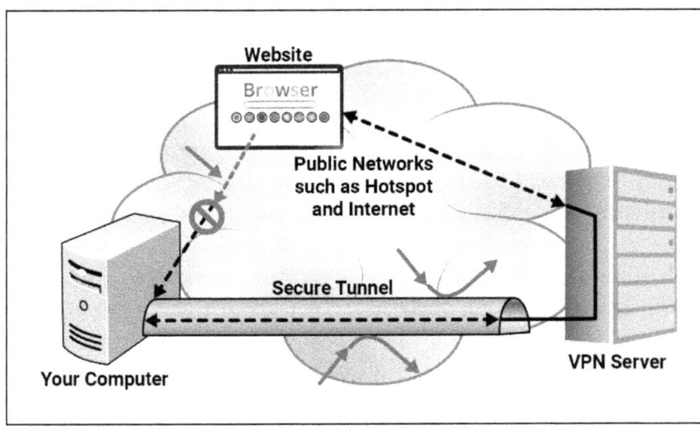

Figure 13-8. A Virtual Private Network

Another way to protect yourself online is through use of a **virtual private network (VPN).** Think of a VPN as a tunnel through which you are sending messages back and forth to a specific destination, in this case a server (Figure 13-8). Your communication is encrypted, and only you and your virtual private server have the keys to decode it. Your IP address also stays hidden. Many businesses connect their employees remotely through a VPN, or you can purchase VPN services from a provider. Using public Wi-Fi is risky, but a VPN will afford some protection from hackers. What it will not do is protect you if a hacker has already compromised the public space's Wi-Fi.

When you upload an app or access a website, you may have an opportunity to configure settings. These settings will be labeled as **security settings** or **privacy settings**, or they may simply be options available when you access your account. For example, Amazon allows you to manage your browsing history by deleting part or all of it. Amazon also allows you to require two-step authentication. Banks and other financial institutions provide many opportunities to configure settings for more secure online transactions. Users can arrange for alerts for transactions over a fixed dollar amount, opt for MFA, and set up alerts for international transactions or fund transfers.

Protecting Privacy

It seems that everyone wants your data today. You enter a store and attempt to make a credit card purchase, and before you tap your card, a salesclerk asks you for your phone number and perhaps your email or other PII. We have become very used to giving out our data for free, pretty much everywhere we go, whether in brick-and-mortar establishments or online. A good policy is to refuse to provide personal information unless it is necessary to complete a transaction. As an example, why give out your phone number when you are making an in-store purchase with a credit card? Ask for a paper receipt if you think you may need to return your purchase. By having your phone number, a hacker can target you with a phone scam or possibly take control of your phone by swapping a SIM card.

All of the data about each of us on the Internet is called our **digital footprint** (Figure 13-9). You have a digital footprint even if you have tried to maintain a low digital profile. If you search on your name with Google, you might find your home address, email address, photos of yourself, your job title, a variety of affiliations, and other personal data and links.

Digital Footprint The trail of data left behind when you use the Internet or digital devices, including information you intentionally share and data collected passively.

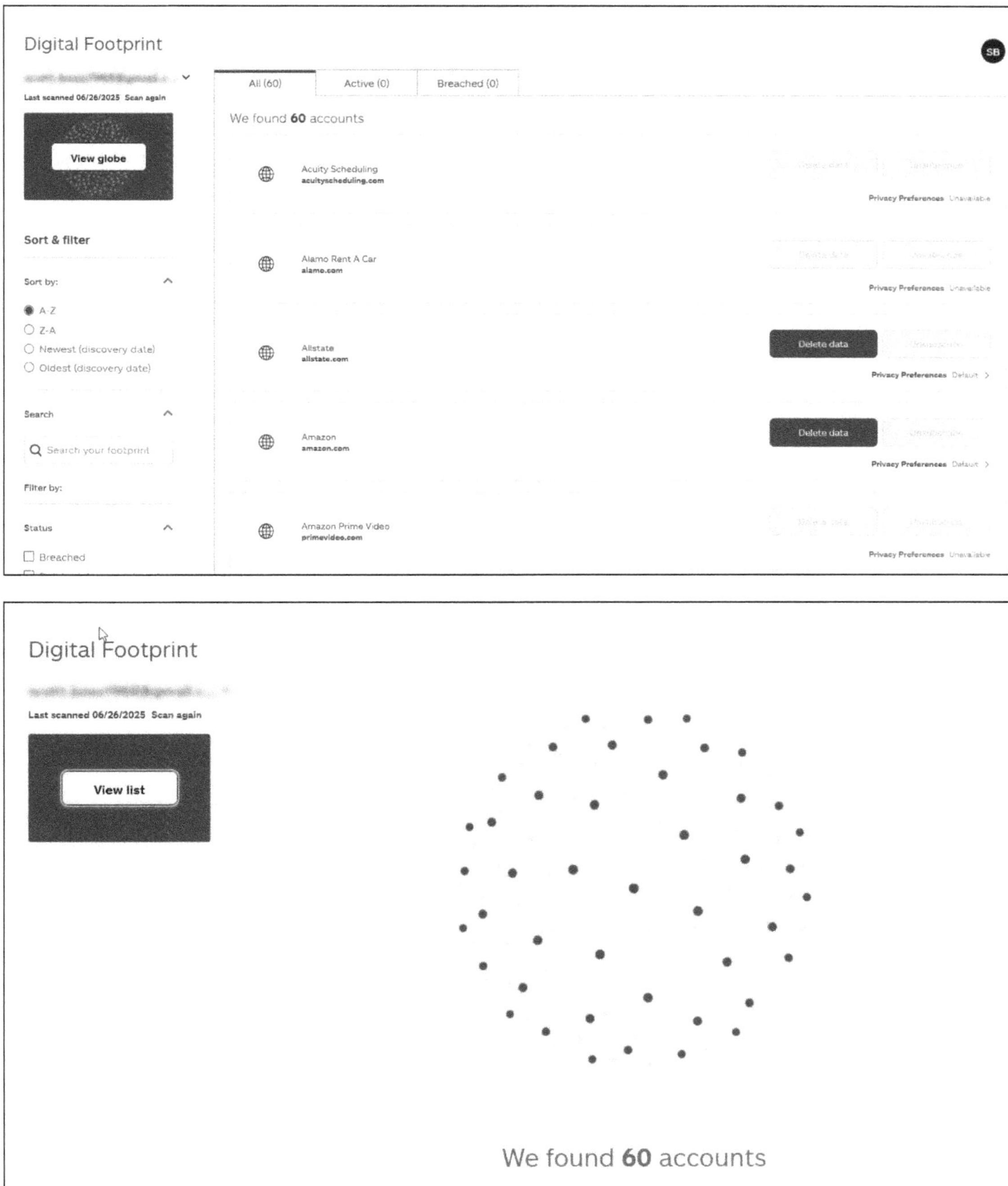

Figure 13-9. Digital Footprint Report

Although it is not likely you can remove all your personal details, you can take steps to reduce the size of your digital footprint, as described in the following case.

CASE IN POINT

Google allows the public to submit requests to remove personal data from search results. To do this, you can visit Google's Support pages and submit a removal request. You can also exert some control over your privacy in Google by managing Activity Controls within the Data and Privacy section of your account. If you have wondered how the ads you see when you conduct searches seem to be tailored just for you, you might consider also turning off the Ad Personalization control.

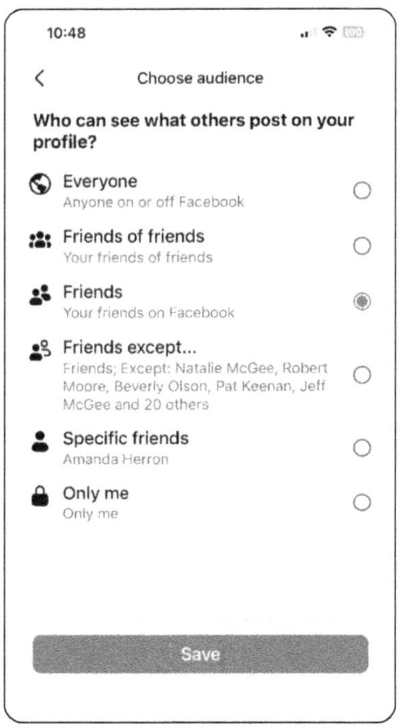

Figure 13-10. Facebook Audience Settings

The age of the Internet has provided many wonderful opportunities for connection. Social media (Facebook, Instagram, etc.) platforms allow us to connect with friends and family who may be physically distant. Grandparents can keep up with their grandchildren's activities, old school roommates can share stories, and you can find people with whom you have lost touch. Unfortunately, social media also has many downsides, of which one is the loss of privacy. We want to share pictures with friends when we are on vacation, but this also lets others know we **are** on vacation (and not in our house), which can lead to several different scams or outright physical burglary. Hackers can use information provided on social media to build profiles, which allow them to engage in pretexting. Taking a personality test on TikTok or a quiz on Facebook can provide a hacker with information that can assist them in phishing scams where they gain trust by seeming to know you. These tests can also help a hacker to answer security questions to gain access to your personal accounts. One cybersecurity expert recommends answering security questions with a nonsensical word that is easy to remember, such as pizza or dodgeball. Why put your mother's real maiden name out there when it could be discovered in a data breach?

Social media companies have steadily been increasing privacy and security options for users over time, though not everyone takes advantage of the features, and the default setting for many of the privacy measures is the least secure option. For example, Meta has a Privacy Center that lets you make choices about your account details and security, your audience, ad preferences, and your information. Figure 13-10 shows one such feature on Facebook.

Credit Bureaus and Identity Theft Protection Providers

Many different service providers can help you protect yourself in cyberspace. **Credit bureaus** or credit reporting agencies collect credit information about individuals and then make that information available to third parties, typically for a fee. A lender will consult a credit bureau when deciding about whether to extend credit. In the United States, the three main credit reporting agencies are Equifax, Experian, and TransUnion. Credit bureaus hold vast amounts of PII, which can make them a target for hackers.

Credit Bureaus Organizations that collect and maintain credit information about individuals and provide it to lenders, employers, and others.

CASE IN POINT

In 2017, Equifax suffered a colossal data breach, in which hackers obtained private records for more than 140 million Americans and millions of British citizens. It remains one of the largest instances of identity theft. Equifax was highly criticized for not doing more to protect its data and for several missteps following the breach. As an example, the company initially offered free data protection to those impacted, but only for one year, after which the victim would need to pay for it.

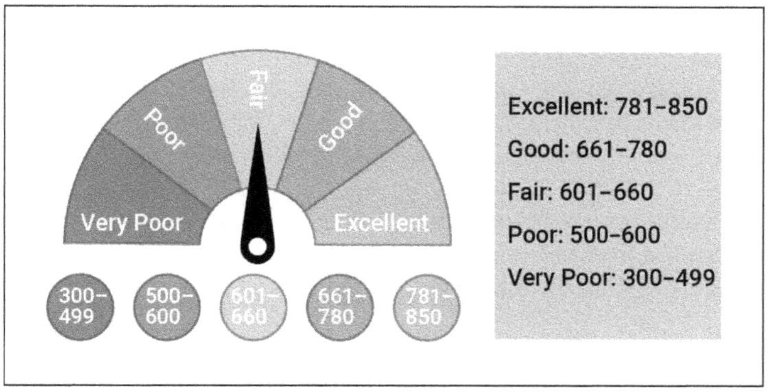

Figure 13-11. FICO Score Ratings

Credit bureaus also use the data they gather to assign **Fair Isaac Corporation (FICO) scores.** FICO scores have a range up to 850, with intervals designating a level of credit risk. US consumers are entitled to receive one free credit report annually. These reports can be useful to you, as they show your FICO score and your current debt. Figure 13-11 shows approximate credit ratings for different FICO scores.

The US **Fair Credit Reporting Act** regulates credit bureaus and reporting. This legislation gives consumers the right to place a security freeze (**credit freeze**) or a fraud alert on their account. Freezing your credit prevents a lender from extending credit in case of identity theft, but to be secure, you will need to initiate a freeze at all three credit bureaus. Freezing your credit can create some delay in legitimate credit needs (loan, new credit card, etc.), but you can lift the freeze if needed. A **fraud alert** is an alternative to a freeze; it requires the credit bureau to verify identity before extending credit.

In addition to working directly with the credit bureaus, many private companies now offer identity theft protection services. You can purchase a package that provides you with credit monitoring, fraud alerts, insurance, and security tools (e.g., antivirus or password manager). The service provider will notify you when there is a data breach at a company that has your PII and when your personal data are found on the dark web.

Fair Isaac Corporation (FICO) Scores Credit scores used to assess a consumer's creditworthiness, created by the Fair Isaac Corporation.

Fair Credit Reporting Act A US federal law, enacted in 1970, that regulates how consumer credit information is collected, shared, and used.

Credit Freeze A security measure that restricts access to your credit report, making it more difficult for identity thieves to open accounts in your name.

Fraud Alert A notice placed on a credit report to warn creditors that the consumer may be a victim of fraud or identity theft.

CASE IN POINT

LifeLock is one of the largest providers of identity theft protection. The company offers different plan levels for individuals or families. The basic plan provides identity and Social Security number alerts, credit monitoring, limited stolen funds reimbursement, and security tools for a limited number of devices. Higher level plans offer additional features, such as alerts on crimes in your name and "Buy Now Pay Later" alerts.

Staying Informed

You can stay up to date on cybersecurity issues by subscribing to a newsfeed or taking training. Several organizations provide free online email subscriptions to cybersecurity news. A few examples are newsletters from Recorded Future, Dark Reading, and *The Hacker News*.

If you are interested in learning more about cybersecurity and ways to protect your accounts or your home network, YouTube has videos available that address these issues.

Figure 13-12. Dark Reading Daily Newsletter

> **EXAMPLE**
>
> **You can take full courses in cybersecurity on YouTube, or you can watch videos from a range of experts. One such expert is Jeff Crume, Distinguished Engineer at IBM. He offers videos on topics that include cybersecurity and AI, Cybersecurity Trends for 2025, Passwords vs. Passkeys, and a 2025 Threat Intelligence Index.**

Many employers offer cybersecurity training to their employees, with some companies having required security training. These training programs all discuss basic cyber hygiene. Your company may also offer warnings about new types of scams.

> **CASE IN POINT**
>
> **Kevin Mitnick was a former hacker who reformed and started several cybersecurity training companies. Prior to his death in 2023, he was Chief Hacking Officer at KnowBe4. The company was founded in 2010 as an Internet Security Awareness Training provider. The company's tagline, "Human Error. Conquered," describes the company's mission. KnowBe4 has continuously expanded its product offering and can provide online training and security products to businesses of all sizes.**

What to Do When You Have Been Hacked

You might realize you have been hacked because you received an alert or are unable to access your accounts, or your device appears to be frozen. Take a breath before you act, and do some research on another device to try to gauge the extent of the problem. Sometimes you have not actually been hacked at all but are receiving a phishing message to that effect, advising you to click on a link or call a number for help. If your computer screen says you have been hacked and offers a number to call for help—ignore it. Hackers use this ploy to act as if they are the "good guys" and then find ways to swindle you out of your money.

A stolen credit card is usually easily remedied. Check with your card provider to identify fraudulent transactions and dispute them. You should also cancel your card and request a new one.

If your identity has been stolen, or you have been defrauded online, you should take the following steps:

- Banks/Creditors
  - Notify your financial institutions and creditors.
  - Contact one or more credit bureaus and place a fraud alert. You might also freeze your credit.
- Personal Steps
  - Change your passwords.
  - Check your credit report.
  - Close any new accounts opened in your name.
  - Remove/dispute any fraudulent transactions in your accounts.
  - Contact utility and Internet providers.
  - Replace any stolen identification.
- Government
  - File a report with the Federal Trade Commission (FTC).
  - File a police report.
  - Contact the Internet Crime Complaint Center (www.ic3.gov)

Summary

This chapter dealt with the thorny issue of computer fraud and crimes against individuals. We noted that vulnerability is the primary characteristic of victims. The chapter described in some detail some common methods hackers use to attack individuals in cyberspace. They include bank account, credit card, and identity theft; investment, online shopping, and romance scams; and wire fraud. You may or may not be able to protect yourself against the most sophisticated cybercriminal, but practicing good personal cyber hygiene can make it more difficult for you to be a target. Best practices include appropriate password management, use of MFA, managing privacy and security settings, enabling banking alerts, and installing security software. You can stay informed of current scams to avoid through various newsfeeds and employer alerts. The chapter concluded with advice on actions to take if you do become a victim of cybercrime.

Chapter Review

Key Terms

Antispyware Applications
Antivirus Software
Biometric Authentication
Credit Bureaus
Credit Freeze
Cyber Hygiene
Digital Footprint
Digital Native
Fair Credit Reporting Act
Fair Isaac Corporation (FICO) Scores
Firewall
Fraud Alert
Identity Theft
Inherent Authentication
Internet Crime Complaint Center (ic3.gov)
Investment Scams
Keylogger
Man-in-the-Middle (MitM) Attack
Multifactor Authentication (MFA)
Passphrase
Password
Password Management Software
Ponzi Scheme
Pretexting
Privacy Settings
Romance Scam
Security Settings
SIM Swap
Skimmer
Virtual Private Network (VPN)
Wire Transfer Fraud/Scheme

Review Questions

1. Why is it important to understand cybersecurity risks at the individual level in addition to understanding the risk to organizations?
2. Compare and contrast the types of digital scams. Which one is the most difficult to detect? Why?
3. Why is it important to practice good cyber hygiene?
4. What are the general practices of good cyber hygiene?
5. How can you stay informed about current cyber scams? Is this important?
6. What should you do when either you have been hacked or your PII/PHI has been compromised?

Multiple Choice

Identify the choice that best completes the statement or answers the question.

1. What is a common reason for a bad actor to assume someone else's identity?
   a. To improve their credit score
   b. To obtain financial resources
   c. To enhance their social media presence
   d. To gain access to government healthcare
   e. To gain employment under a false identity
2. What traditional scam has found new life in the digital environment, often promoted through social media, email, and text messaging?
   a. Identity theft
   b. Ponzi schemes
   c. Insurance fraud
   d. Phishing attacks
   e. Credit card fraud
3. What is the purpose of a password manager?
   a. To monitor and report on your browsing habits
   b. To encrypt your entire Internet traffic
   c. To generate and store passwords for various online accounts
   d. To block websites that do not require passwords
   e. To increase the speed of your Internet connection
4. How can hackers use AI in the context of phishing?
   a. Physically interact with victims
   b. Automatically delete suspicious emails
   c. Restrict the sending of emails to international recipients
   d. Train users to recognize phishing attempts through repetitive quizzes
   e. Craft convincing phishing messages that mimic legitimate sources
5. Which population segment is described as potentially more vulnerable to cyber scams due to their digital habits?

   a. Younger adults, who may be overconfident and less concerned about privacy
   b. Tech professionals, who are overconfident in their security measures
   c. Children, who have extensive access to digital banking
   d. Middle-aged adults, who are completely unfamiliar with technology
   e. Elderly people, because they use technology the most
6. How might taking a personality test on a social media platform like TikTok or Facebook compromise your security?
   a. It increases the quality of ads you receive.
   b. It contributes to academic research on personality.
   c. It can assist hackers in phishing scams by providing them with personal information.
   d. It helps social media platforms improve user experience.
   e. It gives you insights into your own personality traits.
7. What is a SIM swap scam?
   a. A hacker exchanges SIM cards at your local bank branch.
   b. A hacker replaces your bank's SIM card with a faulty one.
   c. A hacker physically swaps your ATM card with another.
   d. A hacker convinces your network provider to transfer your phone number to their SIM card.
   e. A hacker swaps your online banking password through a mobile app.
8. What should raise a red flag for job seekers about a potential employment scam?
   a. The job requires relevant experience.
   b. The job requires sending money in advance.
   c. The job has flexible hours.
   d. The job is with a well-known company.
   e. The job allows for telecommuting.
9. What does placing a security freeze on your credit report do?
   a. It prevents lenders from extending credit in your name.
   b. It prevents any access to your credit report.
   c. It automatically increases your credit score.
   d. It updates your personal information automatically.
   e. It alerts you to any changes in your credit status.

Case Studies

Personal Privacy Check—Natalie Magee

Natalie Magee has been working for some time in cybersecurity, and she has come to understand the risks associated with data breaches and other cyber incidents. She worries that her own online data is not safe and that her identity could be stolen or her PII could be used in a harmful way. Natalie is a digital native, and she has been active on social media. Her new awareness of potential threats due to her activity on social media platforms has encouraged her to review the privacy settings on all of her online accounts.

Case Requirements

1. Take an inventory of your own personal online presence. What social media accounts do you have? What news feeds do you subscribe to?
2. Review the privacy policies and privacy settings on the accounts you listed in Requirement 1. Had you previously changed the default settings? Describe two settings on one of the platforms that you can use to make your online activity more private.

Lost Savings—Mary Cho

Mary Cho is a recent widow. Her husband, Bruce, was a professor of finance at a local private university. He typically managed all their financial transactions and investments. Following his death, Mary has had to get up to speed quickly. One decision she made was to purchase a vacation condo together with her sister. She wired $200,000 to her sister's bank as part of the purchase.

The day after this wire transfer, Mary opened her computer to find that her computer screen was frozen. A number to call for help was on the screen, and Mary called it. A helpful person, Carlos Rodriguez, answered and said he was with Microsoft Security. Carlos provided an ID number and a case number. He said he had a supervisor at Mary's bank, John Allen, who would be in touch. Mary received a call from John, who said he could help her protect her money from the hackers. He called her daily over the next month, advising her on transfers from her savings and retirement accounts. He also advised her to open an account in Coinbase, where the transactions could flow through.

John repeatedly told Mary that she should trust no one but him. As she emptied her accounts, no one at her bank or retirement account company questioned her withdrawals. Over the course of the next month, Mary transferred about $500,000. She became increasingly worried that she might not be doing the right thing, so she finally contacted a friend who works in cybersecurity. At that point, the friend had to tell her that she had actually been transferring money to the hackers themselves and advised her that Carlos and John were scamming her.

Case Requirements

1. Why do you think Mary was a good target for cyber threat actors?
2. What should Mary have done when her computer was frozen?
3. Using AI or other online research tools, assess the likelihood of recouping some or all the funds lost in such an attack.
4. Identify all the red flags (signs of fraud) in this scenario.

Exercises

13-1. Social media offers us many opportunities to enhance our relationships with others. There are many concerns, however, that these platforms offer opportunities for hackers to learn about us and use that information to commit cybercrimes. Required:

1. Review your social media accounts. Do you provide any clues to hackers that might enable them to scam you or one of your online relationships? List some examples. If not, what do you purposely avoid doing when on social media?
2. Check your privacy and security settings to ensure that they are offering you the protection you need. If not, which settings should you, and will you, change? If your settings already protect you, which ones do you consider the most important?

13-2. Most of us have many online accounts and find passwords difficult to manage. Cybercriminals who successfully steal passwords that you use for one website may try that password at other sites. This is known as credential stuffing. For example, if you use the same password at multiple online retailers, and one of them is hacked and your PII is stolen, including that password, the cybercriminal will visit other retailers to see if they can access accounts using the same password. Required:

1. Make a list on paper of all the accounts you use that require a password and what the password is. Do you believe that you have a good password policy?
2. Examine your list for any passwords that you use for multiple accounts. Did you find any? Does this concern you?
3. Do you plan to change your passwords to unique ones for each account or acquire a password manager. Why or why not?
4. Not a question but, please destroy your paper list in a way that it cannot be reassembled!

13-3. Threat actors have many ways to target individuals. Obtain a current year threat advisory report online from a cybersecurity company and answer the following questions. Required:

1. What are the three most popular online scams threat actors perpetrated that year (e.g., investment scam, romance scam)?
2. Does the report note any trends or changes from prior years?
3. Does the report note the characteristics of the targeted population (e.g., age)? What group appears to be most vulnerable?

13-4. The best way to protect yourself is by adopting good cyber hygiene practices. Online tools are available for you to check your personal cyber hygiene. Required:

1. Use an AI tool to ask for five or more best cyber hygiene practices. What are they? Do you engage in each of them?
2. Search the Internet for an online cyber hygiene quiz. There are several available (e.g., https://dartmsp.com/cyber-hygiene-quiz/). Take the quiz and evaluate your personal cyber hygiene.
3. If you are going to do one more thing to improve your cyber hygiene, what will it be?

References

Apple Inc. (2024, December 9). *About Face ID advanced technology*. https://support.apple.com/en-us/102381

Baluch, A. (2023, February 8). 35 e-commerce statistics of 2023. *Forbes*. https://www.forbes.com/advisor/business/ecommerce-statistics/

Federal Bureau of Investigation. (2024). *Internet crime report 2023*. https://www.ic3.gov/Media/PDF/AnnualReport/2023_IC3Report.pdf

Fletcher, E. (2023, February 9). *Romance scammers' favorite lies exposed*. Federal Trade Commission. https://www.ftc.gov/news-events/data-visualizations/data-spotlight/2023/02/romance-scammers-favorite-lies-exposed

Foisy, N. (2025, January 22). What will quantum computing mean for passwords and encryption? *Compass IT Compliance Blog*. https://www.compassitc.com/blog/what-will-quantum-computing-mean-for-passwords-and-encryption

Frew, P., & Petrino, G. (2025, January 27). *62 million Americans experienced credit card fraud last year*. Security.org. Retrieved September 29, 2025, from https://www.security.org/digital-safety/credit-card-fraud-report/

Layden, L. (2018, October 27). Wire fraud: This scam could cost home buyers and sellers a lot of money. *Naples Daily News*. https://www.naplesnews.com/story/money/business/local/2018/10/13/real-estate-wire-fraud-form-cybercrime-intercepts-money-transfers-home-sales/1569672002/

Rohde, M., & Ianzito, C. (2021, September 28). *Woman who lost everything in a romance scam finds new purpose*. AARP. https://www.aarp.org/money/scams-fraud/widow-online-romance-scam/

Shadel, D. (2021, December 1). Revealed! Who gets scammed. *AARP Magazine, December 2021/January 2022*.

US Attorney's Office, District of Massachusetts. (2023, March 21). *Money launderer sentenced for $8 million romance scam fraud scheme*. https://www.justice.gov/usao-ma/pr/money-launderer-sentenced-8-million-romance-scam-fraud-scheme

US Federal Trade Commission. (2025, March 10). *New FTC data show a big jump in reported losses to fraud to $12.5 billion in 2024*. https://www.ftc.gov/news-events/news/press-releases/2025/03/new-ftc-data-show-big-jump-reported-losses-fraud-125-billion-2024

Glossary

Acceptable Risk The level of risk that is deemed tolerable in an organization's operations, given the expected benefits and controls.

Acceptable Use Policy A set of rules that outline how a service or technology can be used.

Access Control Lists Lists used to control who can access a network or system resource and what actions they can perform.

Access Control Policy A set of rules that determines who is allowed to access specific data, applications, or resources and under what conditions.

Accounting Information System (AIS) A structure that an organization uses to collect, store, manage, process, retrieve, and report its financial data.

Active Cyber Defense Law (Japan) An amendment to existing cybersecurity legislation that shifts from caution and restraint to proactive cyber defense. It empowers government agencies to take anticipatory actions against foreign cyber threats while instituting oversight mechanisms to uphold civil liberties and constitutional limits.

Activity-Based Costing (ABC) A costing method that assigns overhead and indirect costs to related products and services. This cost accounting method recognizes the relationship between costs, overhead activities, and manufactured products, assigning indirect costs to products less arbitrarily than traditional costing methods.

Administrative Controls The creation and enforcement of security policies, rules, procedures, and guidelines that are set by management.

Advanced Persistent Threats (APTs) A long-term targeted cyberattack in which the attacker gains unauthorized access to a network and remains undetected for an extended period.

Adware Software that automatically displays or downloads advertising material when a user is online.

Air Gap A security measure in which a computer or network is physically isolated from other networks, especially the Internet.

American Data Privacy and Protection Act (ADPPA) A proposed US federal privacy law that aims to establish comprehensive national standards for how organizations collect, process, and share personal data.

Annualized Loss Expectancy (ALE) A quantitative risk management metric used to estimate the expected monetary loss from a specific risk over the course of a year.

Anonymous (Hacker Group) An organized group of political activists who often engage in hacktivism. Anonymous hackers first came together on 4chan, a private Internet chat room. "Anons," as Anonymous hackers are known, frequently conceal their faces in public with Guy Fawkes masks.

Antispyware Applications Software that detects and removes spyware, a type of malicious software that gathers information about a person without their knowledge.

Antivirus Software A computer program used to prevent, detect, and remove malware. Antivirus software was originally developed to detect and remove computer viruses, hence the name.

Artificial Intelligence (AI) The simulation of human intelligence in machines that are programmed to think and learn like humans.

Asset Identification The process of identifying and documenting all the valuable assets within an organization that need protection from risks.

Assurance The level of confidence an auditor provides that information they have examined is free from mate**rial misstatement and conforms to applicable standards.**

Asymmetric Encryption An encryption method that uses a pair of keys (public and private) to encrypt and decrypt data.

Attack Signatures Patterns or sequences of data that are indicative of a specific type of attack or malicious activity.

Attack Surface The set of points on the boundary of a system, a system element, or an environment where an attacker can try to enter, cause an effect on, or extract data from, that system, system element, or environment.

Attack Vectors The methods or pathways that attackers use to breach a system's defenses.

Attestation A formal statement that confirms the accuracy or authenticity of a company's processes or reports, often in the context of compliance.

Attribute-Based Access Control Access decisions are based on attributes (user, resource, environment) and policies.

Audit Log Injection A type of security vulnerability in which an attacker inserts malicious or deceptive content into an application's audit or security logs. The goal is usually to manipulate, hide, or falsify log records.

Authentication The process of verifying the identity of a user or system.

Authorization The process of determining which users or systems are allowed to access certain resources.

Automated Vulnerability Scanning A process that automatically scans a system or network for security vulnerabilities.

Availability The assurance that systems and information will be operational and accessible within the required time frame, even during disruptions or high demand.

Basic Act on Cybersecurity (Japan) Japan's foundational cybersecurity law, enacted in November 2014. It provides the legal and policy framework for protecting Japan's national security, economy, and society against cyber threats.

Big Data Extremely large and complex data sets that are difficult to process, store, or analyze using traditional data processing tools.

Biometric Authentication A method of verifying identity using unique biological characteristics, such as fingerprint scanning, facial recognition, or retina scans.

Bitcoin A decentralized digital currency that operates without a central bank, based on blockchain technology.

Blockchain A distributed ledger technology used to record transactions in a secure, transparent, and immutable manner.

Blue Screen of Death (BSOD) An error screen displayed on Windows computers following a fatal system error.

Botnet A network of compromised computers that are remotely controlled by a hacker, often used for large-scale cyberattacks.

Bots A bot is a software program that operates on the Internet and performs repetitive tasks. While some bot traffic is from good bots, bad bots can have a huge negative impact on a website or application.

Breach An incident where unauthorized access to data, applications, services, networks, or devices occurs.

Bridge A network device that connects multiple network segments, allowing them to function as a single network.

Bridge Letter A written statement that a service organization (or its auditor) provides to bridge the gap between the end date of the latest SOC report and the user entity's financial reporting period or audit timeline.

Brute Force Attack Attempts to gain unauthorized access to a system by systematically trying all possible combinations of letters, numbers, and symbols for passwords or encryption keys.

Budapest Convention The first international treaty aimed at combating cybercrime through the coordination of national laws, investigative techniques, and international cooperation.

Bus Network Topology All devices share a single communication line (backbone).

Business Continuity Planning (BCP) The creation of a comprehensive framework to maintain or quickly resume critical business functions in the event of a crisis, such as a natural disaster, cyberattack, pandemic, or system failure.

Business Email Compromise Similar to spear phishing, where a scammer gains access to a trusted person's email account and requests actions in the name of the account owner, usually a boss or manager.

California Consumer Privacy Act (CCPA) A state-level data privacy law that gives California residents significant rights over their personal information and imposes transparency and compliance obligations on certain businesses.

Cardholder Data (CHD) Any personally identifiable information associated with a credit card holder, such as the card number, expiration date, and service code.

Carve-Out Method An approach used in SOC 1 or SOC 2 audits when a service organization relies on a subservice organization but does not include that organization's controls in the scope of its own SOC report.

Center for Strategic and International Studies (CSIS) A nonpartisan think tank in Washington, DC, that provides research and policy recommendations on international security, foreign policy, economics, and technology.

Certified Ethical Hacker (CEH) A professional certification provided by the International Council of E-Commerce Consultants (EC-Council) that validates an individual's skills in ethical hacking.

Certified Information Systems Auditor (CISA) A certification for professionals responsible for ensuring that an organization's IT systems are effectively monitored, managed, and protected.

Certified Information Systems Manager (CISM) A certification for professionals who manage, design, and oversee an enterprise's information security program.

Certified Information Systems Security Professional (CISSP) A globally recognized certification in the field of IT security, offered by the International Information System Security Certification Consortium (ISC2).

Certified Risk in Information Systems and Control (CRISC) A certification for professionals who manage enterprise IT risk and implement and maintain information systems controls.

CIA Triad A model designed to guide policies for information security within an organization, consisting of three key principles: confidentiality, integrity, and availability.

CIA Triad A model designed to guide policies for information security within an organization, consisting of three key principles: confidentiality, integrity, and availability.

Cipher An algorithm for encrypting and decrypting information.

CIS Critical Security Controls A risk management model made up of a prescriptive, prioritized, and simplified set of best practices that can be used to strengthen a cybersecurity posture.

Cloud Deployment Model How a cloud computing solution is delivered to the end user, and how that end user can access the resources and tools provided.

Cloud Jacking The act of taking over or compromising a cloud computing environment or resource.

Coaxial Cable A type of electrical cable, consisting of a central conductor, an insulating layer, a metallic shield, and an outer insulating layer, used for transmitting data.

Cold Site An alternate business location where the contracted facility has no active equipment, requiring extensive setup before use.

Cold Site An alternate business location where the contracted facility has no active equipment, requiring extensive setup before use.

Colorado Privacy Act A state-level law that grants Colorado residents specific rights over their personal data and imposes obligations on businesses that collect or process this data.

Common Vulnerabilities and Exposures (CVE) A database of publicly known cybersecurity vulnerabilities and exposures.

Common Vulnerability Scoring System (CVSS) A free and open industry standard for assessing the severity of computer system security vulnerabilities.

Communications Protocol A formal description of the formats and rules for digital messages.

Compensating Control Supplementary measures that are implemented to protect against identified risks or threats when primary controls are ineffective, infeasible, or inefficient.

Complementary Subservice Organization Controls (CSOCs) Specific controls that a subservice organization is expected to implement and operate effectively to support the overall system of controls described in a SOC 1 or SOC 2 report. These controls are excluded from testing by auditors using the carve-out method but are included in testing using the inclusive method.

Complementary User Entity Controls (CUECs) Controls that a service organization assumes its customers (user entities) will implement for the service organization's system of controls to operate effectively.

Compliance Costs Expenses that a company incurs to adhere to industry regulations and government legislation.

Compliance Risk The potential for legal or regulatory sanctions, financial loss, or damage to reputation that an organization may suffer as a result of its failure to comply with laws, regulations, and standards.

CompTIA Security+ A global certification that validates the baseline skills necessary to perform core security functions and pursue an IT security career.

Computer Fraud and Abuse Act (CFAA) A US federal law that criminalizes unauthorized access to computer systems and related offenses.

Computer Misuse Act (UK) A law in the United Kingdom that makes unauthorized access to computer systems a criminal offense.

Computer Security Incident Response Team (CSIRT) A group within an organization that responds to security incidents and breaches.

Confidential Data Information that is meant to be kept secret and is protected from unauthorized access.

Confidentiality The assurance that sensitive data is not disclosed to unauthorized parties, whether intentionally or accidentally.

Control Criteria Standards or benchmarks used to evaluate the design and effectiveness of internal controls in an organization. They define what "good" controls should look like and provide a framework for assessing whether controls are achieving their intended purpose.

Controlled Unclassified Information Information that requires safeguarding or dissemination controls but is unclassified.

Cookies Small text files stored on a user's computer by websites they visit, used to remember login details, preferences, and other information.

Corrective Control A control put in place to correct the damage done as a result of an incident that has already occurred.

Covered Entities (HIPAA) The organizations and individuals directly subject to HIPAA's Privacy, Security, and Breach Notification Rules.

CPA Evolution An initiative by the AICPA and NASBA to reshape the CPA licensure model to meet the needs of the changing profession, including technology and cybersecurity skills.

Credential Stuffing A cyberattack where attackers use stolen username/password combinations to gain unauthorized access to multiple accounts.

Credit Bureaus Organizations that collect and maintain credit information about individuals and provide it to lenders, employers, and others.

Credit Freeze A security measure that restricts access to your credit report, making it more difficult for identity thieves to open accounts in your name.

Credit Monitoring A service that tracks changes in your credit report to alert you of potential fraud or identity theft.

Cross-Scripting (XSS) A type of security vulnerability typically found in web applications, where attackers inject malicious scripts into content from otherwise trusted websites.

Cryptocurrency A digital or virtual currency that uses cryptography for security and operates independently of a central authority.

Cyber Espionage The act of using cyber techniques to gain unauthorized access to confidential information, typically for political or economic gain.

Cyber Hygiene The practices that users of computers and other devices follow, and the steps they take, to maintain system health and improve online security. These practices are often part of a routine to ensure the safety of identity and other details that could be stolen or corrupted.

Cyber Incident Reporting for Critical Infrastructure Act of 2022 (CIRCIA) A US law that requires critical infrastructure entities to report cyber incidents to the federal government.

Cyber Kill Chain A framework developed by Lockheed Martin to understand and defend against cyberattacks.

Cyber Mercenaries Individuals or groups hired to conduct cyberattacks on behalf of others for financial or political gain.

Cyber Risk Insurance A type of commercial coverage that protects businesses from financial losses caused by data breaches and other types of cyberattacks. It covers computer-related risks that are not covered by your general liability or commercial property insurance.

Cyber Risk Quantification The process of measuring the financial impact of cyber risks on an organization.

Cyber Threat Map Also known as a cyberattack map. A real-time map of the computer security attacks that are going on at any given time.

Cyber Value at Risk (VaR) A quantitative risk management approach that applies the financial value at risk (VaR) concept to cybersecurity. It estimates the potential financial loss an organization could face from cyber incidents within a defined time horizon and at a specified confidence level.

Cyberbullying The use of digital platforms to harass, threaten, or humiliate someone.

Cybercriminal An individual, group, or organization that uses computers, networks, or digital technologies to commit illegal activities for personal, financial, political, or malicious gain.

Cybersecurity The practice of protecting systems, networks, and programs from digital attacks, damage, or unauthorized access.

Cybersecurity Act (EU) A law that strengthens the EU Agency for Cybersecurity (ENISA) and establishes a cybersecurity certification framework for products and services.

Cybersecurity and Infrastructure Security Agency (CISA) A US federal agency responsible for protecting the nation's critical infrastructure from physical and cyber threats.

Cybersecurity Audit A formal process for reviewing and assessing an organization's security policies, procedures, and controls.

Cybersecurity Awareness Training Education programs designed to teach employees and users how to protect themselves and the organization from cyber threats.

Cybersecurity Insurance A type of insurance designed to protect businesses from Internet-based risks and risks related to information technology infrastructure and activities.

Cybersecurity Law of the People's Republic of China China's foundational law for regulating cybersecurity, data protection, and Internet governance enacted in 2017.

Cyberwarfare The use of digital attacks by a nation-state or organization to disrupt the activities or infrastructure of another entity.

Cyberweapon A software tool or system designed specifically for the purpose of causing harm in cyberspace, often used in cyberwarfare.

Dark Web A part of the Internet that is not indexed by traditional search engines and requires specific software, configurations, or authorization to access. It is often associated with illegal activities.

Data Raw facts, figures, or characters that are collected, stored, and processed by computer systems, often in digital format.

Data at Rest Digital information that is stored on a physical medium and is not actively moving across networks or being processed.

Data Broker A company or individual that collects, aggregates, processes, and sells/licenses personal and behavioral data regarding individuals or organizations. Most do not have a direct relationship with the data subject.

Data Classification The process of categorizing data based on its sensitivity and the level of security controls needed.

Data Dictionary A centralized repository that contains definitions, descriptions, and metadata about the data elements in a database or information system.

Data in Motion Digital information that is actively moving from one location to another, whether across the Internet or a private network or between devices and applications.

Data Lake A storage repository that holds a vast amount of raw data in its native format until it is needed.

Data Life Cycle The series of stages through which data passes, including creation, storage, usage, sharing, archiving, and deletion, to ensure proper management, security, and value extraction.

Data Loss Prevention (DLP) A set of tools, policies, and processes designed to detect, prevent, and respond to the unauthorized access, transmission, or leakage of sensitive data.

Data Mart A subset of a data warehouse, often oriented to a specific business line or team.

Data Mining The process of discovering patterns, correlations, trends, or useful information in large data sets using statistical, algorithmic, and/or machine learning techniques.

Data Mirroring The process of copying data from one location to another in real time to ensure data redundancy and high availability.

Data Normalization The process of organizing data in a relational database to reduce redundancy and improve data integrity by dividing large tables into smaller, related ones and defining relationships between them.

Data Replication The process of having data in multiple places that are continually synced to enhance access.

Data Replication Method The technique used to copy and maintain data in multiple locations to ensure consistency, availability, fault tolerance, or performance across systems or networks.

Data Security Law (China) A Chinese law to regulate data processing activities (including collection, storage, use, processing, transmission, provision, and disclosure of data) with the goals of protecting national security and the public interest, strengthening data security governance, and promoting the development and use of data in a lawful and safe manner.

Data Security Policy A formal document that outlines an organization's guidelines, rules, and standards for managing and protecting sensitive data assets.

Data Warehouse A system used for reporting and data analysis, storing current and historical data from multiple sources.

Database An organized collection of structured information, or data, typically stored electronically in a computer system.

Database Schema The logical structure that defines how data is organized in a database. It describes the tables, fields (columns), data types, relationships, indexes, views, procedures, and constraints that make up the database.

Deep Web The part of the Internet that is not visible to users without a credential, meaning it requires a login.

Deepfake A piece of synthetic media in which a person's likeness is replaced with another's or is manipulated in other ways, often used maliciously to create fake videos or audio clips.

Default Settings Preconfigured settings or values that are applied to a device or software when it is first installed or reset.

Detective Control A control put in place to detect and alert about an incident that has already occurred.

Dictionary Attack A method of breaking a password or encryption by systematically entering every word in a predefined list of likely passwords.

Differential Backup A process that backs up only the files that have changed since the last full backup.

Digital Certificate An electronic document used to prove the ownership of a public key, issued by a trusted certificate authority.

Digital Footprint The trail of data left behind when you use the Internet or digital devices, including information you intentionally share and data collected passively.

Digital Native A person who has been exposed to digital technology from an early age and is generally comfortable using digital tools such as computers, smartphones, the Internet, and social media.

Digital Operational Resilience Act (DORA) (EU) A law that provides resources to organizations in the financial sector of the EU, specifically in the area of risk management and communications technologies.

Digital Signature A cryptographic technique used to verify the authenticity and integrity of digital messages or documents.

Direct Costs A price that can be directly tied to the production of specific goods or services. A direct cost can be traced to the cost object, which can be a service, product, or department.

Discretionary Access Control The owner of a resource decides who can access it and what they can do.

Disinformation False or misleading information spread deliberately to deceive.

Distributed Denial of Service (DDoS) A cyberattack in which multiple compromised systems are used to flood a targeted system with traffic, overwhelming it and causing it to be unavailable.

Distributed Energy Resources (DERs) Small-scale electricity generation or storage technologies that are located close to where electricity is used rather than at a large, centralized power plant.

DNS (Domain Name Server) Tunneling A digital exploit that hides commands and data in DNS inquiries and responses.

Domain Name A human-readable address, such as www.example.com, that is mapped to an IP address and used to access a website on the Internet.

Domain Name System (DNS) The system that translates domain names (e.g., www.example.com) into IP addresses that computers use to communicate over the Internet.

Double Extortion A ransomware attack where attackers not only encrypt the victim's data but also threaten to release it publicly unless a ransom is paid.

Dumpster Diving The practice of searching through trash to find sensitive information that has been discarded by employees/private citizens.

Electronic Communications Privacy Act (ECPA) A US federal law that protects wire, oral, and electronic communications while they are being made, in transit, and when stored on computers.

Encryption The process of converting data into a coded format to prevent unauthorized access.

Endpoint Any device that connects to a network, such as a computer, smartphone, or server.

Endpoint Detection and Response (EDR) Technology that continuously monitors endpoints for evidence of threats and performs automatic actions to help mitigate them.

Enterprise Resource Planning (ERP) Integrated software systems that help manage core business processes like finance, HR, and supply chain operations.

Enterprise Risk Management A structured, organization-wide approach to identifying, assessing, managing, and monitoring risks that may affect a company's ability to achieve its objectives.

Environmental, Social, and Governance (ESG) Criteria that measure an organization's impact on environmental sustainability, social responsibility, and governance practices.

Ethernet A common technology for wired local area networks (LANs), allowing devices to communicate over a network.

ETL Software Software that can extract, transform, and load big data for analysis.

European Union Agency for Cybersecurity (ENISA) The EU's official agency responsible for cybersecurity.

Exfiltrate To steal and transfer data out of a system or network without authorization. It's the digital equivalent of secretly smuggling information out of a secure facility.

Exploit Code or techniques used to take advantage of a vulnerability to perform unauthorized actions.

External Auditor An independent third-party professional or firm hired to examine an organization's financial statements and issue an opinion on whether they are fairly presented in accordance with accounting standards.

Factor Analysis of Information Risk (FAIR) A risk management framework used to understand, analyze, and quantify information risk in financial terms.

Failover The amount of time it takes to switch from production systems to backup systems.

Fair Credit Reporting Act A US federal law, enacted in 1970, that regulates how consumer credit information is collected, shared, and used.

Fair Isaac Corporation (FICO) Scores Credit scores used to assess a consumer's creditworthiness, created by the Fair Isaac Corporation.

Family Educational Rights and Privacy Act (FERPA) A US federal law that protects the privacy of student education records. It applies to all schools and educational institutions that receive funding from the US Department of Education.

Federal Trade Commission (FTC) A US government agency responsible for protecting consumers and ensuring a competitive market by preventing unfair, deceptive, or fraudulent practices.

Fiber Optic Cables High-speed data transmission cables that use light to transmit information over long distances with minimal loss.

File Transfer Protocol (FTP) A standard network protocol used to transfer files between computers over the Internet.

Financial Reporting A systematic process of recording and representing a company's financial data.

Firewall A security system that monitors and controls incoming and outgoing network traffic based on predetermined security rules.

Form 8-K A report that publicly traded companies must file with the SEC to announce major cybersecurity events that shareholders should know about.

Form 10-K A comprehensive annual report filed by publicly traded companies in the United States that details financial performance and other relevant data.

Form 10-Q A quarterly financial report filed by publicly traded US companies, providing a snapshot of their financial performance.

Fraud Alert A notice placed on a credit report to warn creditors that the consumer may be a victim of fraud or identity theft.

Full Backup The most basic and comprehensive backup method, where a duplicate of all data is sent to another location for storage.

General Data Protection Regulation (GDPR) A regulation in EU law governing data protection and privacy for all individuals within the European Union and the European Economic Area. It also addresses the transfer of personal data outside the EU and EEA.

Generative AI Artificial intelligence that can create new content, such as text, images, or music, based on patterns learned from existing data.

GLBA Privacy Rule A federal regulation under the GLBA of 1999 that requires financial institutions to protect the privacy of consumers' nonpublic personal information (NPI) and provide them with clear notices of their privacy practices.

Global Threat Landscape The evolving array of risks and malicious activities that impact cybersecurity, geopolitical stability, economies, and societies worldwide. It encompasses cyber threats, terrorism, disinformation campaigns, climate-related risks, and emerging technologies that may be exploited for harm.

Gramm-Leach-Bliley Act (GLBA) A US law requiring financial institutions to explain their information-sharing practices and protect sensitive data.

Hacker Someone or some group that uses their technical skills to gain authorized or unauthorized access to systems, networks, or data.

Hacktivist An individual or group that uses hacking techniques to promote political or social agendas.

Hardened (System, Network, Etc.) A system that has been configured and secured to minimize vulnerabilities and reduce the risk of unauthorized access or compromise.

Health Insurance Portability and Accountability Act (HIPAA) A US law designed to provide privacy standards to protect patients' medical records and other health information.

Hidden Backdoor Program Vulnerabilities (Backdoor) Security weaknesses in software that allow unauthorized access to a system, often intentionally left by developers for maintenance purposes.

HIPAA Breach Notification Rule A federal regulation under the Health Insurance Portability and Accountability Act (HIPAA) that requires covered entities and their business associates to notify affected individuals, the US Department of Health and Human Services (HHS), and, in some cases, the media when there is a breach of unsecured protected health information (PHI).

HIPAA Privacy Rule A core part of the Health Insurance Portability and Accountability Act of 1996 (HIPAA) that sets national standards for how protected health information (PHI) is used and disclosed.

HIPAA Security Rule A part of the Health Insurance Portability and Accountability Act of 1996 (HIPAA) that sets national standards for protecting electronic protected health information (ePHI), complementing the HIPAA Privacy Rule.

Homeland Security Act of 2002 The act that created the Department of Homeland Security (DHS) by consolidating 22 diverse organizations to help prevent terrorist attacks in the United States, reduce the vulnerability of the United States to terrorist attacks, and minimize damage and assist in recovery from attacks that do occur.

Hot Site An alternate business location where the target environment is already up and running and can be immediately activated by a failover.

Hub A network device that connects multiple computers or devices, broadcasting data to all connected devices.

Hybrid Cloud A combination of private and public clouds.

Hypertext Transmission Protocol (HTTP) A protocol used for transmitting web pages over the Internet.

Hypertext Transmission Protocol Secure (HTTPS) A secure version of HTTP that uses encryption (SSL/TLS) to protect data exchanged over the Internet.

Identity Theft Identity theft and identity fraud are terms used to refer to all types of crime in which someone wrongfully obtains and uses another person's personal data.

Identity Theft Protection Services and tools that help protect individuals from identity theft, such as credit monitoring and identity recovery assistance.

Incident An event that compromises the confidentiality, integrity, or availability of an information system or the data it processes.

Incident Response Plan (IRP) A set of procedures and guidelines that an organization follows when responding to a cybersecurity incident.

Incident Response Policy A formal document that outlines the procedures and responsibilities an organization follows to detect, respond to, and manage cybersecurity incidents.

Incident Simulation A strategic exercise that replicates potential cyberattacks or security incidents to assess and improve an organization's ability to detect, respond to, and mitigate threats in a controlled environment.

Inclusive Method An approach used in SOC 1 or SOC 2 audits when a service organization includes the controls of a subservice organization within the scope of its own SOC examination. The auditor(s) will therefore evaluate and test the subservice organization's controls as if they were part of the primary organization's system.

Incremental Backup A process that backs up all files that have changed since the last backup occurred.

Indirect Costs Any cost not directly identified with a single, final cost objective but identified with two or more final cost objectives or an intermediate cost objective.

Industrial Control Systems (ICS) Systems used to monitor and control industrial processes, such as manufacturing, power generation, or water treatment.

Information Data that have been processed or organized in a way that provides meaning or value to the user.

Information Security Management System (ISMS) A framework of policies and procedures used to manage an organization's information security risks.

Information Sharing Analysis Centers (ISACs) Groups that share threat intelligence to protect critical infrastructure.

Information Sharing and Analysis Organizations (ISAOs) Groups that voluntarily share threat intelligence within industries.

Information Technology (IT) Auditor A professional who evaluates and tests an organization's technology infrastructure, systems, controls, and processes.

Infrastructure The fundamental physical and organizational systems and facilities needed for the operation of a society, business, or service. It provides the backbone that supports economic activity, public services, and daily life.

Infrastructure as a Service (IaaS) An architecture for providing virtualized computing resources over the Internet.

Inherent Authentication Authentication based on characteristics that are intrinsic to the user, such as biometrics (e.g., fingerprint scanning or facial recognition).

Intangible Assets Nonphysical assets such as intellectual property, goodwill, or brand value.

Integrity The assurance of accuracy, consistency, and trustworthiness of data and systems throughout their lifecycle. It ensures that information is not altered, deleted, or corrupted—whether accidentally or maliciously—without proper authorization or detection.

Intellectual Property (IP) Creations of the mind, such as inventions, literary works, designs, or symbols, that are legally protected from unauthorized use.

Internal Auditor A professional within an organization who provides independent, objective assurance and consulting services designed to evaluate and improve the effectiveness of risk management, internal controls, and governance processes.

Internal Control Over Financial Reporting (ICFR) The policies and procedures a company puts in place to reasonably ensure that its financial statements are accurate, complete, reliable, and prepared in accordance with the applicable accounting standards.

Internal Controls Processes used in an organization that ensure the integrity of the organization's records, systems, and processes as well as ensure regulatory compliance.

Internal Data Data generated and used within an organization, not intended for public distribution.

International Federation of Accountants (IFA) A global organization for the accountancy profession.

Internet A global network of interconnected computers that communicate using standard protocols to share information and services.

Internet Crime Complaint Center (ic3.gov) An initiative run by the FBI that serves as the central hub for the public to report cyber-enabled crimes.

Internet of Things (IoT) A network of physical objects embedded with sensors, software, and other technologies to connect and exchange data with other devices and systems over the Internet.

Internet Protocol (IP) A standard protocol for transmission of data from source to destinations in packet-switched communications networks and interconnected systems of such networks.

Intrusion Detection and Prevention Software (IDPS) Hardware and/or software tools that detect and respond to suspicious activities on a network or system, combining both detection and prevention functionalities.

Investment Scams Fraud schemes where attackers trick people into investing money in fake or nonexistent opportunities.

ISO/IEC 27005 An international standard that provides guidelines for information security risk management.

Isolated Network A full network or segment of a network that has no connection with other sections of an organization's network.

Israeli Computers Law of 1995 A law that criminalizes unlawful computer-related misconduct, hacking, or making malicious use of computers and computer programs.

IT Application Controls Automated controls embedded within software applications that ensure the accuracy, completeness, validity, and authorization of data processing specific to a business process (e.g., payroll, invoicing, purchasing). They rely on strong IT general controls to function effectively.

IT General Controls Foundational controls that apply to all systems, applications, and data across an organization's IT environment. They ensure the reliability, integrity, and security of information systems that support business processes and financial reporting.

Japanese Act on Prohibition of Unauthorized Access (UCAL) A Japanese law that criminalizes unauthorized access to computer systems and networks.

Keylogger Software or hardware devices that track and record a user's keystrokes.

Knowledge Information prepared, sorted, and organized so that it can be used in high-level decision-making processes to set policy or strategy.

Least Privilege An information security concept in which a user is given the minimum level of access—or permissions—needed to perform their job functions.

Legacy Hardware Older, possibly superseded hardware still being used by an organization.

List Broker A marketing intermediary who rents, sells, or manages access to contact lists on behalf of clients seeking to reach specific target audiences.

Local Area Network (LAN) A network that connects computers within a limited area, such as a home, school, or office building.

Logical Access Control Tools and/or protocols used to identify, authenticate, authorize, and hold accountable individuals trying to access a system.

Login Name A unique identifier that a user enters to gain access to a system, network, or application.

Machine Learning A subset of artificial intelligence that involves training algorithms to learn from and make predictions based on data.

Malware Malicious software designed to damage, disrupt, or gain unauthorized access to a computer system.

Management Representation Letter A formal, written statement provided by an organization's management to their external auditor at the end of an audit engagement. It confirms that management has fulfilled its responsibilities and provided the auditor with accurate, complete, and truthful information necessary for the audit.

Mandatory Access Control Access is controlled by system-enforced rules based on classifications (e.g., Top Secret, Confidential). Users cannot change permissions.

Man-in-the-Middle (MitM) Attack A cyberattack in which an attacker secretly intercepts and relays communication between two parties who believe they are directly communicating with each other.

Materiality The term that determines whether a cyber incident is significant enough that it must be reported to regulators, investors, or stakeholders. Just like in financial reporting, it is not only about the size of the financial loss but also about the nature, timing, and context of the incident.

Metadata Structured information that describes the characteristics, origin, structure, and usage of data.

Modem A device that converts digital data into a format suitable for transmission over a communication medium, such as a telephone line or cable system.

Monte Carlo Simulation A mathematical technique used to model the probability of different outcomes when there is uncertainty in the input variables. Instead of producing a single outcome, it runs a large number of trials (often thousands or millions), each time using randomly generated values within defined ranges, to estimate the distribution of possible results.

Moral Hazard The risk that a party insulated from risk may behave differently from how they would if they were fully exposed to the risk.

Multifactor Authentication (MFA) A security process that requires users to provide two or more authentication factors to verify their identity.

NERC Critical Infrastructure Protection (CIP) Reliability Standards A set of cybersecurity and physical security requirements developed by the North American Electric Reliability Corporation (NERC). They are designed to protect the bulk electric system (BES) in North America from cyber threats, physical attacks, and other disruptions.

Network and Information Security (NIS2) Directive (EU) A European Union directive aimed at improving the cybersecurity of network and information systems across the EU.

Network and Information Security Regulations (UK) UK regulations designed to enhance cybersecurity by requiring organizations to improve their network and information systems.

Network Client A software or hardware component that accesses a service made available by a server.

Network Detection and Response (NDR) Security solutions that use machine learning and behavioral analytics to detect and respond to threats on a network.

Network Interface Card (NIC) A hardware component that allows a computer to connect to a network.

Network Management Software An application or set of applications that lets network engineers manage a network's independent components.

Network Operating System Software that manages network resources and allows communication between computers and devices on a network.

Network Protocols Rules and conventions for communication between network devices.

Network Segmentation The practice of dividing a computer network into smaller, isolated sections (segments) to improve security, performance, and manageability.

Network Topology The arrangement of network elements, including nodes and connecting lines, that determines the structure and performance of a network.

Next Generation (Next-Gen) Antivirus Software Antivirus software that uses a combination of artificial intelligence, behavioral detection, machine learning algorithms, and exploit mitigation, so known and unknown threats can be anticipated and immediately prevented.

Nist SP800-30 A NIST special publication that provides guidelines for conducting risk assessments for information systems.

Operating System The software that manages a computer's hardware and software resources, providing services to applications and users.

Operational Information Security Risk The risk of loss resulting from inadequate or failed processes, systems, human errors, or external events relating to information security.

Operational Loss A financial loss that arises from the failure of internal processes, people, or systems, or from external events rather than from market movements or credit risk.

Operational Vulnerabilities Weaknesses in an organization's processes or systems that can be exploited to cause operational failures or security breaches.

Opinion (Auditor) The formal conclusion that an independent auditor provides after evaluating a company's financial statements and related disclosures.

Parsing Errors Errors that occur when data is improperly interpreted or processed by a program or system.

Passphrase A longer, more secure version of a password, typically composed of multiple words or a sentence.

Password A secret word or phrase used to authenticate a user and provide access to a system or resource.

Password Management Software Software that helps users generate, store, and manage complex passwords securely.

Patch An update to software.

Patch Management The process of managing updates to software, especially security software, to protect systems from vulnerabilities.

Payload The part of malicious software (malware) or exploit code that actually delivers the intended harmful action once the attack successfully compromises a system.

Penetration (PEN) Testing A security exercise where a cybersecurity expert attempts to find and exploit vulnerabilities in a computer system. The purpose of this simulated attack is to identify any weak spots in a system's defenses.

Personal Information Protection Law (PIPL) China's first comprehensive law specifically focused on protecting personal information.

Personally Identifiable Information (PII) Information that, when used alone or with other relevant data, can identify an individual.

Petty Cybercriminals Individuals who engage in minor cybercrimes, often for personal gain or mischief.

Phishing A social engineering technique that attempts to acquire sensitive data through a fraudulent solicitation in email or on a website, in which the perpetrator masquerades as a legitimate business or reputable person.

Physical Controls The measures that organizations take to physically protect their assets.

Piggybacking Same as Tailgating. A physical security breach where an unauthorized person gains access to a restricted area by following an authorized individual.

Platform as a Service (PaaS) An architecture that provides a platform allowing customers to develop, run, and manage applications.

Ponzi Scheme A fraudulent investment scheme in which returns are paid to earlier investors using the capital of new investors rather than from legitimate profit earned by the operation of a business.

Port Scanning The act of systematically scanning a computer's ports to find vulnerabilities.

Pretexting The use of a fabricated story, or pretext, to gain a victim's trust and trick or manipulate them into sharing sensitive information, downloading malware, sending money, or otherwise harming themselves or the organization for which they work.

Preventive Control A control put in place with the intention to avoid an incident or prevent it from occurring.

Privacy The right of individuals to control access to their personal information, communications, and activities. It encompasses the ability to determine what data are collected, how they are used, who can access them, and for how long.

Privacy Policy A statement by an organization that outlines how it collects, uses, and protects personal data.

Privacy Settings User-configurable settings that determine how personal data is shared and managed in an application or service.

Private Cloud Cloud infrastructure operated solely for a single organization.

Programmable Logic Controllers (PLCs) Industrial digital computers used to control machinery and processes, such as in manufacturing environments.

Protected Health Information (PHI) Any information about health status, provision of healthcare, or payment for health care that can be linked to an individual.

Protection of Privacy Law (Israel) An Israeli law that regulates the collection, processing, and dissemination of personal data to protect individual privacy.

Public Cloud Cloud services offered over the public Internet and available to anyone.

Public Data Information that is freely available to the public and does not require any specific access controls.

Public Relations (PR) The set of techniques and strategies related to managing how information about an individual or company is disseminated to the public, and especially the media.

Qualitative Risk Assessment A method of evaluating risks based on subjective judgment and non-numeric criteria, such as severity or likelihood.

Quantitative Risk Assessment A method of evaluating risks using numerical data, typically assigning values to probability and impact.

Quantum Computing A theoretical field of computing that uses the principles of quantum mechanics—the physics governing atoms and subatomic particles—to process information to solve certain classes of problems (e.g., factoring large numbers) much faster than classic computers.

Quishing Short for QR code phishing. A type of phishing attack where cybercriminals use malicious QR codes to trick victims into visiting fraudulent websites, downloading malware, or revealing sensitive information.

Racketeer Influenced and Corrupt Organizations Act (RICO) A US federal law designed to combat organized crime by allowing prosecution of all individuals involved in a corrupt organization.

Radio Waves Electromagnetic waves used for transmitting data wirelessly, including in communication technologies such as Wi-Fi and radio.

Ransomware Malicious software that encrypts a victim's data and demands payment for the decryption key.

Ransomware as a Service (Raas) A hacker business model in which cybercriminals lease ransomware tools and infrastructure to other attackers for a share of the ransom profits.

Red Team A group that attempts a physical or digital intrusion against an organization.

Relational Schema The formal structure of a relational database, describing how data is organized into tables (relations) and how those tables relate to each other.

Remote Access Policy A written document containing the guidelines for connecting to an organization's network from outside the office.

Remote Access Trojan (RAT) A Trojan malware type that allows the attacker remote access to and control of the infected system.

Remote Code Execution (RCE) A cyberattack where an attacker can execute code on a remote system.

Reputational Risk The potential for any event, controllable or otherwise, to damage your organization's reputation.

Residual Risk The remaining risk after all mitigation efforts have been implemented.

Restricted Data Information that is highly sensitive and subject to strict access controls to protect its confidentiality.

Ring Network Topology Devices form a circular chain; data travels in one or both directions.

Risk The possibility that an event, action, or condition will occur that adversely affects objectives, assets, or outcomes.

Risk Appetite The amount of risk that an organization is willing to accept in pursuit of its objectives.

Risk Tolerance The acceptable level of variation in performance or outcomes that an organization (or individual) is willing to withstand in pursuit of its objectives.

Risk-Adaptive Access Control A type of control that considers contextual risk and adjusts access dynamically.

Role-Based Access Control A type of control where permissions are assigned to roles (e.g., "Manager," "HR") and users inherit permissions through their role(s).

Romance Scam A type of fraud where an attacker feigns romantic interest to gain the victim's trust and steal money or personal information.

Rootkit Trojan A Trojan malware type that operates at a deep level within the system, often evading detection by standard security measures.

Router A device that directs data traffic between different networks, such as from a local network to the Internet.

Rule-Based Access Control A type of control that uses predefined rules to grant/deny access, often based on conditions like time of day or IP address.

Sandbox An isolated environment used for testing and analyzing software without affecting the rest of the system.

Sarbanes-Oxley Act (SOX) A US federal law that sets requirements for public company boards, management, and public accounting firms to protect investors from fraudulent financial reporting.

Script Kiddies Inexperienced hackers who use existing scripts or tools to launch attacks without understanding the underlying technology.

SEC Cybersecurity Breach Disclosure Rules SEC guidelines requiring public companies to disclose their cybersecurity risk management, strategy, governance, and incidents.

SEC Public Company Cybersecurity Risk Management, Strategy, Governance, and Incident Disclosure SEC guidelines requiring public companies to disclose their cybersecurity risk management, strategy, governance, and incidents.

Secure Boot A security feature that ensures a computer only boots using software that is trusted by the manufacturer or owner.

Secure by Design A principle that emphasizes building security into the design and development of systems rather than adding it later.

Secure Configuration The practice of configuring systems and software securely to reduce vulnerabilities.

Security Certificates Digital certificates used to verify the identity of websites or systems and to encrypt communications.

Security Logs Records of events and activities related to the security of a system, used for monitoring, analysis, and auditing.

Security Settings Configurations that control access to resources and determine how security features are applied in a system.

Separation of Duties An organizational principle that divides tasks and responsibilities among multiple individuals to reduce the risk of fraud or error.

Server A system that responds to requests from clients by delivering specific functions, such as file storage, websites, applications, or databases.

Service Auditor An independent auditor who evaluates and reports on the controls of a service organization.

Service Organization Controls (SOC) Reports Reports that provide information about the controls at a service organization, including SOC 1, SOC 2, and SOC 3; SOC for Cybersecurity; and SOC for Supply Chain reports.

Short Message Service (SMS) Attacks Same as "Smishing." A type of phishing attack carried out via SMS text messages, aiming to deceive users into providing sensitive information.

Shoulder Surfing A social engineering technique where an attacker observes someone's private information by either watching them directly or through visual aids (cameras, mirrors, etc.) without the target realizing it.

SIM Swap A type of fraud in which an attacker transfers a victim's phone number to a new SIM card to gain access to their accounts.

Simple Mail Transfer Protocol (SMTP) A standard protocol used to send and receive email messages between servers.

Single Sign-On (SSO) An authentication process where the user authenticates once to gain access to all of the systems for which they are authorized based on their defined role.

Skimmer A malicious device or software that secretly captures payment card data, either through physical attachments on

ATMs/POS machines or code injected into online checkout pages.

Smart Card Any ID card embedded with a microchip that allows secure processing and control over physical access, cryptographic keys, or digital signatures.

Smishing Derived from "SMS" and "phishing," is a type of cybercrime that uses deceptive text messages to manipulate victims into divulging sensitive information.

Snowflake Schema A type of database schema that organizes data into multiple related tables to reduce redundancy.

SOC 2 Description Criteria The required elements that a service organization must include when describing its system in a SOC 2 report.

SOC 2 Readiness Assessment A pre-audit evaluation designed to help an organization prepare for a SOC 2 examination. It identifies gaps in its security, privacy, and internal control practices before the formal audit begins.

SOC (System and Organization Controls) Engagement A third-party audit that a CPA or accounting firm performs to evaluate and report on the controls of a service organization—typically related to security, availability, processing integrity, confidentiality, or privacy of systems and data.

Social Engineering The manipulation of individuals into divulging confidential information or performing actions that compromise security.

Software as a Service (SaaS) A cloud computing service model that delivers software applications over the Internet on a subscription basis.

Software Patch A piece of code released by developers to update or fix issues in a software program, such as bugs, security vulnerabilities, or performance problems.

Software Vulnerabilities Weaknesses or flaws in software that can be exploited by attackers to gain unauthorized access or cause damage.

SOX Section 404 A provision of the Sarbanes-Oxley Act of 2002 (SOX) that requires public companies in the United States or companies that trade in the United States to establish, assess, and report on the effectiveness of their internal control over financial reporting (ICFR).

Spear Phishing A targeted phishing attack aimed at a specific individual or organization, often using personalized information to make the attack more convincing.

Spyware Malicious software that secretly monitors and collects information about a user's activities.

SQL Injection A type of attack where malicious SQL code is inserted into a database query, allowing the attacker to gain unauthorized access to data.

Stack Overflow A type of software bug that occurs when a program tries to use more memory than is allocated, which can lead to crashes or vulnerabilities.

Star Network Topology All devices connect to a central hub or switch.

Star Schema A type of database schema that organizes data into a central fact table connected to multiple dimension tables.

Statement on Auditing Standards (SAS) Official guidelines issued by the Auditing Standards Board (ASB) of the American Institute of Certified Public Accountants (AICPA). The SAS provides rules and procedures auditors must follow when conducting audits of companies in the United States.

Storage Media The physical materials or devices used to store digital data in a computer system. It allows data to be saved, retrieved, and transferred.

STRIDE Threat Model A developer-focused model to identify and classify threats under six types of attacks—spoofing, tampering, repudiation, information disclosure, denial of service (DoS), and elevation of privilege.

Structured Query Language (SQL) A standard programming language used to create, read, update, and delete data in relational databases.

Subsequent Event An event that occurs after the formal end of the audit testing period but before the audit opinion is issued that may require disclosure or adjustment of the management statements.

Subservice Organization A service organization used by another service organization to perform some of the services provided to user entities.

Supervisory Control and Data Acquisition (SCADA) A system used for monitoring and controlling industrial processes and infrastructure.

Supply Chain The network of individuals, organizations, resources, activities, and technologies involved in the creation and distribution of a product or service—from raw materials to final delivery to the customer.

Supply Chain Attack A type of cyberattack where attackers target less-secure elements within an organization's supply chain to compromise the larger network. These attacks can occur in both software and hardware supply chains.

Surface Web The portion of the Internet that is indexed by standard search engines and accessible to the general public.

Switch A network device that connects multiple devices and manages data traffic between them.

Symmetric Encryption An encryption method that uses a single shared key to encrypt and decrypt data.

SYN Flood A type of denial-of-service (DoS) attack that targets the TCP three-way handshake used to open connections. The attacker overwhelms a server with so many half-open TCP connection requests (SYN packets) that the server runs out of resources for legitimate clients — causing slowdowns or service outages.

Syntax or Logic Error Mistakes in a program's code, in either its structure (syntax) or logic, that prevent it from functioning correctly.

System Information and Event Management (SIEM) A solution that provides real-time analysis of security alerts generated by network hardware and applications.

System Misconfiguration Incorrect or suboptimal settings in a system that can lead to vulnerabilities or performance issues.

Tabletop Exercise An informal, discussion-based session in which a team discusses their roles and responses during an emergency, walking through one or more example scenarios.

Tactics, Techniques, and Procedures (TTPs) The methods and strategies that threat actors use to carry out cyberattacks.

Tailgating Same as Piggybacking. A physical security breach where an unauthorized person gains access to a restricted area by following an authorized individual.

Tangible Assets Physical items of value that an organization owns, such as buildings, equipment, and inventory.

Technical Controls The measures that organizations put in place to safeguard their digital assets.

Testing Matrix A structured tool used in audits, risk assessments, or compliance reviews to organize and track the testing of internal controls. It helps auditors or control testers clearly map what is being tested, how it's being tested, and what the results are.

Third-Party Coverage A type of insurance policy that the insured (first party) purchases from the insurance company (second party) for protection against the claims of another (third party).

Threat Any circumstance or event with the potential to cause harm to an information system.

Threat Hunters Security professionals who proactively search for cyber threats both within and outside an organization's network.

Threat Index A measure used to quantify the level of threat that various cyber threats pose.

Threat Intelligence Evidence-based knowledge about past, current, and future threats, including information on threat actors and their motives.

Threat Landscape The overall view of the threats facing an organization, including the types and sources of threats.

Threat Maps Also known as a cyberattack map, a real-time map of computer security attacks that are going on at any given time. Can be at the company, regional, national, or worldwide level.

Threat Modeling A structured process for identifying, evaluating, and addressing potential security threats.

Threat-Risk Pairing The process of linking identified threats to specific risks to assess their potential impact.

TOR (The Onion Router) A technology that uses layered encryption to route Internet traffic through multiple volunteer-run servers (nodes), hiding the user's identity, location, and activity.

Transmission Control Protocol (TCP) A communication protocol that establishes a reliable connection between two systems and ensures the correct delivery of data packets.

Transmission Media The physical or wireless medium through which data is transmitted, such as fiber optics, twisted pair cables, or radio waves.

Trojan A type of malware that disguises itself as legitimate software or code to trick users into installing it on their systems.

Trolling The deliberate act of posting provocative, offensive, misleading, or disruptive messages online with the intent to provoke emotional reactions, cause conflict, or derail conversations.

Twisted Pair Cables A type of wiring where two conductors are twisted together to reduce electromagnetic interference, commonly used in telecommunication networks.

UN Convention Against Cybercrime A multilateral treaty currently under negotiation with the United Nations. It is intended to create a globally agreed legal framework for combating cybercrime and promoting international cooperation.

Uniform Resource Locator (URL) The web address used to access resources on the Internet, typically in the format https://www.example.com, for example.

Universal Serial Bus (USB) A standard for connecting peripherals such as keyboards, mice, and external drives to computers.

US Privacy Act of 1974 A federal law that governs how federal government agencies collect, maintain, use, and share personally identifiable information (PII) about individuals.

Virginia Consumer Data Protection Act (VCDPA) A state-level data privacy law enacted in March 2021 that establishes privacy rights for Virginia residents and compliance obligations for businesses that process their personal data.

Virtual Private Network (VPN) A technology that creates a secure, encrypted connection over a less-secure network, such as the Internet.

Virus A computer program that spreads by first infecting files or the system areas of a computer or network router's hard drive and then making copies of itself.

Vishing Short for voice phishing, refers to fraudulent phone calls or voice messages designed to trick victims into providing sensitive information or access.

Vulnerability A specific weakness in a system that a threat can exploit.

Warm Site An alternate business location where the target environment has some active equipment, requiring minimal setup before use.

Web Application Threats Cyber threats specifically targeting web-based applications, often through techniques like SQL injection, cross-site scripting (XSS), or session hijacking.

Web Browser Software used to access and navigate the World Wide Web on the Internet.

Website Errors Technical issues or malfunctions that prevent a website from functioning as intended, often caused by coding errors, server problems, or connectivity issues.

Whaling A type of phishing attack that targets high-level executives or other important individuals in an organization, often involving personalized and sophisticated tactics.

Wide Area Network (WAN) A network that spans a large geographical area, connecting multiple local area networks (LANs) across cities, regions, or countries.

Wiper Malware Malicious software designed to delete or "wipe" data from a system, rendering it irrecoverable.

Wire Fraud A criminal act involving the use of telecommunications or the Internet to defraud someone of money or property.

Wire Transfer Fraud/Scheme Fraudulent schemes that trick individuals or organizations into transferring money to the scammer's account.

Wire Transfer Scams Scams where attackers trick individuals or businesses into sending money through wire transfers to accounts controlled by the fraudsters.

Wired Equivalent Privacy (WEP) An outdated and vulnerable security protocol for wireless networks, replaced by WPA, WPA2, and WPA3.

Wireless Protected Access (WPA, WPA2, and WPA3) Security protocols used to secure wireless networks, with WPA2 offering stronger encryption than WPA and WPA3 stronger than WPA2.

With-and-Without Method Also called the "incremental cash flow" or "differential income" method, a common approach in intellectual property (IP) valuation, particularly for patents, trademarks, and technology assets.

Workstation A personal computer or terminal connected to a network, used for professional or technical tasks such as software development or data analysis.

World Wide Web (WWW) A system of interlinked hypertext documents and resources that can be accessed via the Internet using a web browser.

Worm Malicious software that replicates itself and spreads to other systems without user intervention.

Zero Trust A security model that assumes no implicit trust and requires verification for every access request.

Zero-Day Vulnerability An undiscovered flaw in an application or operating system, a gap in security for which there is no defense or patch because the software maker does not know it exists—they have had "zero days" to prepare an effective response.

Index

Note: Page numbers in **bold** refer to terms defined in the margins. Numbers in *italics* refer to figures and tables.

www.ingramcontent.com/pod-product-compliance
Ingram Content Group UK Ltd.
Pitfield, Milton Keynes, MK11 3LW, UK
UKHW051453290926
15471UKWH00008B/1180

9 781958 303443